Serono Symposia USA
Norwell, Massachusetts

PROCEEDINGS IN THE SERONO SYMPOSIA USA SERIES

Continued after Index

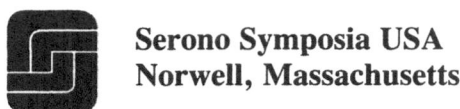

Serono Symposia USA
Norwell, Massachusetts

Eli Y. Adashi   Michael O. Thorner
Editors

The Somatotrophic Axis and the Reproductive Process in Health and Disease

With 92 Figures

Springer-Verlag
New York Berlin Heidelberg London Paris
Tokyo Hong Kong Barcelona Budapest

Eli Y. Adashi, M.D.
Department of
  Obstetrics and Gynecology
University of Maryland
  School of Medicine
Baltimore, MD 21201
USA

Michael O. Thorner, M.B., D.Sc.
Division of Endocrinology and
  Metabolism
Department of Medicine
University of Virginia
  Health Sciences Center
Charlottesville, VA 22908
USA

ISBN-13:978-1-4612-7567-1   e-ISBN-13:978-1-4612-2518-8
DOI: 10.1007/978-1-4612-2518-8

Proceedings of the Symposium on the Somatotrophic Axis and the Reproductive Process in Health and Disease, sponsored by Serono Symposia USA, Inc., held October 28 to 31, 1993, in Baltimore, Maryland.

For information on previous volumes, please contact Serono Symposia USA, Inc.

Library of Congress Cataloging-in-Publication Data
The somatotrophic axis and the reproductive process in health and disease/Eli Y. Adashi, Michael O. Thorner, editors.
     p.   cm.
   "Proceedings of the Symposium on the Somatotrophic Axis and the Reproductive Process in Health and Disease, sponsored by Serono Symposia USA, Inc., held October 28 to 31, 1993, in Baltimore, Maryland"—T.p. verso.
   Includes bibliographical references and index.
   ISBN-13:978-1-4612-7567-1   e-ISBN-13:978-1-4612-2518-8
   DOI: 10.1007/978-1-4612-2518-8

   1. Somatotropin—Physiological effect—Congresses.   2. Endocrine gynecology—Congresses.
I. Adashi, E.Y.   II. Thorner, Michael O.   III. Serono Symposia USA, Inc.   IV. Symposium on the Somatotrophic Axis and the Reproductive Process in Health and Disease (1993: Baltimore, Md.)
   [DNLM: 1. Somatotropin—congresses.   2. Ovulation—physiology—congresses.
3. Reproduction—physiology—congresses.   4. Puberty—metabolism—congresses.
WK 515 S6935 1995]
QP572.S6S65   1995
612.6—dc20   94-39174

Printed on acid-free paper.

© 1995 Springer-Verlag New York, Inc.
Softcover reprint of the hardcover 1st edition 1995

Production coordinated by Marilyn Morrison and managed by Francine McNeill; manufacturing supervised by Jacqui Ashri.
Typeset by Best-set Typesetter Ltd., Hong Kong.

9 8 7 6 5 4 3 2 1

ISBN-13:978-1-4612-7567-1

SYMPOSIUM ON THE SOMATOTROPHIC AXIS AND THE REPRODUCTIVE PROCESS IN HEALTH AND DISEASE

Scientific Committee

Eli Y. Adashi, M.D.
University of Maryland
School of Medicine
Baltimore, Maryland

Michael O. Thorner, M.B., D.Sc.
University of Virginia
Health Sciences Center
Charlottesville, Virginia

Organizing Secretaries

Bruce K. Burnett, Ph.D.
Leslie Nies
Serono Symposia USA, Inc.
100 Longwater Circle
Norwell, Massachusetts

Preface

For many years now, our understanding of the somatotrophic and reproductive axes has evolved essentially independently, both fields of study reaching a highly advanced, although far from complete, level of understanding. Along the way, however, it became apparent that in some circumstances the reproductive and somatotrophic axes may be interdependent. Inklings to this effect were at times feeble and at other times more convincing. Among those inklings are the clinical recognition by pediatric endocrinologists of the apparent association between isolated GH deficiency and delayed puberty, as well as of the apparent permissive, puberty-promoting property of GH. Equally important is a body of experimental studies establishing the ovary of multiple species as a site of GH reception and action. Arguing against an essential role for GH in the reproductive process is the observation that individuals who have GH resistance of the Laron variety are fertile and that isolated GH deficiency does not constitute an absolute barrier to the attainment of sexual maturation and fertility.

The intraovarian insulin-like growth factor (IGF) hypothesis proposes that IGFs may serve as amplifiers of gonadotropin action. Although the dependence of intraovarian IGFs on systemic GH action has never been unequivocally demonstrated, that leap of faith has often been made. The intraovarian IGF hypothesis serves as the rationale for the adjunctive use of GH in the induction of ovulation. It is in this context that yet another group of studies is accumulating in which the use of GH in both chronic anovulation and assisted reproductive technology is being assessed. This body of work led to a growing recognition that the concurrent use of GH may decrease gonadotropin requirements.

It was against this backdrop that a meeting was convened by Serono Symposia USA to reassess the science related to reproductive and somatotrophic interactions. The meeting was concerned with the critical exploration of the work accomplished thus far and with the possible delineation of alternative, presumably improved approaches along which this line of investigation might proceed. Convened in Baltimore in late October 1993, the international meeting featured a multidisciplinary faculty whose experience covered the spectrum of somatotrophic and reproductive physiology. This volume represents a summary of those proceedings.

The editors are indebted to the dedicated group of contributors, without whose help this compilation could not have been realized. We are especially grateful to Serono Symposia USA and in particular to Leslie Nies and Bruce Burnett for their guidance, considerable financial assistance, and their patience throughout the process of organizing the symposium and the publication of this volume.

ELI Y. ADASHI
MICHAEL O. THORNER

Contents

**Part II.   Pubertal, Menstrual, Gestational,
and Menopausal Adaptation**

Part III.   Growth Hormone: The Ovarian Connection

Contributors

ELI Y. ADASHI, Division of Reproductive Endocrinology, Department of Obstetrics and Gynecology, University of Maryland School of Medicine, Baltimore, Maryland, USA.

JOSEPH A. ALOI, Division of Endocrinology and Metabolism, Department of Medicine, University of Virginia Health Sciences Center, Charlottesville, Virginia, USA.

KLAUS AMBURN, Department of Medicine, Center for Endocrinology, Metabolism and Nutrition, Northwestern University Medical School, Chicago, Illinois, USA.

BRIAN J. AREY, Department of Biochemistry, Molecular Biology, and Cell Biology, Northwestern University, Evanston, Illinois, USA.

C.M. ASPLIN, Division of Endocrinology and Metabolism, Department of Internal Medicine, University of Virginia Health Sciences Center, Charlottesville, Virginia, USA.

GERHARD BAUMANN, Department of Medicine, Center for Endocrinology, Metabolism and Nutrition, Northwestern University Medical School, Chicago, Illinois, USA.

MARC R. BLACKMAN, Division of Endocrinology and Metabolism, Department of Medicine, Johns Hopkins University School of Medicine, Baltimore, Maryland, USA.

ROBERT M. BLIZZARD, Department of Pediatrics, University of Virginia Health Sciences Center, Charlottesville, Virginia, USA.

PAUL A. BOEPPLE, Harvard Medical School and Pediatric and Reproductive Endocrine Units, Massachusetts General Hospital, Boston, Massachusetts, USA.

R.A. BOOTH, JR., Division of Endocrinology and Metabolism, Department of Internal Medicine, University of Virginia Health Sciences Center, Charlottesville, Virginia, USA.

DAVID R. CLEMMONS, Division of Endocrinology, Department of Medicine, University of North Carolina, Chapel Hill, North Carolina, USA.

NANCY E. COOKE, Departments of Medicine and Genetics, University of Pennsylvania School of Medicine, Philadelphia, Pennsylvania, USA.

NORMA DAVILA, Department of Medicine, Center for Endocrinology, Metabolism and Nutrition, Northwestern University Medical School, Chicago, Illinois, USA.

W.S. EVANS, Division of Endocrinology and Metabolism, Department of Internal Medicine, University of Virginia Health Sciences Center, Charlottesville, Virginia, USA.

A.C.S. FARIA, Department of Clinical Medicine, Federal University of Rio de Janeiro, Rio de Janeiro, Brazil.

MARCO FILICORI, Reproductive Endocrinology Center, University of Bologna, Bologna, Italy.

J.K. FINDLAY, Prince Henry's Institute of Medical Research, Clayton, Victoria, Australia.

STEPHEN FRANKS, Department of Obstetrics and Gynaecology, St. Mary's Hospital Medical School, Imperial College of Science, Technology and Medicine, University of London, London, UK.

BRUCE D. GAYLINN, Division of Endocrinology and Metabolism, Department of Medicine, University of Virginia Health Sciences Center, Charlottesville, Virginia, USA.

RANDALL GRIMES, Division of Endocrinology, Diabetes, and Metabolism, Pennsylvania State University College of Medicine, Hershey, Pennsylvania, USA.

DAVID GUTHRIE, U.S. Department of Agriculture, Agricultural Research Service, Beltsville Agricultural Research Center, Beltsville, Maryland, USA.

JOHN G. HADDAD, Division of Endocrinology, Department of Medicine, University of Pennsylvania School of Medicine, Philadelphia, Pennsylvania, USA.

DANIEL HAGEN, Department of Dairy and Animal Science, Pennsylvania State University, University Park, Pennsylvania, USA.

DIANA HAMILTON-FAIRLEY, Department of Obstetrics and Gynaecology, St. Thomas' Hospital Medical School, University of London, London, UK.

JAMES M. HAMMOND, Division of Endocrinology, Diabetes, and Metabolism, Pennsylvania State University College of Medicine, Hershey, Pennsylvania, USA.

MARK L. HARTMAN, Division of Endocrinology and Metabolism, Department of Medicine, University of Virginia Health Sciences Center, Charlottesville, Virginia, USA.

KEN K.Y. HO, Garvan Institute of Medical Research, St. Vincent's Hospital, Sydney, Australia.

A. IRANMANESH, Veterans Affairs Medical Center, Salem, Virginia, USA.

OLLE G.P. ISAKSSON, Research Centre for Endocrinology and Metabolism, Division of Endocrinology, Department of Internal Medicine, Sahlgrenska Hospital, University of Göteborg, Göteborg, Sweden.

JÖRGEN ISGAARD, Research Centre for Endocrinology and Metabolism, Division of Endocrinology, Department of Internal Medicine, Sahlgrenska Hospital, University of Göteborg, Göteborg, Sweden.

HOWARD S. JACOBS, Cobbold Laboratories, University College London Medical School, and The Middlesex Hospital, London, UK.

MICHAEL JOHNSON, Department of Pharmacology, University of Virginia Health Sciences Center, Charlottesville, Virginia, USA.

BEVERLY K. JONES, Department of Genetics, University of Pennsylvania School of Medicine, Philadelphia, Pennsylvania, USA.

JILL A. KANALEY, Division of Endocrinology and Metabolism, Department of Medicine, University of Virginia Health Sciences Center, Charlottesville, Virginia, USA.

JOHN J. KELLY, Department of Medicine, St. George Hospital, Kogarah, Sydney, Australia.

SUSAN E. KIRK, Division of Endocrinology and Metabolism, Department of Medicine, University of Virginia Health Sciences Center, Charlottesville, Virginia, USA.

ANTHONY A. KOSSIAKOFF, Department of Protein Engineering, Genentech, Inc., South San Francisco, California, USA.

GAIL A. LAUGHLIN, Department of Reproductive Medicine, University of California at San Diego, La Jolla, California, USA.

STEPHEN A. LIEBHABER, The Howard Hughes Medical Institute and Departments of Genetics and Medicine, University of Pennsylvania School of Medicine, Philadelphia, Pennsylvania, USA.

ANDERS LINDAHL, Research Centre for Endocrinology and Metabolism, Department of Clinical Chemistry, Sahlgrenska Hospital, University of Göteborg, Göteborg, Sweden.

DANIEL I.H. LINZER, Department of Biochemistry, Molecular Biology, and Cell Biology, Northwestern University, Evanston, Illinois, USA.

PAUL M. MARTHA, JR., Genentech, Inc., South San Francisco, California, USA.

HELEN D. MASON, Department of Obstetrics and Gynaecology, St. Mary's Hospital Medical School, Imperial College of Science, Technology and Medicine, University of London, London, UK.

MOISES MERCADO, Department of Medicine, Center for Endocrinology, Metabolism and Nutrition, Northwestern University Medical School, Chicago, Illinois, USA.

ANITA MISRA-PRESS, Department of Genetics and The Howard Hughes Medical Institute, University of Pennsylvania School of Medicine, Philadelphia, Pennsylvania, USA.

ARLENE J. MORALES, Department of Reproductive Medicine, University of California at San Diego, La Jolla, California, USA.

E. KIRK NEELY, Department of Pediatrics, Stanford University Medical Center, Stanford, California, USA.

CLAES OHLSSON, Research Centre for Endocrinology and Metabolism, Division of Endocrinology, Department of Internal Medicine, Sahlgrenska Hospital, University of Göteborg, Göteborg, Sweden.

SUZAN S. PEZZOLI, Division of Endocrinology and Metabolism, Department of Medicine, University of Virginia Health Sciences Center, Charlottesville, Virginia, USA.

ALAN D. ROGOL, Departments of Pediatrics and Pharmacology, University of Virginia Health Sciences Center, Charlottesville, Virginia, USA.

RON G. ROSENFELD, Department of Pediatrics, Oregon Health Sciences University, Portland, Oregon, USA.

J. ERIC RUSSELL, Departments of Medicine and Genetics, University of Pennsylvania School of Medicine, Philadelphia, Pennsylvania, USA.

ALAN SALZMAN, Department of Medicine, University of Pennsylvania School of Medicine, Philadelphia, Pennsylvania, USA.

SUSAN SAMARAS, Division of Endocrinology, Diabetes, and Metabolism, Pennsylvania State University College of Medicine, Hershey, Pennsylvania, USA.

MELISSA SHAW, Department of Medicine, Center for Endocrinology, Metabolism and Nutrition, Northwestern University Medical School, Chicago, Illinois, USA.

CORINNE M. SILVA, Division of Endocrinology and Metabolism, Department of Medicine, University of Virginia Health Sciences Center, Charlottesville, Virginia, USA.

PETER SÖNKSEN, Department of Medicine, United Medical and Dental School of Guy's and St. Thomas' Hospital, London, UK.

MICHAEL O. THORNER, Division of Endocrinology and Metabolism, Department of Medicine, University of Virginia Health Sciences Center, Charlottesville, Virginia, USA.

R.J. URBAN, Department of Internal Medicine, University of Texas Medical Branch, Galveston, Texas, USA.

MARGRIT URBANEK, Department of Genetics, University of Pennsylvania School of Medicine, Philadelphia, Pennsylvania, USA.

MARY LEE VANCE, Division of Endocrinology and Metabolism, Department of Medicine, University of Virginia Health Sciences Center, Charlottesville, Virginia, USA.

JOHANNES D. VELDHUIS, Division of Endocrinology and Metabolism, Department of Internal Medicine, University of Virginia Health Sciences Center, Charlottesville, Virginia, USA.

KATHERINE VERIKIOU, Department of Medicine, United Medical and Dental School of Guy's and St. Thomas' Hospital, London, UK.

ANDREW J. WEISSBERGER, Department of Medicine, United Medical and Dental School of Guy's and St. Thomas' Hospital, London, UK.

JAMES A. WELLS, Department of Protein Engineering, Genentech, Inc., South San Francisco, California, USA.

ARTHUR WELTMAN, Exercise Physiology Laboratories, Department of Human Services, Curry School of Education, University of Virginia, Charlottesville, Virginia, USA.

DAVINIA M. WHITE, Department of Obstetrics and Gynaecology, St. Mary's Hospital Medical School, Imperial College of Science, Technology and Medicine, University of London, London, UK.

DEBBIE WILLIS, Department of Obstetrics and Gynaecology, St. Mary's Hospital Medical School, Imperial College of Science, Technology and Medicine, University of London, London, UK.

SAMUEL S.C. YEN, Department of Reproductive Medicine, University of California at San Diego, La Jolla, California, USA.

Part I

Growth Hormone
and Its Receptor:
State of the Art

1

Growth Hormone:
A Current Perspective

M.O. THORNER, M.L. HARTMAN, C.M. SILVA, B.D. GAYLINN,
J.A. ALOI, S.E. KIRK, S.S. PEZZOLI, AND M.L. VANCE

It is well recognized that *growth hormone* (GH) acts at multiple peripheral tissues, including, among other tissues, the liver, adipocytes, muscle, and the growth plate. There is good evidence in animals that GH deficiency is associated with combined immune deficiency, which suggests a potential physiological role for GH in the regulation of immune function. However, in humans there is no definitive evidence to indicate such a role for GH in immune modulation. Perhaps this will be the focus of a future meeting. In this volume, we focus our attention on the somatotrophic axis and the reproductive process. This is a new area, and it is our hope that the chapters herein will clarify those issues that have been resolved and focus the attention of researchers on those areas where there are gaps in our knowledge.

GH Secretion

Growth hormone is secreted from the anterior pituitary under the dual control of the hypothalamic peptides *somatostatin* (SRIH) and *growth hormone releasing hormone* (GHRH). GH is secreted in a pulsatile fashion; prominent changes in the pattern of GH secretion occur at different stages of the life cycle (reviewed in 1). One unresolved question that is central to future strategies for GH replacement therapy is whether the pulsatile pattern of GH is important for some or all of the actions of GH in humans.

GH is detectable at the end of the first trimester and reaches a peak of 100–150 µg/L at about 20 weeks of gestation. GH levels decline to about 30 µg/L in cord serum and continue to decline for about 3 months. GH secretion increases to maximal levels during puberty and progressively declines with increasing age. Beyond age 60 GH secretion is similar to that observed in GH-deficient children. This diminished GH secretion associated with aging may reflect an alteration in the release of GHRH and/or SRIH, enhanced sensitivity to *insulin-like growth factor I* (IGF-I) feedback, or decreased somatotroph mass. This hyposomatotropism of aging may contribute to the frailty of aging, which includes decreased muscle and bone mass and increased adiposity. Gonadal steroids have a profound effect on GH secretion, such that premenopausal women have greater GH secretion than age-matched men.

During the day, GH secretion is suppressed by food ingestion and stimulated by exercise. At night, GH secretion is maximal during slow wave sleep. Nutrient deprivation and type I diabetes mellitus are associated with increased GH concentrations, while GH secretion is suppressed in obese subjects. GH-deficient subjects have higher percent body fat and lower *lean body mass* (LBM) than normal subjects. These changes in body composition are reversed by GH administration (reviewed in 2).

GH Receptor

We have developed a homologous assay to study human *GH receptor* (GH-R) signal transduction using the IM-9 human B cell lymphocyte cell line that has receptors for GH, insulin, IGF-I, IGF-II, and glucocorticoids. The GH-R is a member of the cytokine family of receptors, and activation of the GH receptor leads to tyrosine phosphorylation of at least three proteins in the human IM-9 lymphocyte cell line: a 134-kd protein that represents the GH-R itself (3) and a 120-kd protein and 93-kd protein whose natures have yet to be elucidated. Recently, in a 3T3 mouse fibroblast cell line, the 120-kd protein has been shown to be the tyrosine kinase JAK2 (4). The IM-9 cell line has been useful in demonstrating that receptor dimerization is necessary for GH-R signal transduction and that mutant GH molecules with defective site 2 not only are incapable of activating the receptor, but also act as receptor antagonists (Fig. 1.1). In this volume we will read about molecular modeling and peptide engineering studies and the expectations for major advances that will come from these approaches as they relate to the GH-R.

GH Secretagogues

The possibilities for future therapeutic manipulation of the somatotrophic axis are encouraging. There are two major classes of compounds that could serve as GH secretagogues with therapeutic potential. The first is GHRH and its analogs; the second is *GH releasing peptide* (GHRP) and its analogs, including a recently developed nonpeptidal GHRP mimetic.

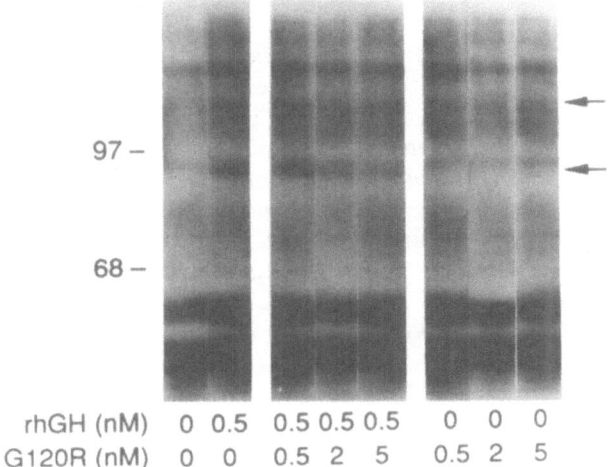

rhGH (nM)   0   0.5   0.5 0.5 0.5   0   0   0
G120R (nM) 0   0   0.5   2   5   0.5   2   5

FIGURE 1.1. Antagonist activity of GH mutant G120R. IM-9 cells were treated for 15 min at 37°C either under control conditions (0) or with rhGH alone at 0.5 nM (left panel); with increasing concentrations (0.5, 2, and 5 nM) of mutant G120R alone (right panel); or with a combination of 0.5 nM of rhGH plus increasing concentrations (0.5, 2, and 5 nM) of G120R mutant (middle panel). Mutant G120R has a normal site 1, but a mutation that inhibits binding to site 2, and therefore does not result in dimerization of the receptor. Lysates were analyzed by denaturing electrophoresis, transferred to nitrocellulose, and probed with an antibody that detects proteins that are tyrosine phosphorylated. Antibody binding was detected using ^{125}I-protein A. Reprinted with permission from Silva, Weber, and Thorner (33), © The Endocrine Society, 1993.

GHRH

GHRH, or agents that act through *GHRH receptors* (GHRH-Rs), holds promise as a therapeutic agent. Ectopic GHRH secretion unequivocally causes acromegaly, thus demonstrating that a GHRH agonist can increase GH secretion to any desired level. In addition, the administration of GHRH in GH-deficient children as a therapeutic agent results in both stimulation of GH secretion and acceleration of growth velocity.

In 1982, following isolation, characterization, and synthesis of human GHRH(1-40)-OH by Dr. Vale's group (5), clinical studies in normal human volunteers were performed at the University of Virginia to determine biological effects and peptide specificity. Six healthy young men were given *intravenous* (IV) human GHRH(1-40)-OH (1 μg/kg) or vehicle (control), and the responses of multiple hormones were determined (6). On the control day, 2 subjects had small spontaneous pulses of GH secretion. Following GHRH administration, all subjects had increased serum GH levels; however, there was considerable variability in the degree of responsiveness. We theorize that this is a result of variable hypothalamic SRIH secretion between subjects. Thus, high hypothalamic SRIH secretion yields low GH response to GHRH; whereas, when SRIH secretion is low, the GH response to GHRH will be greater. There was no change in serum *prolactin* (PRL), cortisol (a reflection of adrenocorticotropin secretion), *luteinizing hormone* (LH), and thyrotropin. Additionally, there was no effect on blood glucose, insulin, pancreatic glucagon, pancreatic polypeptide, cholecystokinin, gastric inhibitory peptide, motilin, or SRIH. Thus, GHRH specifically stimulated GH secretion in normal men.

A dose-response study was performed to determine the range of GHRH doses that would stimulate GH release (7, 8). Normal men were given either vehicle or GHRH at doses ranging from 0.003 to 10 µg/kg as a single IV bolus injection. Doses of 0.3–10 µg/kg resulted in significant stimulation of GH secretion; again, there were variable responses among individuals. Serum IGF-I levels increased 24 h after GHRH administration in 11 of 13 subjects and demonstrated that GH secreted in response to GHRH had a biological effect.

To determine whether GHRH could stimulate linear growth, we studied 24 GH-deficient children for 6 months or longer. GHRH (1–4 µg/kg per dose) was administered *subcutaneously* (sc) every 3 h, or every 3 h overnight only, or by *twice-daily* (b.i.d.) injections. Twenty-one children had an increase in growth rate during GHRH treatment (Fig. 1.2), with a significant correlation between average daily dose and growth velocity ($r = 0.57$, $P = 0.004$) (9). Indeed, many such studies have been performed that demonstrate that GHRH can stimulate growth in some children.

The potential feasibility of administering a sustained-release GHRH preparation or a long-acting GHRH analog was determined by administration of a continuous IV infusion of GHRH (10 ng/kg/min) for 14 days to 5 normal men. As previously observed during 24-h and 6-h infusions, continuous GHRH infusion resulted in an augmentation of pulsatile GH release that was sustained for 14 days. Serum IGF-I concentrations increased in all subjects during the 14-day infusion and declined to pretreatment levels by 2 weeks after discontinuation of the infusion. As in patients with ectopic GHRH secretion, these studies confirmed that constant GHRH exposure results in an enhanced GH secretion that is sustained as long as GHRH levels are elevated.

GHRH-R

The recent cloning of the GHRH-R, together with the clinical data described above, may lead to the development of new GHRH analogs. A representation of the deduced sequence of the human GHRH-R is shown in Figure 1.3. It is the fifth member of a new family of receptors that includes those for secretin, vasoactive intestinal peptide, parathyroid

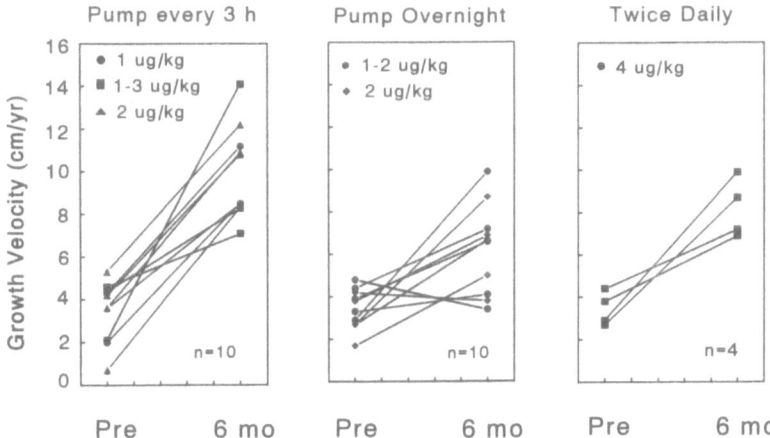

FIGURE 1.2. Effects of 3 different regimens of GHRH therapy on height velocity. The regimens are a pump every 3 h at doses of 1, 2, or 3 µg/kg (left panel); a pump overnight at doses of 1–2 µg/kg or 2 µg/kg (middle panel); and twice-daily injections of 4 µg/kg (right panel). Redrawn with permission from Thorner, Rogol, Blizzard, et al. (9).

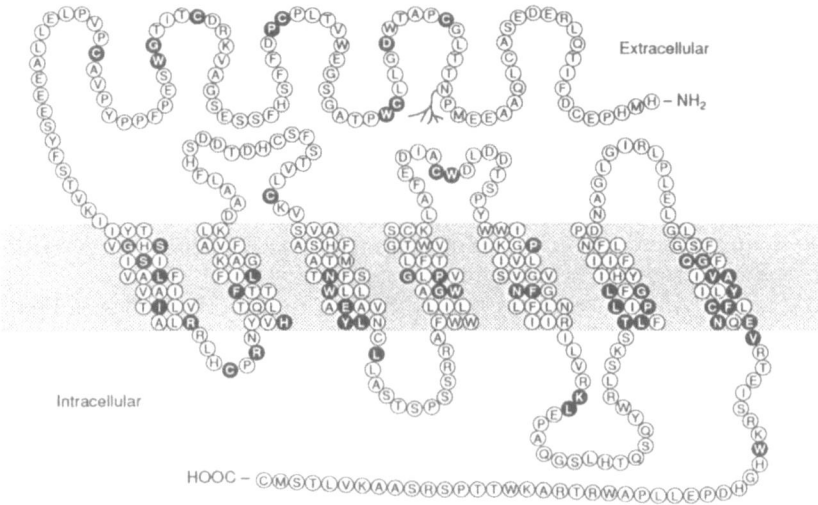

FIGURE 1.3. Cartoon of the human GHRH-R demonstrating predicted transmembrane structure and location of glycosylation site. Shaded amino acids are conserved in all 8 known members of this receptor family (GHRH, VIP, secretin, PACAP, glucagon, GLP-1, PTH, and calcitonin). Data from Gaylinn, Harrison, Zysk, Lyons, Lynch, and Thorner (11). Modified with permission from Thorner (34).

hormone, calcitonin, glucagon, glucagon-like peptide, pituitary adenylate cyclase activating peptide, and, most recently, *corticotropin releasing hormone* (CRH) (10–12). Recently, the molecular basis of a mutant GH-deficient mouse has been identified as a point mutation that altered Asp60 to Gly in the GHRH-R in the little mouse (13, 14). This mutation abolishes the cAMP response to GHRH in cell lines transfected with the mutant receptor, while cell lines transfected with the intact receptor have the typical cAMP response. However, binding studies with this mutant GHRH-R have not been reported.

In conclusion, GHRH potentially can be used to restore or enhance spontaneous GH secretion in children and adults. The primary obstacle is that GHRH (and its analogs) needs to be administered parenterally. In addition, sc administration results in lower bioavailability when compared with IV administration in humans. The development of analogs that have enhanced activity (super analogs), but are not readily metabolized to biologically inactive molecules, continues to be a challenge in this area. Because shortened peptides with simpler structures are less expensive to synthesize, the use of such GHRH analogs will almost certainly become an attractive alternative to the use of native peptide. The cloning of the GHRH-R and the definition of structure-activity relationships, as well as the development of stable cell lines that overexpress the receptor, now make it feasible to design and screen stable molecules (nonpeptides) that not only stimulate the GHRH-R, but are also orally active.

GHRP

The novel synthetic hexapeptide GHRP (His-D-Trp-Ala-Trp-D-Phe-Lys-NH_2) acts through unidentified pathways to stimulate GH secretion in animals and humans. GHRP acts synergistically with GHRH and is orally active in humans (15, 16). However, the bioavailability of GHRP is dramatically reduced when administered orally.

Current evidence supports the hypothesis that GHRP releases GH via nonopiate, non-GHRH, non-SRIH pituitary receptors. GHRP may also act on the hypothalamus since specific binding sites for GHRP are present in both the pituitary gland and the hypothalamus (17). The response to GHRP in rats is abolished by anti-GHRH serum (18, 19), but is enhanced by anti-SRIH serum (18). Similarly, GH responses to GHRP in monkeys are enhanced by propranolol, which inhibits SRIH release (20). These data suggest that GHRP is dependent on endogenous GHRH for its action and does not act by decreasing SRIH secretion. However, continuous infusions of GHRP abolish the cyclical refractoriness in GHRH in rats, strongly suggesting that GHRP antagonizes SRIH action (19).

Recent work has demonstrated that GHRP depolarizes and SRIH hyperpolarizes somatotroph cell membranes (21–23). Using the reverse hemolytic plaque assay, our laboratory has shown that GHRP increases and SRIH decreases the number of somatotrophs secreting GH without affecting the amount of GH secreted per cell (24). This is in contrast to the known effect of GHRH to increase both the number of cells secreting and the amount of GH secreted per cell (25). These data support the hypothesis that GHRP acts via a distinct pituitary receptor to functionally antagonize the action of SRIH. However, GHRP may stimulate GH release in the absence of SRIH since it depolarizes the somatotroph cell membrane. Finally, a recent report provides evidence that GHRP may stimulate the release of GHRH (26).

Although several studies of the short-term, potent, GH-releasing effects of GHRP had been performed in humans, there were few data to show that it could stimulate GH secretion over a prolonged period of time. To determine the effects of a 24-h GHRP infusion, 8 normal young men received 24-h infusions of saline or GHRP (1 µg/kg/h) on 2 occasions. The 24-h infusions were preceded by a 2-h saline infusion as a baseline period and were followed by an IV bolus of GHRP or GHRH (1 µg/kg), concluding with a 2.5-h saline infusion. Serum GH was measured every 10 min throughout the 28.5-h period.

GH secretion was enhanced and remained pulsatile throughout the 24-h GHRP infusions, and GH secretion rates (calculated by deconvolution analysis) were increased 8-fold compared to saline (Fig. 1.4). The number of GH pulses, pulse duration and height, incremental pulse amplitude, interpeak valley concentration, and individual pulse areas (calculated by cluster analysis) were significantly greater during GHRP infusion than during saline infusion. Attributes of pulsatile GH release on the 2 GHRP

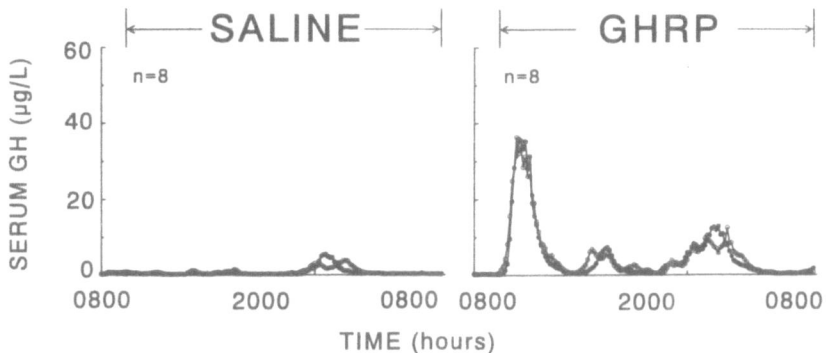

FIGURE 1.4. Mean serum GH concentrations in 8 subjects during 2-h baseline period (0800–1000 h, study day 1) infusions of saline and GHRP (0800–0930, study days 1–2). Each subject received 2 saline and 2 GHRP infusions. Note the highly reproducible pattern of GH release during the 2 GHRP infusions. Each subject responded with an initial large GH pulse. GH release for the balance of the infusion period remained pulsatile. Reprinted with permission from Huhn, Hartman, Pezzoli, and Thorner (27), © The Endocrine Society, 1993.

infusion days were significantly correlated, indicating that enhancement of GH secretion by GHRP is highly reproducible. Mean IGF-I concentrations increased 12% and 22%, respectively, on GHRP infusion days, whereas IGF-I levels declined 18% and 20%, respectively, during saline infusions. GHRP infusion significantly attenuated the GH response to a subsequent GHRP bolus injection. In contrast, peak GH concentrations in response to GHRH were significantly increased after GHRP infusion compared with saline infusion. Twenty-four-hour GHRP infusions augment pulsatile GH release and increase plasma IGF-I concentrations without significant adverse effects. Attenuation of the GH response to a subsequent GHRP bolus is not caused by depletion of pituitary GH since the response to a GHRH bolus was enhanced by prior infusion of GHRP (27).

To date, there have been no long-term studies of GHRP reported in humans; however, animal studies indicate that GHRP enhances GH secretion and promotes body weight gain over a 25-day period (28). There are two recent advances that make GHRP even more interesting. First, several analogs of GHRP (GHRP-1 and GHRP-2) have been developed that have increased potency, although both these peptides have the same disadvantages as GHRP-6. The second is the development of a nonpeptide GHRP mimetic, *L-692,429* ([L]), at Merck Research Laboratories. It is a substituted benzolactam that has in vitro and in vivo activity. GHRP and [L] appear to have a similar specificity of action in normal young men (23, 29).

We have recently studied this compound to determine whether it can enhance GH secretion in 2 distinct populations: the elderly and obese young men. In the initial study, 8 men and 8 women (ages 65–79) received the following on 4 separate occasions: an IV bolus injection of saline, a supramaximal IV dose of GHRH (1 µg/kg), and 15-min infusions of [L] at doses of 0.2 and 0.75 mg/kg. Peak and integrated GH concentrations increased significantly in a dose-dependent manner following [L]. Responses to [L] at either dose were significantly greater than the responses to GHRH. As with GHRP, serum cortisol and PRL concentrations increased following both doses of [L], but to values within the respective normal ranges. A surprising finding was the significant correlation between the integrated GH concentrations after GHRH (1 µg/kg) and the 0.75 mg/kg dose of [L] ($R^2 = 0.61$, $P < 0.0004$) (30).

Since GH secretion is suppressed in obese individuals, we hypothesized that the GH stimulatory effect of [L] would also be inhibited in obese but otherwise healthy male volunteers. Twelve young men (ages 20–30) with *body mass indices* (BMIs) of 30–39 kg/m^2 were studied. Following a 1-h baseline saline infusion, saline, or 0.2-mg/kg [L], or 0.75-mg/kg [L] was infused over 15 min at 0900 h in a 3-period, randomized, crossover design. The mean peak GH-R after saline, 0.2-, and 0.75-mg/kg [L] were 0.5 ± 0.1, 10 ± 2.8, and 16 ± 3.8 µg/L, respectively (31). In a second arm of the study, the effect of 0.2-mg/kg [L] was examined in 8 normal-weight young men (BMI < 28kg/m^2, ages 18–30). Peak GH responses in the fed (30 min after an 817-kilocalorie mixed meal) and fasted states were 18 ± 3.9 and 44 ± 3.9 µg/L, respectively; $P < 0.01$ (31). In an earlier study by Williams et al., peak GH responses in normal-weight and obese subjects to insulin-induced hypoglycemia were 25 ± 3.5 versus 7.5 ± 3.1 µg/L, respectively. The responses to GHRH (1 µg/kg) were 42 ± 7 versus 4 ± 1 µg/L, respectively (32).

In conclusion, the GH response to [L] is reduced by obesity, as it is to all known GH stimuli in the obese. This impaired GH response is similar to that observed in fed normal-weight individuals. In addition, [L] appears to be a more effective GH secretagogue than is GHRH in both elderly subjects and obese young men.

In summary, both GHRH and GHRP stimulate GH secretion. Long-term administration of GHRH enhances GH secretion and can accelerate growth velocity in GH-deficient children. Acutely, GHRP stimulates GH secretion more effectively than GHRH. Continuous administration of either GHRH or GHRP augments pulsatile GH secretion. A nonpeptidal GHRP agonist has been developed that stimulates GH release effectively in obese and older subjects.

Conclusions

We currently stand at a threshold of GH research. Not only have new GH functions been identified—for example, in reproduction and immunology—but molecular modeling and the understanding of GH-R transduction have revealed new possibilities for the development of GH agonists and antagonists for clinical use.

Endogenous GH secretion can be augmented by GH-releasing compounds. The GH secretagogues are clearly functional, and it is only a matter of time before an orally active, stable molecule is developed. Scientific, technological, and pharmacological barriers are rapidly being overcome to allow this development to take place. Such orally active GH secretogogues will offer new options in the management of short stature, adult GH deficiency syndrome, and reproductive disorders and, possibly, in the reversal of aging-associated changes in body composition.

References

1. Hartman ML, Veldhuis JD, Thorner MO. Normal control of growth hormone secretion. Horm Res 1993;40:37–47.
2. Cuneo RC, Salomon F, McGauley GA, Sonksen PH. The growth hormone deficiency syndrome in adults. Clin Endocrinol (Oxf) 1992;37:387–97.
3. Silva CM, Day RN, Weber MJ, Thorner MO. Human growth hormone receptor is characterized as the 134 kDa tyrosine phosphorylated protein

activated by growth hormone treatment in IM-9 cells. Endocrinology 1993; 133:2307–12.

4. Argetsinger LS, Campbell GS, Yang X, et al. Identification of JAK2 as a growth hormone receptor-associated tyrosine kinase. Cell 1993;74:237–44.

5. Rivier J, Spiess J, Thorner M, Vale W. Characterization of a growth hormone-releasing factor from a human pancreatic islet tumour. Nature 1982; 300:276–8.

6. Thorner MO, Rivier J, Spiess J, et al. Human pancreatic growth-hormone-releasing factor selectively stimulates growth-hormone secretion in man. Lancet 1983;1:24–8.

7. Vance ML, Borges JL, Kaiser DL, et al. Human pancreatic tumor growth hormone-releasing factor: dose-response relationships in normal man. J Clin Endocrinol Metab 1984;58:838–44.

8. Evans WS, Vance ML, Kaiser DL, et al. Effects of intravenous, subcutaneous, and intranasal administration of growth hormone (GH)-releasing hormone-40 on serum GH concentrations in normal men. J Clin Endocrinol Metab 1985;61:846–50.

9. Thorner MO, Rogol AD, Blizzard RM, et al. Acceleration of growth rate in growth hormone-deficient children treated with human growth hormone-releasing hormone. Pediatr Res 1988;24:145–51.

10. Mayo KE. Molecular cloning and expression of a pituitary-specific receptor for growth hormone-releasing hormone. Mol Endocrinol 1992;6:1734–44.

11. Gaylinn BD, Harrison JK, Zysk JR, Lyons CE, Lynch KR, Thorner MO. Molecular cloning and expression of a human anterior pituitary receptor for growth hormone-releasing hormone. Mol Endocrinol 1993;7:77–84.

12. Chen R, Lewis KA, Perrin MH, Vale WW. Expression cloning of a human corticotropin-releasing-factor receptor. Proc Natl Acad Sci USA 1993;90: 8967–71.

13. Lin SC, Lin CR, Gukovsky I, Lusis AJ, Sawchenko PE, Rosenfeld MG. Molecular basis of the little mouse phenotype and implications for cell type-specific growth. Nature 1993;364:208–13.

14. Godfrey P, Rahal JO, Beamer WG, Copeland NG, Jenkins NA, Mayo KE. GHRH receptor of little mice contains a missense mutation in the extracellular domain that disrupts receptor function. Nature Genet 1993;4:227–32.

15. Bowers CY, Reynolds GA, Durham D, Barrera CM, Pezzoli SS, Thorner MO. Growth hormone (GH)-releasing peptide stimulates GH release in normal men and acts synergistically with GH-releasing hormone. J Clin Endocrinol Metab 1990;70:975–82.

16. Hartman ML, Farello G, Pezzoli SS, Thorner MO. Oral administration of growth hormone (GH)-releasing peptide (GHRP) stimulates GH secretion in normal men. J Clin Endocrinol Metab 1992;74:1378–84.

17. Codd EE, Shu AYL, Walker RF. Binding of a growth hormone releasing hexapeptide to specific hypothalamic and pituitary binding sites. Neuropharmacology 1989;28:1139–44.

18. Bowers CY, Sartor AO, Reynolds GA, Badger TM. On the actions of the growth hormone-releasing hexapeptide, GHRP. Endocrinology 1991;128: 2027–35.

19. Clark RG, Carlsson LMS, Trojnar J, Robinson ICAF. The effects of a growth hormone-releasing peptide and growth hormone-releasing factor in conscious and anaesthetized rats. J Neuroendocrinol 1989;1:249–55.

20. Malozowski S, Hao EH, Ren SG, et al. Growth hormone (GH) responses to the hexapeptide GH-releasing peptide and GH-releasing hormone (GHRH) in the cynomolgus macaque: evidence for non-GHRH-mediated responses. J Clin Endocrinol Metab 1991;73:314–7.

21. Pong SS, Chaung LYP, Smith RG. GHRP-6 (His-D-Trp-Ala-Trp-D-Phe-Lys-NH2) stimulates growth hormone secretion by depolarization in rat pituitary cell cultures [Abstract]. Prog 73rd meet Endocr Soc, 1991:88.

22. Koch BD, Blalock JB, Schonbrunn A. Characterization of the cyclic AMP-independent actions of somatostatin in GH cells, I. An increase in potassium conductance is responsible for both the hyperpolarization and the decrease in intracellular free calcium produced by somatostatin. J Biol Chem 1988;263: 216–25.

23. Smith RG, Cheng K, Schoen WR, et al. A nonpeptidyl growth hormone secretagogue. Science 1993;260:1640–3.

24. Goth MI, Lyons CE, Canny BJ, Thorner MO. Pituitary adenylate cyclase activating polypeptide, GH-releasing peptide and GH-releasing hormone stimulate GH release through distinct pituitary receptors. Endocrinology 1992;130:939–44.

25. Ho KY, Leong DA, Sinha YN, Johnson ML, Evans WS, Thorner MO. Sex-related differences in GH secretion in rat using reverse hemolytic plaque assay. Am J Physiol 1986;250:E650–4.

26. Dickson SL, Leng G, Robinson ICAF. Systemic administration of growth hormone-releasing peptide activates hypothalamic arcuate neurons. Neurosci Lett 1993;53:303.

27. Huhn WC, Hartman ML, Pezzoli SS, Thorner MO. 24-hour growth hormone (GH)-releasing peptide (GHRP) infusion enhances pulsatile GH secretion and specifically attenuates the response to a subsequent GHRP bolus. J Clin Endocrinol Metab 1993;76:1202–8.

28. Bowers CY, Momany FA, Reynolds GA, Hong A. On the in vitro and in vivo activity of a new synthetic hexapeptide that acts on the pituitary to specifically release growth hormone. Endocrinology 1984;114:1537–45.

29. Gertz BJ, Barrett JS, Eisenhandler R, et al. Growth hormone response in man to L-692,429, a novel nonpeptide mimic of growth hormone releasing peptide (GHRP-6). J Clin Endocrinol Metab 1993;77:1393–7.

30. Aloi JA, Huhn WC, Gertz BJ, et al. GH response to L-692,429, a substituted benzolactam which mimics GHRP, is greater than the response to GHRH in older persons. J Clin Endocrinol Metab 1994.

31. Kirk SE, Aloi JA, Gertz BJ, et al. L-692,429, a substituted benzolactam with GHRP-like activity, stimulates GH secretion in obese young men [Abstract]. Prog 75th meet Endocr Soc, 1993.

32. Williams T, Berelowitz M, Joffe SN, et al. Impaired growth hormone responses to growth hormone-releasing factor in obesity: a pituitary defect reversed with weight reduction. N Engl J Med 1984;311:1403–7.

33. Silva CM, Weber MJ, Thorner MO. Stimulation of tyrosine phosphorylation in human cells by activation of the growth hormone receptor. Endocrinology 1993;132:101–8.

34. Thorner MO. On the discovery of growth hormone-releasing hormone. Acta Paediatr Scand Suppl 1993;388:2–7.

2

Probing and Designing Growth Hormone-Receptor Interactions

JAMES A. WELLS AND ANTHONY A. KOSSIAKOFF

The mechanism through which extracellular signals from hormones trigger intracellular responses is central to understanding how cells and tissues interact. The signaling process is initiated by the binding of polypeptide hormones to cell-surface receptors; however, the molecular basis for binding and receptor activation is not well understood. Through a series of mutational and structural studies, the molecular basis for ligand binding and activation of the growth hormone receptor has been revealed in perhaps greater detail than for any other hormone-receptor system (reviewed in 1–3).

Human growth hormone (hGH) is a 22-kd pituitary hormone that stimulates and regulates the growth of bone, muscle, and cartilage by binding to specific cell-surface receptors (reviewed in 4). Two specific receptors that bind hGH have been identified, activation of which results in distinct pharmacological effects: the *GH receptor* (GH-R) and the *prolactin receptor* (PRL-R). Based on sequence homology of their extra-cellular domains, the GH-R and PRL-R have been categorized as belonging to the hematopoietic receptor superfamily, which includes the endocrine receptors and those for a number of other cytokines (5–7). These receptors have a 3-domain organization: the extracellular portion that binds the activating hormone ligand, a transmembrane segment, and a cytoplasmic domain that is involved in producing the response—the so-called second message—within the cell. The intracellular domain, which generates the cellular signal, is not homologous to any known tyrosine kinase. However, there is recent evidence that it can associate non-covalently with a tyrosine kinase known as *JAK2* (8). Presumably, the binding of hGH to the extracellular domain transmits a change in structure to the intracellular domain that causes, among other things, the activation of JAK2.

Several mechanisms for the activation process of this class of receptors have been proposed (9). The mechanism with most experimental support was based on some type of receptor aggregation. It assumed formation of complexes of hormone-receptor pairs; that is, aggregates of some multiple of hormone-receptor dimers. However, the biochemical data assigning stoichiometry were equivocal, and there was no structural information about the nature of the aggregation.

The focus of our work at Genentech has been to understand at a molecular level how hGH binds and activates hGH-R and hPRL-R. This information has been applied to design GH analogs to better probe its pharmacology and to produce GH antagonists and potentially better GH agonists.

Structure and Function of GH/Receptor Complexes In Vitro

A series of biophysical studies (10, 11) on the complex between hGH and the extracellular domain of its receptor—referred to here as *hGHbp*—established that when hGH binds the hGHbp, it forms a remarkable dimeric complex: 1 hGH molecule is complexed to 2 hGHbps. When crystals of the complex were dissociated and their composition analyzed by HPLC, 1 equivalent of hGH was found along with 2 equivalents of the hGHbp. Gel filtration and titration calorimetry experiments confirmed that hGH binds 2 molecules of the hGHbp in solution as well (10). Homolog-scanning and alanine-scanning mutational analysis (10, 12, 13) elaborated the functional determinants for 2 separate sites on hGH (called site 1 and site 2), as well as a binding site on the hGHbp (14). The mutational studies further showed that the receptor binds these sites on hGH sequentially, first at site 1 and then at site 2, to form the hGH (hGHbp)$_2$ complex. The structural rationale for this binding sequence is discussed below.

A 2.8-Å X-ray structure analysis was performed that revealed the atomic details of the global structure and molecular interactions of the hGH(hGHbp)$_2$ complex (15). As shown in Figure 2.1, hGH is a 4-helix bundle protein, which has been shown to be a common folding motif for hormones in the cytokine family. The hGHbp consists of 2 distinct domains of approximately equal size, each of which contains about 120 residues and has a characteristic immunoglobulin-like fold. The domains are linked together by 4 residues, and the interface surface to the hormone is shared between the 2 domains. Figure 2.1 shows that in the complex, the C-termini of the receptors come into direct contact. Thus, it is believed that once the hGH induces dimerization of the extracellular domains, the transmembrane and intracellular segments of the receptor are brought

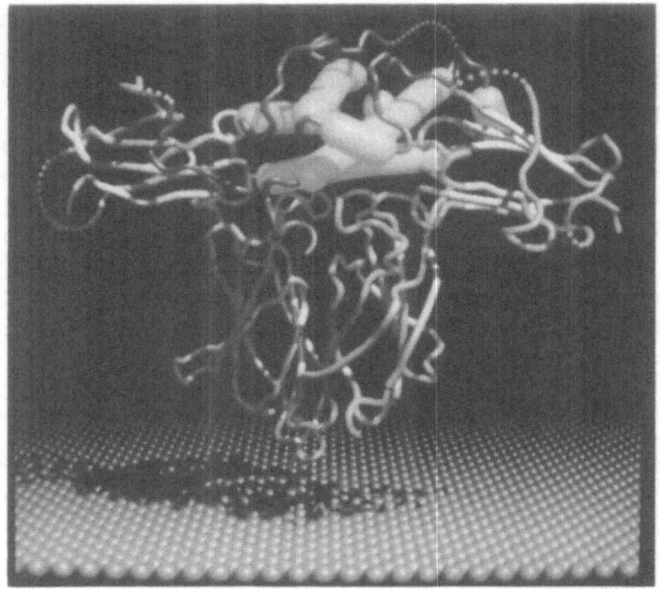

FIGURE 2.1. Structure of the hGH(hGHbp)$_2$ complex. The hGH molecule is represented in white, and the 2 binding proteins are represented in different shadings. The hGH structure is a 4-helix bundle; the α-helices are represented by the thicker tubes. The 2-domain organization of the binding protein is apparent. Each domain is principally made up of an extensive 7-stranded β-sheet and contains about 120 amino acids. The 2 domains are linked together by a 4-residue segment. The dotted lines represent regions of the polypeptide chain that are disordered and not visible in the electron density maps. Reproduced, with permission, from the Annual Review of Biophysics and Biomolecular Structure, Volume 22, © 1993, by Annual Reviews, Inc.

together so that they may associate and somehow activate JAK2 and/or other intracellular signaling components.

The high-resolution structure showed that the hormone binds to essentially the same sites on the two receptors. It is the difference in conformation of several surface loops of the receptor(s) that allows binding to the 2 independent sites on the hormone. Although the hormone-binding epitope of the two receptors encompasses essentially the same residues, the surface areas defining the 2 interfaces are substantially different. For receptor 1, which binds to hGH at site 1 (Fig. 2.1), the buried surface area is approximately 1250 Å^2, whereas for receptor 2 the area is 850 Å^2. The topography of the hGH molecule at each site is also quite different. At binding site 1, hGH has a large cavity into which the contacting side chains of the hGHbp insert; at binding site 2, the surfaces of both the hGH and receptor are relatively flat and appear not to be as complementary.

The differences in the surface interactions are presumed to play an active role in the sequential binding of receptors to the hormone. The binding sequence results because the association energy that is available for site 2 is not, in itself, sufficient to support binding. However, in the case of site 1 binding, the larger surface area and topography present for the hGH-hGHbp1 association are sufficient to allow the formation of a tight 1:1 complex. Binding of the second receptor is synergized by the initial 1:1 complex. The X-ray structure shows that there is a significant interaction surface (500 Å²) between the two bound receptors in the complex between their C-terminal domains. What appears to happen is that once the 1:1 complex is formed, the second receptor can use the binding energy derived from its interaction with the hormone at site 2, as well as its interaction with the first bound receptor.

Given that the receptors do have surfaces through which they interact, is it possible that they can dimerize to produce a biological signal in the absence of hormone? The answer is no, and nature has designed into the receptors a form of inhibition. The structure of the 1:1 complex was also solved in this lab, and there was only one significant difference between this structure and the hormone and the equivalent part of the 2:1 complex (De Vos, unpublished results). That difference was contained within the exact set of residues that formed the interface between the two receptors. In the 1:1 complex this region was highly disordered. It is generally recognized that there are energetic difficulties with trying to form a stable association between 2 disordered segments of polypeptide chain. Apparently, once the interfacing segments of hGHbp1 and hGHbp2 are in close proximity, the side-chain complementary is strong enough that the residues order in productive binding conformations.

How does hGH bind to the PRL-R? Are the same surface regions used for the interaction? Homolog- and alanine-scanning mutagenesis indicated that although the functional epitope defining the hGH-hPRLbp interface was not identical to the hGH-hGHbp case, it was very close, and the same mode of binding was confirmed (16). Figure 2.2 shows the sequence alignment of the hGH and hPRL extracellular domains; the residues involved in the points of contact between the hormone and the receptors are marked. Five of the 6 sites of contact have residues with reasonable sequence homology; the tryptophanes at 104 and 169 play an especially important role in the binding of both receptors. The 6th site, 217–221, shows considerable variation between the two receptors. In fact, there is a residue deletion in this segment. It had been determined previously that the binding of hGH to hPRL-R involved zinc binding (17), and the structure clearly shows that it is at this 6th site where the hPRL sequence provides a site for the zinc coordination (Somers, De Vos, Kossiakoff, unpublished results).

```
humghr    1 F S G S E A T A A I L S R A P W S L Q S V N P G L K T N S S K E P K F T K C R S P E R E T F S C H W
humprlr   1 - - - - - - - - - - - - - - - - - - - - - - - - - - Q L P P G K P E I F K C R S P N K E T F T C W W

humghr   51 T D E V H H G T K N L G P I Q L F Y T R R N T Q E W T Q E W K E C P D Y V S A G E N S C Y F N S S F
humprlr  25 R P G T D G G L P - - T N Y S L T Y - H R E G E T L M H - - - E C P D Y I T G G P N S C H F G K Q Y

humghr  101 T S I W I P Y C I K L T S - - - - N G G T V D E K C F S V D E I V Q P D P P I A L N W T L L N V S L
humprlr  69 T S M W R T Y I M M V N A T N Q M G S S F S D E L Y V D V T Y I V Q P D P P L E L A V E V K Q P - -

humghr  147 T G I H A D I Q V R W E A P R N A D I Q K G W M V L E Y E L Q Y K E V N E T K W K M M D P I L T T S
humprlr 117 E D R K P Y L W I K W S P P T L I D L K T G W F T L L Y E I R L K P E K A A E W E I H F A G Q Q T E

humghr  197 V P V Y S L K V D K E Y E V R V R S K Q R N S G N Y G E F S E V L Y V T L P Q M S Q F T C E E D F Y
humprlr 167 F K I L S L H P G Q K Y L V Q V R C K - P D H G Y W S A W S P A T F I Q I P - - S D F T M N D T
```

FIGURE 2.2. Sequence alignment of the extracellular domains of hGH-R (humghr) and hPRL-R (humprlr). Sequence identities are noted in boxes. Residues contacting the hGH molecule are underlined.

Comparison of the X-ray structure and the mutational analysis revealed a surprising finding. The hGH epitope defined by the contact surface established by the X-ray structure contained the mutagenic epitope, but it extended over a substantially larger surface area (18). This indicated that a significant percentage of the contact residues provides no binding energy to the formation of the complex. In an attempt to gain insight into the source of this finding, we used a biochip device called BIAcore™ (from Pharmacia) to investigate the hormone binding mechanism in detail.

The kinetics and affinities for alanine mutations at all contact residues between site 1 and receptor 1 have been measured (18). These studies showed that of 30 contact residues, 15 have virtually no effect on binding when converted to alanine. Of the remaining 15, 7 can account for >80% of the binding free energy. Thus, the functional epitope (defined by alanine scanning) is considerably smaller than the contact epitope (as defined by X-ray crystallography). The alanine substitutions had a much more dramatic effect on off rate than on rate, suggesting that association is a multistep process. Moreover, converting the specific basic residues to alanines decreased on rate, while converting some acidic residues to alanine increased on rate. This indicates that electrostatic effects are important to steer the hormone to the receptor.

Correlations with In Vivo Studies

Studies on whole cells show that receptor dimerization is required to activate the receptor and induce cell growth (19–23). For example, antibodies to the hGH-R (19) or PRL-R (20), but not their corresponding *fragment antigen binding* (FAB) fragments, can activate these receptors. Based on the sequential receptor dimerization mechanism, it is predicted that the dose-response curve should be bell-shaped. This is because high concentrations of hGH should antagonize the receptor by saturation of cellular receptors as 1:1 complexes with hGH so that spare receptors are not available to dimerize. Indeed, this was found for both the hGH-R (19) and PRL-R (21). Further evidence for this dimerization mechanism has been provided in both IM-9 lymphocytes (22, 23) and rat preadipocytes (23). Studies of hGH mutants on these cells have shown that reducing the affinity of either site 1 or site 2 reduces the potency of the analog in these cell-based assays (19, 21–23).

Engineering New Properties into hGH

Based on the differences found in the way that hGH binds to the hGH-R and PRL-R in site 1, it has been possible to design receptor-selective variants of hGH (18). Analogs have been produced that bind tightly to either the PRL-R or hGH-R or to neither. These have been used to probe some of the pharmacological effects of the hormone. For example, hGH is known to activate human neutrophils, and Fu and coworkers (24), using receptor-selective hGH analogs, showed that this is likely the result of binding to a PRL-like receptor. By using such analogs, Feldman et al. (25) showed that the GH-R is involved in mammary development in the rat.

From the structure on the complex, it has been possible to build potent antagonists to the hGH-R and PRL-R by making hGH analogs that bind at site 1 but not at site 2 (19, 21). The GH antagonists have been shown to be effective in antagonizing cells containing transfected receptors, as well as resident ones (21–23). It has been possible to engineer nonbinding analogs of hGH, such as hPRL (26) or hPL (27), to bind to the hGH-R by incorporating into them hGH site 1 binding determinants. These also turn out to be antagonists because they lack the site 2 determinants. The engineered antagonists may be clinically useful for treating pituitary tumors that cause excess secretion of hGH and PRL, as in acromegaly and hyperprolactinemia.

Extremely high affinity variants of hGH at site 1 have been produced using a technique we call *phage display* (28, 29). In this method the gene for hGH was inserted into the aminoterminal domain of the geneIII protein of the filamentous phage M13. GeneIII is normally displayed in 5 copies at one end of the virion and functions to attach the phage to its host, *E. coli*. Phage particles displaying the fusion protein contain the DNA for that protein packaged within. We mutated the hGH gene at 20 different residues in site 1 and allowed the variants to bind to beads containing immobilized hGHbp. Tighter-binding variants could be isolated by binding to these beads. After many selections and recombinations, a variant of hGH was isolated that bound 400 times tighter to the hGHbp. When this was incorporated into the mutant that disrupts binding to site 2, a more potent receptor antagonist was produced.

Conclusions

Determination of the molecular basis for hormone action is fundamental to understanding signal transduction. This pharmacologic research leads naturally to the generation of improved hormone agonists and antagonists that may be of clinical use. Ultimately, we believe such studies will be instrumental in the rational design of small molecule mimics of peptide hormones.

Acknowledgments. We are grateful to our colleagues and Genentech for their continued and enthusiastic support and to Bart De Vos and Will Somers for making unpublished data available.

References

1. Wells JA, De Vos AM. Structure and function of human growth hormone: implications for the hematopoietins. Annu Rev Biophys Biomol Struct 1993; 22:329–51.
2. Wells JA, Cunningham BC, Fuh G, et al. The molecular basis for growth hormone-receptor interactions. Recent Prog Horm Res 1993;48:253–75.
3. De Vos AM, Kossiakoff AA. Receptor action and interaction. Curr Opin Struct Biol 1992;2:852–8.
4. Isaksson O, Eden S, Jansson JO. Mode of action of growth hormone on target cells. Annu Rev Physiol 1985;47:483–99.
5. Leung DW, Spencer SA, Cachianes G, et al. Growth hormone receptor and serum binding protein: purification, cloning and expression. Nature 1987; 330:537–43.
6. Boutin JM, Edrey M, Shirota M, et al. Identification of a cDNA encoding a long form of prolactin receptor in human hepatoma and breast cancer cells. Mol Endocrinol 1989;3:1455–61.
7. Bazan JF. Structural design and molecular evolution of a cytokine receptor superfamily. Proc Natl Acad Sci USA 1990;87:6934–8.
8. Argetsinger LS, Campbell GS, Yang X, Witthuhn BA, Silvennoinen IJ, Carter-Su C. Identification of JAK2 as a growth hormone receptor-associated tyrosine kinase. Cell 1993;74:237–44.
9. Ullrich A, Schlessinger J. Signal transduction by receptors with tyrosine kinase activity. Cell 1990;61:203–12.
10. Cunningham BC, Ultsch M, De Vos AM, Mulkerrin MG, Clauser KR, Wells JA. Dimerization of the extracellular domain of the human growth hormone receptor by a single hormone molecule. Science 1991;254:821–5.
11. Ultsch M, De Vos AM, Kossiakoff AA. Crystals of the complex between human growth hormone and the extracellular domain of its receptor. J Mol Biol 1991;222:865–8.
12. Cunningham BC, Jhurani P, Ng P, Wells JA. Receptor and antibody epitopes in human growth hormone identified by homolog-scanning mutagenesis. Science 1989;243:1330–6.
13. Cunningham BC, Wells JA. High-resolution epitope mapping of hGH-receptor interactions by alanine-scanning mutagenesis. Science 1989;244: 1081–5.
14. Bass SH, Mulkerrin MG, Wells JA. A systematic mutational analysis of hormone-binding determinants in the human growth hormone receptor. Proc Natl Acad Sci USA 1991;88:4498–502.
15. De Vos AM, Ultsch M, Kossiakoff AA. Human growth hormone and extra-cellular domain of its receptor: crystal structure of the complex. Science 1992;255:306–12.
16. Cunningham BC, Wells JA. Rational design of receptor-specific variants of human growth hormone. Proc Natl Acad Sci USA 1991;88:3407–11.
17. Cunningham BC, Bass S, Fuh G, Wells JA. Zinc mediation of the binding of human growth hormone to the human prolactin receptor. Science 1990;250: 1709–12.
18. Cunningham BC, Wells JA. Comparison of a structural and functional epitope. J Mol Biol 1993;233:554–63.

19. Fuh G, Cunningham BC, Fukunaga R, Nagata S, Goeddel DV, Wells JA. Rational design of potent antagonists to the human growth hormone receptor. Science 1992;256:1677–80.
20. Elberg G, Kelly PA, Djiane J, Binder L, Gertler A. Mitogenic and binding properties of monoclonal antibodies to the prolactin receptor in Nb2 rat lymphoma cells: selective enhancement by anti-mouse IgG. J Biol Chem 1990;265:14770–6.
21. Fuh G, Colosi P, Wood WI, Wells JA. Mechanism-based design of prolactin receptor antagonists. J Biol Chem 1993;268:5376–81.
22. Silva CM, Weber MJ, Thorner MJ. Stimulation of tyrosine phosphorylation in human cells by activation of the growth hormone receptor. Endocrinology 1992;132:101–8.
23. Ilondo MM, Damholt AB, Cunningham BC, Wells JA, Shymko RM, De Meyts P. Receptor dimerization determines the effects of growth hormone in primary rat adipocytes and cultured human IM-9 lymphocytes. Endocrinology 1994;134:2397–403.
24. Fu Y-K, Arkins S, Fuh G, et al. Growth hormone augments superoxide anion secretion of human neutrophils by binding to the prolactin receptor. J Clin Invest 1992:451–7.
25. Feldman M, Ruan W, Cunningham BC, Wells JA, Kleinberg DL. Evidence that the growth hormone receptor mediates differentiation and development of the mammary gland. Endocrinology 1994.
26. Cunningham BC, Henner DJ, Wells JA. Engineering human prolactin to bind to the human growth hormone receptor. Science 1990;247:1461–5.
27. Lowman HB, Cunningham BC, Wells JA. Mutational analysis and protein engineering of receptor-binding determinants in human placental lactogen. J Biol Chem 1991;266:10982–8.
28. Lowman HB, Bass SH, Simpson N, Wells JA. Selecting high-affinity binding proteins by monovalent phage display. Biochemistry 1991;30:10832–8.
29. Lowman HB, Wells JA. Affinity maturation of human growth hormone by monovalent phage display. J Mol Biol 1993;234:564–78.

3

Human Growth Hormone Binding Proteins: Regulation and Physiological Significance

Gerhard Baumann, Moises Mercado, Norma Davila, Melissa Shaw, and Klaus Amburn

Circulating *growth hormone binding proteins* (GHBPs) have recently been recognized in a variety of species. Their discovery has added a new dimension to the physiology of *growth hormone* (GH) action. The human high-affinity GHBP, first described in 1986 (1, 2), was subsequently found to represent the soluble ectodomain of the *GH receptor* (GH-R) (3). This property has attracted considerable attention among basic scientists and clinicians alike. Another circulating GHBP, of lower affinity and probably not related to the GH-R, has received less scrutiny and remains only partially characterized (4). Comprehensive reviews of the GHBPs have recently appeared (5, 6); this chapter summarizes salient features and focuses on aspects that are new and potentially relevant to reproduction. For full citation the reader is referred to the mentioned reviews.

The high-affinity human GHBP is a 60-kd, ~246-amino acid, single-chain glycoprotein. It corresponds to the extracellular domain of the GH-R and probably arises from the receptor by proteolytic cleavage. Neither the precise site of cleavage nor the enzyme responsible is known. None of the classical proteases appear to be involved. The sulfhydryl group of a free cysteine in the region of cleavage may play an important role in proteolysis (7). It is not clear whether cleavage is linked to GH binding, although this seems unlikely based on the lack of a temporal relationship between GH pulses and changes in GHBP in vivo (8, 9). The human GHBP is highly specific for human GH, has a rapid association rate, a slower dissociation rate, and circulates in blood at about nanomolar concentrations. These functional properties allow the GHBP to both efficiently bind and release GH in the circulation in vivo, resulting in a dynamic equilibrium with, on the average, about 45% of circulating GH complexed with the GHBP (10). A recombinant form of this GHBP has been characterized in great structural detail (11, 12).

The low-affinity GHBP is not related to the GH-R, as judged by a variety of criteria. It is probably heterogeneous; components of 100, 165, and 174 kd have been described (4, 13). Further evidence for heterogeneity rests with the fact that it contains a component/binding site that is specific for the 20,000 variant of human GH (14). Little is known about the binding kinetics of this GHBP; it has been estimated that about 5%–8% of circulating 22,000 GH is bound to this GHBP (15). However, 20,000 GH seems to be primarily bound to the low-affinity GHBP, with approximately 25% of circulating 20,000 GH bound (14).

Serum levels of the high-affinity GHBP are developmentally up-regulated during childhood. They are very low in fetal life and at birth, increase rapidly during the first few years and more slowly in later childhood and adolescence, and remain relatively constant during adult life (15–17). This pattern coincides with the onset of GH responsivity after birth; it probably mirrors what occurs with GH-Rs in tissues. There is no effect of puberty per se on GHBP levels in either sex (15, 18). The normal range of GHBP levels is very wide, with up to 10-fold differences among individuals (13, 15, 16). Women have slightly higher GHBP levels than men.

Oral, but not transdermal, estrogen treatment results in an increase of serum GHBP (19); testosterone treatment results in a decrease (20). Ovulation induction with gonadotropin results in the up-regulation of serum GHBP, in parallel with estradiol and progesterone levels (21). Pregnancy is attended by a mild increase in the 1st trimester, with return to nongravid levels during the remainder of pregnancy (15, 21). The GHBP is equally effective in binding the *placental variant of hGH* (hGH-V) as it is in binding pituitary GH (22), consistent with the biological efficacy of hGH-V as a somatogen (23). The hGH-V is the predominant GH in the maternal circulation during the later stages of pregnancy. The regulation of GHBP levels by GH in humans is not fully understood, with some studies showing GH-dependent up-regulation (20, 24) and others showing no GH dependence (25, 26). Of note is that acromegaly is associated with low to low-normal GHBP levels (26–28).

Serum GHBP levels are decreased in a variety of GH-resistant states, such as Laron syndrome, pygmy dwarfism, malnutrition, fasting, insulin-dependent diabetes, liver cirrhosis, uremia, and hypothyroidism (reviewed in 5, 6). Conversely, GHBP is high in obesity, a condition of enhanced GH responsivity. Based on these associations and direct evidence in animals, it is reasonable to assume that GHBP levels in serum reflect GH-R concentration in tissues. A particularly compelling study has shown that the serum GHBP level is highly correlated with the biochemical and growth response to a given dose of exogenous GH (25). Clearly, there appears to be a link between GHBP levels and GH action or responsivity. This may be explained by the parallelism between GHBP and GH-Rs. It may also be due to the direct effect of GHBP on GH clearance, which prolongs the bioavailability of GH (29).

GHBP has been found in ovarian follicular fluid; its concentration correlates with that in serum (30). Follicular fluid GHBP concentration is also positively correlated with estradiol levels in either serum or follicular fluid. It is not clear whether the GHBP is locally produced or transferred from blood into follicular fluid. Human milk also contains a binding protein that binds GH and *prolactin* (PRL). This binding protein is distinct from the serum GHBP as assessed by a variety of functional, immunochemical, and structural characteristics (31). Its precise nature is presently unknown.

The physiological significance of the GHBP is far from clear. Although much has been learned about its structure, function, and origins, it is not known whether the GHBP is an essential component of the GH axis, nor whether it plays an active role in GH action. Serum levels appear to be positively correlated with GH action in most, but not all, circumstances. The GHBP prolongs GH half-life in blood, but also inhibits GH binding to receptors through competition for ligand (32). The net effect on GH action in vivo can be null (33), enhancing (34), or inhibitory (35). These apparent discrepancies may be due in part to differential effects at different GHBP levels. Thus, very high GHBP levels may inhibit GH action because of ligand trapping, whereas lower levels may enhance GH action through stabilization of circulating GH concentrations. Virtually all studies to date relate to GHBP levels in blood. It should be recognized that GHBP in the interstitium is perhaps more important for GH action at the local level, yet there is presently little information about GHBP in that compartment. Most studies have focused on the high-affinity GHBP, and the role of the low-affinity GHBP has been largely ignored. For these and other reasons, our understanding of the ultimate physiological significance of the GHBPs remains incomplete.

Acknowledgments. This work was supported in part by NIH Grants DK-38128 and DK-45265 (G.B.), a Young Investigator Award from the American Diabetes Association (M.M.), and a grant from the Fondo de Investigación de la Seguridad Social Española (N.D.).

References

1. Baumann G, Stolar MW, Amburn K, Barsano CP, DeVries BC. A specific growth hormone-binding protein in human plasma: initial characterization. J Clin Endocrinol Metab 1986;62:134–41.
2. Herington AC, Ymer S, Stevenson J. Identification and characterization of specific binding proteins for growth hormone in normal human sera. J Clin Invest 1986;7:1817–23.

3. Leung DW, Spencer SA, Cachianes G, et al. Growth hormone receptor and serum binding protein: purification, cloning and expression. Nature 1987; 330:537–43.

4. Baumann G, Shaw MA. A second, lower affinity growth hormone-binding protein in human plasma. J Clin Endocrinol Metab 1990;70:680–6.

5. Baumann G. Minireview: growth hormone-binding proteins. Proc Soc Exp Biol Med 1993;202:392–400.

6. Mercado M, Baumann G. Growth hormone-binding proteins. Endocrinologist 1993;3:268–77.

7. Trivedi B, Daughaday WH. Release of growth hormone binding protein from IM-9 lymphocytes by endopeptidase is dependent on sulfhydryl group inactivation. Endocrinology 1988;123:2201–6.

8. Snow KJ, Shaw MA, Winer LM, Baumann G. Diurnal pattern of plasma growth hormone-binding protein in man. J Clin Endocrinol Metab 1990;70: 417–20.

9. Carlsson LMS, Rosberg S, Vitangcol RV, Wong WLT, Albertsson-Wikland K. Analysis of 24-hour plasma profiles of growth hormone (GH)-binding protein, GH/GH-binding protein complex, and GH in healthy children. J Clin Endocrinol Metab 1993;77:356–61.

10. Baumann G, Vance ML, Shaw MA, Thorner M. Plasma transport of human growth hormone in vivo. J Clin Endocrinol Metab 1990;71:470–3.

11. Bass SH, Mulkerrin MG, Wells JA. A systematic mutational analysis of hormone-binding determinants in the human growth hormone receptor. Proc Natl Acad Sci USA 1991;88:4498–502.

12. De Vos AM, Ultsch M, Kossiakoff T. Human growth hormone and extracellular domain of its receptor: crystal structure of the complex. Science 1992; 255:306–12.

13. Tar A, Hocquette JF, Souberbielle JC, Clot JP, Brauner R, Postel-Vinay M-C. Evaluation of the growth hormone-binding proteins in human plasma using high pressure liquid chromatography. J Clin Endocrinol Metab 1990; 71:1202–7.

14. Baumann G, Shaw MA. Plasma transport of the 20,000 dalton variant of human growth hormone (20K): evidence for a 20K-specific binding site. J Clin Endocrinol Metab 1990;71:1339–43.

15. Baumann G, Shaw MA, Amburn K. Regulation of plasma growth hormone-binding proteins in health and disease. Metabolism 1989;38:683–9.

16. Daughaday WH, Trivedi B, Andrews BA. The ontogeny of serum GH binding protein in man: a possible indicator of hepatic GH receptor development. J Clin Endocrinol Metab 1987;65:1072–4.

17. Silbergeld A, Lazar L, Erster B, Keret R, Tepper R, Laron Z. Serum growth hormone binding protein activity in healthy neonates, children and young adults: correlation with age, height and weight. Clin Endocrinol (Oxf) 1989; 31:295–303.

18. Massa G, Bouillon R, Vanderschueren-Lodeweyckx M. Serum levels of growth hormone-binding protein and insulin-like growth factor-I during puberty. Clin Endocrinol (Oxf) 1992;37:175–80.

19. Weissberger AJ, Ho KKY, Lazarus L. Contrasting effects of oral and transdermal routes of estrogen replacement therapy on 24-hour growth hormone (GH) secretion, insulin-like growth factor I, and GH-binding protein in postmenopausal women. J Clin Endocrinol Metab 1991;72:374–81.

20. Postel-Vinay M-C, Tar A, Hocquette J-F, et al. Human growth hormone (GH)-binding proteins are regulated by GH and testosterone. J Clin Endocrinol Metab 1991;73:197–202.
21. Blumenfeld Z, Barkey RJ, Youdim MBH, Brandes JM, Amit T. Growth hormone (GH) binding protein regulation by estrogen, progesterone, and gonadotropins in human: the effect of ovulation induction with menopausal gonadotropins, GH and gestation. J Clin Endocrinol Metab 1992;75:1242–9.
22. Baumann G, Dávila N, Shaw MA, Jay R, Liebhaber S, Cooke NE. Binding of human growth hormone-variant (hGH-V; placental GH) to growth hormone binding protein in human plasma. J Clin Endocrinol Metab 1991; 73:1175–9.
23. MacLeod JN, Worsley I, Ray J, Friesen HG, Liebhaber SA, Cooke NE. Human growth hormone-variant is a biologically active somatogen and lactogen. Endocrinology 1991;128:1298–302.
24. Hochberg Z, Barkey RJ, Even L, Peleg I, Youdim MBH, Amit T. The effect of human growth hormone therapy on GH binding protein in GH-deficient children. Acta Endocrinol (Copenh) 1991;125:23–7.
25. Martha PM Jr, Reiter EO, Dávila N, Shaw MA, Holcombe JH, Baumann G. Serum growth hormone (GH)-binding protein/receptor: an important determinant of GH responsiveness. J Clin Endocrinol Metab 1992;75:1464–9.
26. Ho KY, Valiontis E, Waters MJ, Rajkovic IA. Regulation of growth hormone binding protein in man: comparison of gel chromatography and immunoprecipitation methods. J Clin Endocrinol Metab 1993;76:302–8.
27. Amit T, Ishshalom S, Glaser B, Youdim MBH, Hochberg Z. Growth hormone-binding protein in patients with acromegaly. Horm Res 1992;37: 205–11.
28. Mercado M, Carlsson L, Vitangcol R, Baumann G. Growth hormone-binding protein determination in human plasma: a comparison of immunofunctional and growth hormone-binding assays. J Clin Endocrinol Metab 1993;76: 1291–4.
29. Baumann G, Amburn KD, Buchanan TA. The effect of circulating growth hormone-binding protein on metabolic clearance, distribution and degradation of human growth hormone. J Clin Endocrinol Metab 1987;64:657–60.
30. Amit T, Dirnfeld M, Barkey RJ, et al. Growth hormone-binding protein (GH-BP) levels in follicular fluid from human preovulatory follicles: correlation with serum GH-BP levels. J Clin Endocrinol Metab 1993;77:33–9.
31. Mercado M, Baumann G. A growth hormone-binding protein in human milk [Abstract]. Prog 74th meet Endocr Soc, 1992:225.
32. Mannor DA, Winer LM, Shaw MA, Baumann G. Plasma growth hormone binding proteins: effect on growth hormone binding to receptors and on growth hormone action. J Clin Endocrinol Metab 1991;73:30–4.
33. Mannor DA, Shaw MA, Winer LM, Baumann G. Circulating growth hormone-binding protein inhibits growth hormone (GH) binding to GH receptors but not in vivo GH action [Abstract]. Clin Res 1988;36:870A.
34. Clark RG, Cunningham B, Moore JA, et al. Growth hormone binding protein enhances the growth promoting activity of GH in the rat [Abstract]. Prog 73rd meet Endocr Soc, 1991:1611.
35. Rieu M, Le Bouc Y, Villares SM, Postel-Vinay M-C. Familial short stature with very high levels of growth hormone binding protein. J Clin Endocrinol Metab 1993;76:857–60.

4

Ovarian Prolactin Receptors and Their Placental Ligands

Daniel I.H. Linzer and Brian J. Arey

Prolactin (PRL), which is closely related to *growth hormone* (GH), has a diverse array of physiological effects, including, for example, the regulation of reproduction, mammary development and lactation, metabolism, immune response, osmotic balance, and mating and maternal behavior. These effects of PRL are initiated by binding of the hormone to a cell-surface receptor that is a member of the cytokine receptor superfamily (1). At least three factors contribute to the complexity of PRL action: (i) the expression of multiple forms of the *PRL receptor* (PRL-R), (ii) the presence and action of these receptors in distinct cell backgrounds, and (iii) the synthesis of several different receptor ligands.

In the mouse at least 4 forms of the PRL-R are synthesized, 3 proteins with relatively short cytoplasmic domains and 1 with a long cytoplasmic region (2, 3). In contrast, only 1 short and 1 long receptor form have been identified in other species (1). The ability of the long receptor form to transduce PRL binding into an intracellular signaling cascade has been demonstrated in model cell systems (4), but the roles of the short receptor forms are unknown. Possibly, these short receptor forms may connect to different signaling pathways or they may target bound hormone to distinct intracellular fates.

As a reproductive hormone, PRL has direct effects on both the primary reproductive organs (ovary and testis) and the secondary organs, such as the mammary glands. Within the ovary PRL-R expression has been detected by immunocytochemistry, radioligand binding, and in situ hybridization in multiple cell types, including granulosa, luteal, thecal, and interstitial cells (3, 5–10). These cellular targets of PRL in the ovary are components of various structures, including small preantral follicles, mature preovulatory follicles, and corpora lutea.

The effects of PRL on these different cell types are numerous (Fig. 4.1). In granulosa and luteal cells, PRL acts to increase progesterone

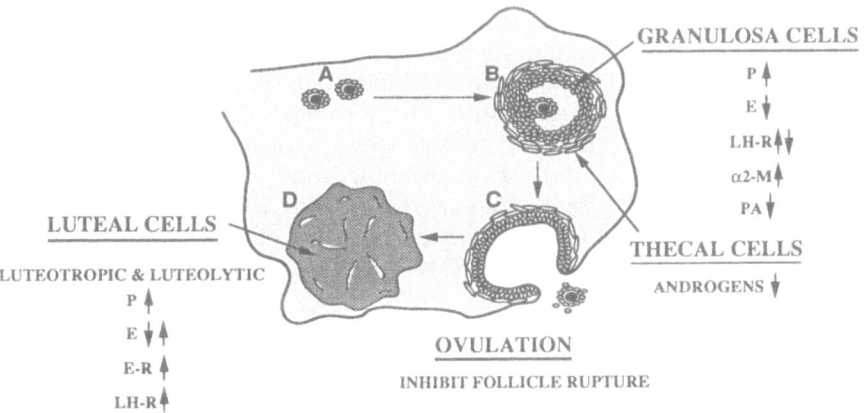

FIGURE 4.1. Summary of PRL's actions on the ovary. PRL-R expression occurs in small, preantral follicles (A), large, preovulatory follicles (B), ovulatory follicles (C), and corpora lutea (D), as described in reference 3. The effects of PRL on these cell types and structures are varied and include the regulation of the levels of progesterone (P), estradiol (E) and its receptor (E-R), LH-R, α2 macroglobulin (α2-M), and plasminogen activator (PA), as described in references 11–34.

production by increasing cholesterol uptake and the amount of progesterone synthetic enzymes and by decreasing progesterone-modifying enzymes (11–18). Under most conditions PRL inhibits androgen synthesis in thecal cells and estradiol production in granulosa and luteal cells, although PRL can also stimulate estradiol secretion during pregnancy (19–25). In addition to regulating hormone synthesis, PRL also affects the levels of hormone receptors, typically acting to increase and maintain *luteinizing hormone receptor* (LH-R) numbers on granulosa and luteal cells, but PRL has also been reported to cause a decrease in LH-R levels (15, 26–32). Finally, PRL influences the synthesis of proteins involved in proteolysis and tissue remodeling, including plasminogen activator and α2 macroglobulin, that in turn can influence ovulation and corpus luteum formation (33, 34).

The expression of 4 distinct receptor forms in the mouse liver and ovary (3) suggests that each of these forms may have a specific role in PRL physiology. For example, even though all 4 mRNAs are coexpressed in most cell types in the mouse ovary, we have found that 1 of the short receptor forms—designated PRL-R$_S$2—is uniquely expressed in the granulosa cells of atretic follicles at early to midpregnancy (3). Similarly, we have observed distinct expression patterns of the long and 1 short PRL-R mRNA in the rat ovary (10). On proestrous morning, both of these mRNAs are expressed at similar levels in the granulosa cells of large preovulatory follicles. In contrast, the short PRL-R mRNA is the predominant form detected in the thecal cell layer of these follicles, whereas the long PRL-R mRNA is the major species found in the granulosa cells of small preantral follicles in the center of the ovary. Thus, the individual receptor forms may provide specific functions in the processes of follicular early development, maturation, and atresia.

PRL is not the only ligand that can bind to and activate the PRL-R. In rodents placental hormones designated *placental lactogen I* (PL-I) and *II* (PL-II) both bind to the PRL-R with high affinity and have similar bioactivity compared to PRL (35, 36). Similarly, placental proteins in the PRL/GH family are synthesized in both ruminants and primates, including humans (36). Gene mapping and sequence comparison of these placental hormones in rodents, ruminants, and primates suggest that they may have arisen from independent duplications of the PRL or GH gene during evolution, rather than by the retention of a gene that was expressed in the placenta of a common ancestral mammal. The independent selection for the expression of placental hormones related to PRL and GH in at least three mammalian lineages suggests that these proteins provide an important reproductive advantage.

PRL, PL-I, and PL-II form a triad of hormones that are expressed sequentially during rodent gestation. Surges of PRL secretion occur at 12-h intervals during the first half of gestation, but PRL release then decreases significantly (37). Concurrent with the decrease in PRL release is a marked elevation in PL-I synthesis and in its accumulation in the maternal serum to levels approaching 10 μg/mL (38). The presence of PL-I in the serum is transient, however, and within 2–3 days PL-I has been replaced by PL-II as the major lactogenic hormone in the circulation. This latter hormone continues to be produced at high levels during the remainder of gestation (39).

An unanswered question is what effects PL-I and PL-II might have that would not be achieved simply by the continued presence of circulating PRL. One possibility is that the extremely high level of PL-I in the circulation at midgestation serves as a gestational synchronizer. In this model occupancy of most or all of the cell-surface PRL-R molecules at midgestation would simultaneously trigger physiological changes in the many target tissues, thereby coordinating their actions during the latter half of pregnancy and in preparation for the postpartum period (Fig. 4.2).

An alternative hypothesis is that the high levels of PL-I and PL-II, coupled with their distinct amino acid sequences compared to PRL, enable these placental hormones to be delivered into compartments that PRL is unable to enter. The ability of PL-I and PL-II to be transported across barriers at a sufficiently high concentration to act on potential target tissues may be important, for example, in the maternal brain and in the developing fetus (Fig. 4.3). Voogt and colleagues have demonstrated that experimental delivery of PL-I to the rat hypothalamus can down-

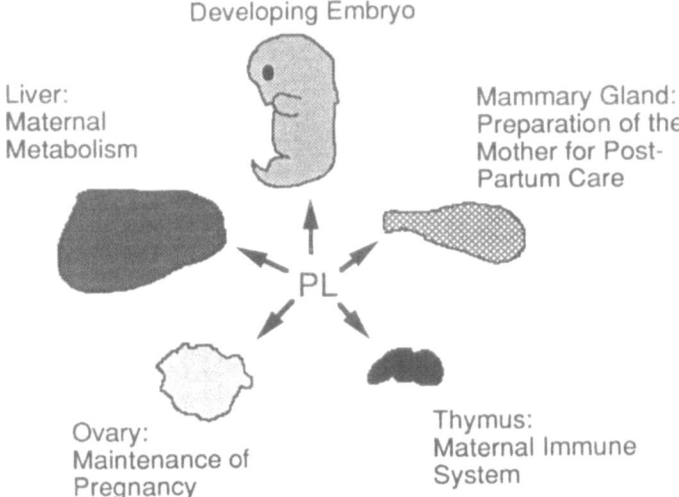

FIGURE 4.2. Hypothesis of PL action: synchronization of gestational events. One possible role of PL, especially for PL-I that is present at high levels transiently at midpregnancy, is to trigger simultaneous alterations in numerous target organs. This might coordinate maternal and fetal physiology during gestation and in preparation for the postpartum period. Some of the likely targets are the maternal liver, mammary gland, ovary, T lymphocytes, and the developing embryo.

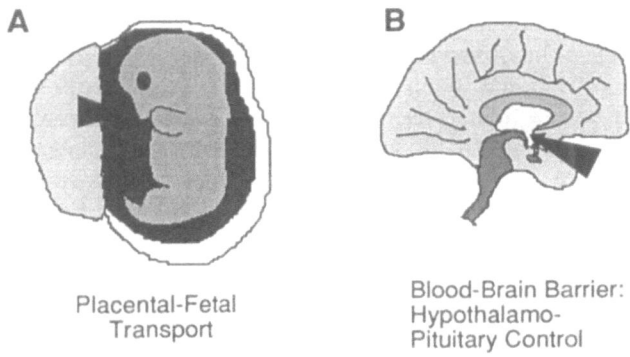

FIGURE 4.3. Hypothesis of PL action: crossing barriers. Another possible explanation for the synthesis of PL-I and PL-II is that the unique amino acid sequences and high circulating levels of these proteins enable them to enter compartments such as the brain and the fetus at sufficiently high levels to have physiological effects. PL-I has been reported to alter hypothalamic control of pituitary PRL release in the rat, as described in references 40–42. PL-II has been detected in the fetal circulation in the mouse, as described in reference 43.

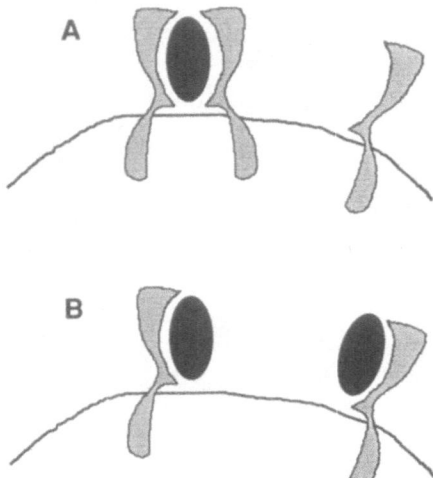

FIGURE 4.4. Model of PRL-R interaction with its ligand. Based on the structure of the GH/GH-R complex, as described in references 47 and 48, it is likely that PRL activates the PRL-R by forming a 1 ligand:2 receptor complex (54), as shown in *A*. However, when the circulating ligand concentration is much higher than that of cell-surface receptor or when modified ligands are present that have only a single functional binding site for the PRL-R, as described in reference 54, receptors would form predominantly 1:1 inactive complexes, as shown in *B*.

regulate PRL release from the pituitary through a dopaminergic signaling pathway (40–42). The structure and concentration of PL-I in the serum may therefore enable enough of this protein to cross the blood-brain barrier to cause the observed decrease in PRL secretion at midpregnancy.

Similarly, PL-II has been detected in the fetal circulation at levels significantly lower than those found in the maternal serum (43). Thus, PL-II may act on fetal cells expressing the PRL-R; for example, in the fetal lung where PRL has been shown to regulate surfactant production (44, 45). The ability of PL-II to enter the fetal circulation apparently depends on both its high level of expression and its structure since another member of the PRL/GH family in the mouse, proliferin-related protein, is also secreted from the placenta at high levels and with a time course similar to that of PL-II, but cannot be detected in the fetal circulation (46).

Based on studies of GH (47, 48), PRL, PL-I, and PL-II are each predicted to have 2 sites for receptor interaction, thereby providing a means to dimerize and activate cell-surface PRL-R proteins. At very high ligand concentrations at midpregnancy, though, it is possible that most of the available PRL-R molecules would be occupied by PL-I, leaving few free PRL-R molecules to add to the complex; 1:1 inactive complexes of

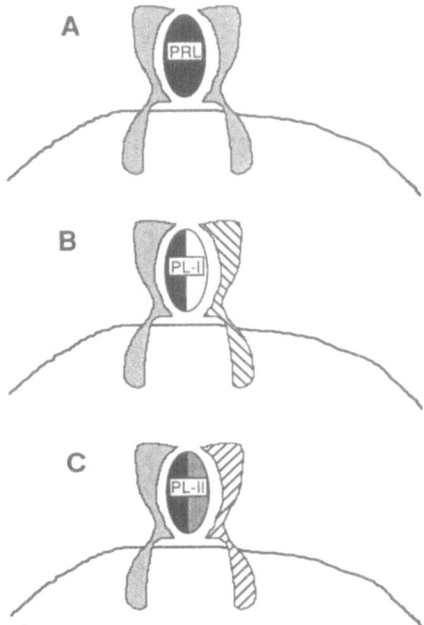

FIGURE 4.5. Hypothesis of PL action: unique functions. The finding that PRL, PL-I, and PL-II bind the PRL-R with high affinity, as described in references 50 and 51, indicates that these 3 proteins all must have at least 1 domain capable of binding the PRL-R. Once bound to a single PRL-R molecule, though, it is possible that the specificity of a second binding domain within PRL, PL-I, and PL-II differs among these hormones. For example, PRL might bind a second molecule of the PRL-R (A), whereas the second binding site on PL-I (B) and PL-II (C) might bring different membrane proteins into the complex, thereby creating heteromeric complexes with potentially distinct activities.

the PRL-R and PL-I might then predominate over active 2:1 PRL-R:ligand complexes (Fig. 4.4). Thus, in this model the high level of PL-I is predicted to result in decreased signaling through the PRL-R.

The requirement for a second PRL-R ligand gene might therefore be explained by the need to have quite different modes of regulation of hormone levels, with the PRL gene subject to the limited expression evident in the adult and the PL-I gene able to provide maximum expression in the pregnant animal. This hypothesis seems unlikely, though, for at least two reasons. First, if an increased PRL serum concentration is all that is needed, a simpler mechanism might be to up-regulate PRL synthesis and secretion from the pituitary during pregnancy. Alternatively, several genes utilize more than one transcriptional regulatory region to

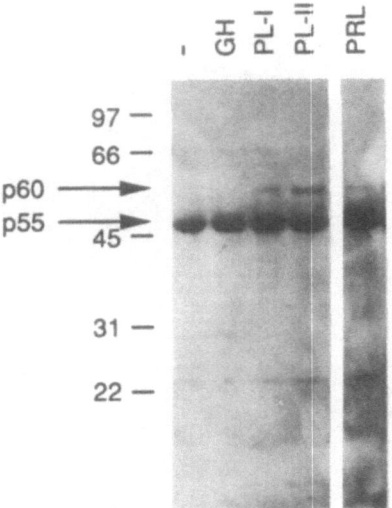

FIGURE 4.6. Hormonal stimulation of protein tyrosine phosphorylation. To test the hypothesis described in Fig. 4.5, cultures of primary mouse ovarian granulosa cells were incubated for 5 min without hormone (−), or with 1-µg/mL bovine GH, mouse PL-I, mouse PL-II, or ovine PRL. Bovine GH binds the GH-R but not the PRL-R, whereas the other 3 hormones bind the PRL-R but not the GH-R. Proteins released by cell lysis were immunoprecipitated with a monoclonal anti-body against phosphotyrosine, fractionated by gel electrophoresis, transferred to nitrocellulose, and detected with the monoclonal antibody against phospho-tyrosine. Phosphorylation of a 60-kd protein (p60) is induced by PRL, PL-I, and PL-II, but not by GH.

allow for different levels of expression in different cell types. Thus, another mechanism for achieving high PRL-R ligand concentrations at midgestation might have been the evolution of a placental-specific transcriptional control region in the PRL gene. A second limitation of this hypothesis is that it fails to explain the requirement for the expression of a PRL-R ligand that is distinct from PRL.

Once bound to the PRL-R, it is also possible that PL-I and PL-II initiate distinct biological effects. Each of these placental hormones shares only 40%−50% amino acid sequence identity with PRL (49). These sequence differences do not alter the ability of these proteins to bind to the PRL-R with high affinity (50, 51), suggesting (by analogy to the interaction of GH with the GH-R) that binding site 1 within each of these hormones is intact. However, once bound to the PRL-R, PL-I and PL-II may present a low-affinity binding site 2 different from that of PRL, thereby bringing a second cell-surface protein other than a PRL-R into a heteromeric complex with the PRL-R (Fig. 4.5). This could result in the

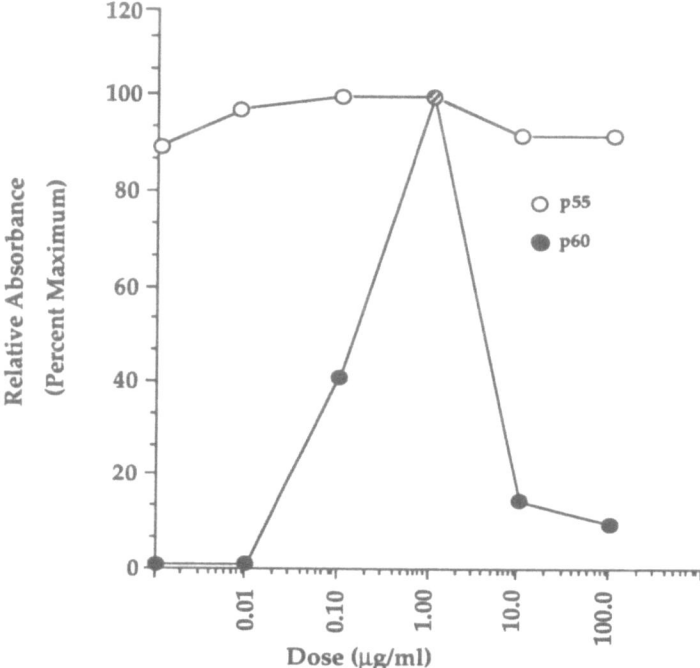

FIGURE 4.7. Induction of p60 phosphorylation by increasing concentrations of PRL. Primary mouse ovarian granulosa cells were incubated for 5 min with increasing doses of PRL (from 0 to 100 µg/mL). Tyrosine phosphorylation of p60—and, for comparison, a second protein of 55-kd (see Fig. 4.6)—was quantified by densitometric analysis of experimental results obtained as described in Fig. 4.6.

activation of signaling pathways different from those initiated by PRL-R dimers. Although this hypothesis is attractive in providing a rationale for the expression of PRL-R ligands with significantly different amino acid sequences, no evidence for this hypothesis has been obtained to date. Indeed, all available data indicate that PRL, PL-I, and PL-II trigger identical biological responses (36, 43).

To test if PRL, PL-I, and PL-II all induce the identical molecular events in a reproductive target tissue, primary cultures of mouse and rat ovarian granulosa cells were prepared and exposed to hormone. Cell lysates were then prepared and analyzed for rapid changes in tyrosine phosphorylation of specific proteins since PRL has been shown to induce this type of modification in Nb2 lymphoma cells (52, 53). As can be seen in Figure 4.6, the three lactogenic hormones induced the tyrosine phosphorylation of a 60-kd protein (hereafter referred to as *p60*); in contrast, ovine GH (which does not bind to the PRL-R) failed to induce p60 phosphorylation. Thus, the placental proteins and PRL appear to have the same activity in this bioassay, suggesting that the hypothesis diagrammed in Figure 4.5 is not correct, at least in this cell background.

Quantification of p60 phosphorylation in response to an increasing PRL dose revealed the pattern shown in Figure 4.7. Instead of a plateau a peak concentration was detected; increasing PRL concentration beyond this peak value resulted in decreased p60 phosphorylation. These data are consistent with the model presented in Figure 4.4. Furthermore, since PL-I concentrations approach 10 μg/mL in the maternal serum, a level above the peak concentration for PRL-induced p60 phosphorylation, PL-I in vivo might actually be occupying granulosa cell PRL-Rs primarily in 1:1 complexes, thereby decreasing PRL-R signaling in these cells.

A phosphoprotein of 60 kd was not reported to be induced by PRL in Nb2 cells (52, 53). Thus, the signaling pathways activated by PRL in granulosa and T lymphoma cells may have distinct components. One potentially important difference between these systems is that PRL acts as a mitogen on Nb2 cells, but not on primary granulosa cells. Thus, the distinct effects invoked by PRL in different cell types may depend on the cellular composition of protein kinases and phosphatases, as well as the substrates for these signaling enzymes. This variation in signaling might then explain, at least in part, the complexity of PRL regulation of fetal, maternal, and adult physiology.

Acknowledgments. This work was supported by NIH Grant HD-P01-21921. B.A. is a postdoctoral trainee on the Reproductive Biology Training Grant. D.L. is the recipient of an American Cancer Society faculty research award.

References

1. Kelly PA, Djiane J, Postel-Vinay M-C, Edery M. The prolactin/growth hormone receptor family. Endocr Rev 1991;12:235–51.
2. Davis JA, Linzer DIH. Expression of multiple prolactin receptors in mouse liver. Mol Endocrinol 1989;3:674–80.
3. Clarke DL, Linzer DIH. Changes in prolactin receptor expression during pregnancy in the mouse ovary. Endocrinology 1993;133:224–32.
4. Lesueur L, Edery M, Ali S, Paly J, Kelly PA, Djiane J. Comparison of long and short forms of the prolactin receptor on prolactin-induced milk protein gene transcription. Proc Natl Acad Sci USA 1991;88:824–8.
5. Midgley AR. Autoradiographic analysis of gonadotropin binding to rat ovarian tissue sections. Adv Exp Biol Med 1973;36:365–78.
6. Costlow ME, McGuire WL. Autoradiographic localization of the binding of ^{125}I-labelled prolactin to rat tissues in vitro. J Endocrinol 1977;75:221–6.

 7. Dunaif AE, Zimmerman EA, Friesen HG, Frantz AG. Intracellular locali-
    zation of prolactin receptor and prolactin in the rat ovary by immunocyto-
    chemistry. Endocrinology 1982;110:1465–71.
 8. Oxberry BA, Greenwald GS. An autoradiographic study of the binding of
    ^{125}I-labeled follicle-stimulating hormone, human chorionic gonadotropin and
    prolactin to the hamster ovary throughout the estrous cycle. Biol Reprod
    1982;27:505–16.
 9. Oxberry BA, Greenwald GS. Autoradiographic analysis of proestrous
    changes in the binding of ^{125}I-labeled prolactin to the hamster ovary. Biol
    Reprod 1982;29:1255–63.
10. Clarke DL, Arey BJ, Linzer DIH. Prolactin receptor expression in the ovary
    during the rat estrous cycle. Endocrinology 1993;133:2594–603.
11. Lamprecht SA, Lindner HR, Strauss JF. Induction of 20α-hydroxysteroid
    dehydrogenase in rat corpora lutea by pharmacological blockade of pituitary
    prolactin secretion. Biochim Biophys Acta 1969;187:133–43.
12. Armstrong DT, Miller LS, Knudsen KA. Regulation of lipid metabolism and
    progesterone production in rat corpora lutea and ovarian interstitial elements
    by prolactin and luteinizing hormone. Endocrinology 1969;85:393–401.
13. Armstrong DT, Knudsen KA, Miller LS. Effects of prolactin upon choles-
    terol metabolism and progesterone biosynthesis in corpora lutea of rats hypo-
    physectomized during pseudopregnancy. Endocrinology 1970;86:634–41.
14. Behrman HR, Orczyk GP, MacDonald GJ. Prolactin induction of enzymes
    controlling luteal cholesterol turnover. Endocrinology 1970;87:1251–6.
15. Gibori G, Richards JS. Dissociation of two distinct luteotropic effects of
    prolactin: regulation of luteinizing hormone-receptor content and proges-
    terone secretion during pregnancy. Endocrinology 1978;102:767–74.
16. Jones PBC, Valk CA, Hsueh AJW. Regulation of progestin biosynthetic
    enzymes in cultured rat granulosa cells: effects of prolactin, β₂-adrenergic
    agonist, human chorionic gonadotropin and gonadotropin releasing hormone.
    Biol Reprod 1983;29:572–85.
17. Fortune JE, Vincent SE. Prolactin modulates steroidogenesis by rat granulosa
    cells: effects on progesterone. Biol Reprod 1986;35:84–91.
18. Martel C, Labrie C, Dupont E, et al. Regulation of 3β-hydroxysteroid dehy-
    drogenase/Δ5-Δ4 isomerase expression and activity in the hypophysectomized
    rat ovary: interactions between the stimulatory effect of human chorionic
    gonadotropin and the luteolytic effect of prolactin. Endocrinology 1990;
    127:2726–37.
19. Magoffin DA, Erickson GP. Prolactin inhibition of luteinizing hormone-
    stimulated androgen synthesis in ovarian interstitial cells cultured in defined
    medium: mechanism of actions. Endocrinology 1982;111:2001–7.
20. Wang C, Hsueh AJW, Erickson GF. Prolactin inhibition of estrogen produc-
    tion by cultured rat granulosa cells. Mol Cell Endocrinol 1980;20:135–44.
21. Dorrington J, Gore-Langton RE. Prolactin inhibits oestrogen synthesis in the
    ovary. Nature 1981;290:600–2.
22. Wang C, Chan V. Divergent effects of prolactin on estrogen and proges-
    terone production by granulosa cells of rat graafian follicles. Endocrinology
    1982;110:1085–93.
23. Tsai-Morris CH, Ghosh M, Hirshfield AN, Wise PM, Brodie AMH. Inhibi-
    tion of ovarian aromatase by prolactin in vivo. Biol Reprod 1983;29:342–6.

24. Hickey GJ, Oonk RB, Hall PF, Richards JS. Aromatase cytochrome P450 and cholesterol side-chain cleavage cytochrome P450 in corpora lutea of pregnant rats: diverse regulation by peptide and steroid hormones. Endocrinology 1989;125:1673–82.
25. Krasnow JS, Hickey GJ, Richards JS. Regulation of aromatase mRNA and estradiol biosynthesis in rat ovarian granulosa and luteal cells by prolactin. Mol Endocrinol 1990;4:13–21.
26. Grinwich DL, Hichens M, Behrman HR. Control of the LH receptor by prolactin and prostaglandin F2α in rat corpora lutea. Biol Reprod 1976; 14:212–8.
27. Richards JS, Williams JJ. Luteal cell receptor content for prolactin (PRL) and luteinizing hormone (LH): regulation by LH and PRL. Endocrinology 1976;99:1571–81.
28. Behrman HR, Grinwich DL, Hichens M, MacDonald GJ. Effects of hypophysectomy, prolactin, and prostaglandin F2α on gonadotropin binding in vivo and in vitro in the corpus luteum. Endocrinology 1978;103:349–57.
29. Holt JA, Richards JS, Midgley AR Jr, Reichert LE Jr. Effect of prolactin on LH receptor in rat luteal cells. Endocrinology 1976;98:1005–13.
30. Casper RF, Erickson GF. In vitro heteroregulation of LH receptors by prolactin and FSH in rat granulosa cells. Mol Cell Endocrinol 1981;23:161–71.
31. Lane TA, Chen TT. Heterologous down-modulation of luteinizing hormone receptors by prolactin: a flow cytometry study. Endocrinology 1991; 128:1833–40.
32. Piquette GN, LaPolt PS, Oikawa M, Hsueh AJW. Regulation of luteinizing hormone receptor messenger ribonucleic acid levels by gonadotropins, growth factors, and gonadotropin-releasing hormone in cultured rat granulosa cells. Endocrinology 1991;128:2449–56.
33. Yoshimura Y, Maruyama K, Shiraki M, Kawakami S, Fukushima M, Nakamura Y. Prolactin inhibits plasminogen activator activity in the preovulatory follicles. Endocrinology 1990;126:631–6.
34. Gaddy-Kurten D, Richards JS. Regulation of α2-macroglobulin by luteinizing hormone and prolactin during cell differentiation in the rat ovary. Mol Endocrinol 1991;5:1280–91.
35. Soares MJ, Faria TN, Roby KF, Deb S. Pregnancy and the prolactin family of hormones: coordination of anterior pituitary, uterine, and placental expression. Endocr Rev 1991;12:402–23.
36. Ogren L, Talamantes F. Prolactins of pregnancy and their cellular source. Int Rev Cytol 1988;112:1–65.
37. Barkley MS, Radford GE, Geschwind II. The pattern of plasma prolactin concentration during the first half of mouse gestation. Biol Reprod 1978; 19:291–6.
38. Ogren L, Southard JN, Colosi P, Linzer DIH, Talamantes F. Mouse placental lactogen I: RIA and gestational profile in maternal serum. Endocrinology 1989;125:2253–7.
39. Soares MJ, Talamantes F. Placental lactogen secretion in the mouse: in vitro responses and ovarian and hormonal influences. J Exp Zool 1985;234:97–104.
40. Tomogane H, Arbogast LA, Soares MJ, Robertson MC, Voogt JL. A factor(s) from a rat trophoblast cell line inhibits prolactin secretion in vitro and in vivo. Biol Reprod 1993;48:325–32.

41. Arbogast LA, Soares MJ, Tomogane H, Voogt JL. A trophoblast-specific factor(s) suppresses circulating prolactin levels and increases tyrosine hydroxylase activity in tuberoinfundibular dopaminergic neurons. Endocrinology 1992;131:105–13.
42. Mathiasen JR, Tomogane H, Voogt JL. Serotonin-induced decrease in hypothalamic tyrosine hydroxylase activity and corresponding increase in prolactin release are abolished at midpregnancy and by transplants of rat choriocarcinoma cells. Endocrinology 1992;131:2527–32.
43. Talamantes F, Ogren L. The placenta as an endocrine organ. In: Knobil E, Neill JD, eds. The physiology of reproduction. New York: Raven Press, 1988:2093–144.
44. Mendelson CR, Snyder JM. Role of prolactin, cortisol, and insulin in the regulation of surfactant synthesis by the human fetal lung. Pediatr Pulmonol 1985;1:S91–8.
45. Schellenberg JC, Liggins GC, Manzai M, Kitterman JA, Lee CCH. Synergistic hormonal effects on lung maturation in fetal sheep. J Appl Physiol 1988;65:94–100.
46. Lopez MF, Ogren L, Linzer DIH, Talamantes F. Pituitary-placental interaction during pregnancy: regulation of prolactin-like proteins. Endocr J 1993;1: 513–8.
47. Cunningham BC, Ultsch M, De Vos AM, Mulkerrin MG, Clauser KR, Wells JA. Dimerization of the extracellular domain of the human growth hormone receptor by a single hormone molecule. Science 1991;254:821–5.
48. De Vos AM, Ultsch M, Kossiakoff AA. Human growth hormone and extracellular domain of its receptor: crystal structure of the complex. Science 1992;255:306–12.
49. Colosi P, Talamantes F, Linzer DIH. Molecular cloning and expression of mouse placental lactogen I complementary deoxyribonucleic acid. Mol Endocrinol 1987;1:767–76.
50. Harigaya T, Smith WC, Talamantes F. Hepatic placental lactogen receptors during pregnancy in the mouse. Endocrinology 1988;122:1366–72.
51. MacLeod KR, Smith WC, Ogren L, Talamantes F. Recombinant mouse placental lactogen-I binds to lactogen receptors in mouse liver and ovary: partial characterization of the ovarian receptor. Endocrinology 1989; 125:2258–66.
52. Rillema JA, Campbell GS, Lawson DM, Carter-Su C. Evidence for a rapid stimulation of tyrosine kinase activity by prolactin in Nb2 rat lymphoma cells. Endocrinology 1992;131:973–5.
53. Rui H, Djeu JY, Evans GA, Kelly PA, Farrar WL. Prolactin receptor triggering: evidence for rapid tyrosine kinase activation. J Biol Chem 1992; 267:24076–81.
54. Fuh G, Colosi P, Wood WL, Wells JA. Mechanism-based design of prolactin receptor antagonists. J Biol Chem 1993;268:5376–81.

5

Regulatory Actions of Testosterone on Pulsatile Growth Hormone Secretion in the Human: Studies Using Deconvolution Analysis

J.D. Veldhuis, A. Iranmanesh, A.D. Rogol, and R.J. Urban

Sex steroid hormones profoundly influence multiple endocrine axes, including not only the gonadotropic but also the *growth hormone* (GH) axis (1). In the rat intact male and female animals exhibit strikingly different time profiles of pulsatile GH release (2–4). These sexually dimorphic temporal patterns of GH release appear to convey significant growth-stimulating information to target tissues (5–11). Indeed, pulsatile infusions of GH in the female rat will induce increased somatic growth rates approaching those in male rats when the patterned infusions mimic the high-amplitude ultradian rhythmicity of the normally growing male animal. In addition, experimental simulation of male and female patterns of GH delivery in hypophysectomized animals will induce hepatic enzymes in a sexually dimorphic manner. Thus, the specific gender-defined mode of pulsatile GH secretion strongly influences the nature and magnitude of target tissue responses.

Conversely, the steroid hormone milieu within the animal controls the pattern of GH secretion (12, 13). For example, estrogen exposure results in a low-amplitude, high-baseline profile of GH release as observed in the intact female rodent, whereas androgen administration under appropriate experimental conditions reproduces the occasional high-amplitude volleys of GH release typical of the intact male animal (12, 14, 15). In brief, in the experimental animal steroid hormones on the one hand can influence the pulsatile mode of GH secretion; on the other hand, the pulsatile mode of GH secretion can control the target tissue response to this somatotrophic hormone.

In the human blood-sampling studies consisting of observations at 10- or 20-min intervals over 24 h tend to demonstrate a lower-amplitude, higher-frequency pattern of episodic GH secretion in women of reproductive age and an apparently lower detectable GH pulse frequency with higher-amplitude events in healthy young men (16–20). Although these studies have not yet been repeated using ultrasensitive GH assays combined with high-intensity blood sampling (e.g., 5-min blood sample withdrawal), one investigation has used an increased sensitivity *immunoradiometric assay* (IRMA) in men and women following 20-min blood sampling.

In other studies, varying estrogen concentrations within the course of the normal menstrual cycle were associated with substantial amplification (late follicular phase) and diminution (early follicular phase) in the height of GH release episodes in healthy women (20). Moreover, the administration of small oral doses of estrogen (100 ng/kg ethinyl estradiol daily) in prepubertal girls with ovarian dysgenesis induces a 2- to 3-fold amplification of GH secretory burst amplitude and mass, as assessed by deconvolution analysis or by discrete peak detection methods (21, 22). Thus, whether evaluated indirectly within the normal menstrual cycle when increased serum estradiol concentrations correlate with heightened GH pulse amplitude or in the estrogen-deprived prepubertal girl treated with small amounts of exogenous estrogen, estrogenic steroids appear to preferentially augment the amplitude of serum GH concentration pulses. Deconvolution analysis (23, 24) further shows that estrogen specifically increases the mass of GH secreted per burst without changing the calculated GH secretory burst duration, the GH burst frequency, or the half-life of endogenous GH.

In healthy boys the onset of increased androgen secretion in puberty is accompanied by a consistent (2- to 3-fold) increase in the mean serum GH concentration (Fig. 5.1). Notably, sampling at 20-min intervals over 24 h in a cohort of 44 boys evaluated cross-sectionally at various stages of normal puberty revealed a highly specific increase in the serum GH peak amplitude, as assessed by such discrete peak methods as cluster analysis (25, 26). More mechanistic analysis by deconvolution techniques demonstrated that the increase in peak serum GH concentrations resulted from an amplified mass and maximal GH secretory rate attained within each discrete release episode, with no alteration in the duration of the calculated secretory burst, the frequency of pulsatile GH secretion, or the apparent half-life of endogenous GH (27). The administration of testosterone by parenteral injection to boys with constitutionally delayed puberty recapitulated the GH responses observed during spontaneous healthy pubertal progression; that is, a doubling of serum GH concentration peaks due to a doubling or tripling of the mass of GH secreted per burst (28). These experimental observations in boys are consistent with the inference that androgen can positively regulate GH secretory burst mass or amplitude.

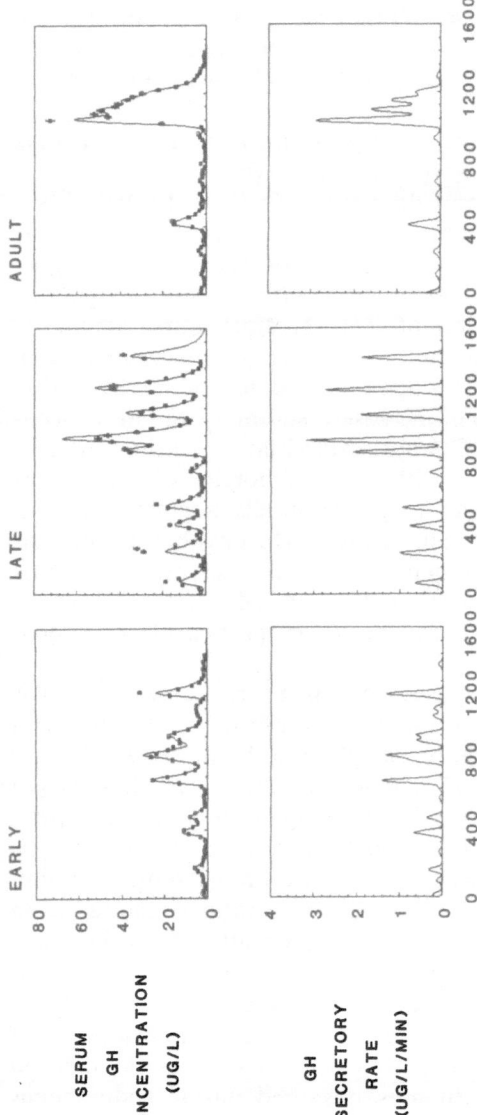

FIGURE 5.1. Amplified mode of pulsatile GH secretion in healthy pubertal boys as estimated by deconvolution analysis of 24-h serum GH concentration profiles. Blood samples were collected at 20-min intervals for 24 h in 44 boys at different stages of healthy puberty. Data from 3 boys are illustrated, as described in reference 27. The upper panels depict the serial serum GH concentration measurements over time and the deconvolution-predicted fits of the observed data. Below the fitted profiles are the deconvolution-calculated GH secretory rates (lower panels) that consist of punctuated bursts of GH secretion, as described in references 23, 24, and 33. As summarized in the text, mid and late puberty are accompanied by a 2- to 3-fold increase in the mass of GH secreted per burst, with no change in the computed half-life of endogenous GH, the GH secretory burst frequency, or the duration of secretory events. Reprinted with permission from Martha, Goorman, Blizzard, Rogol, and Veldhuis (27), © The Endocrine Society, 1992.

The mechanisms subserving androgen's stimulation of pulsatile GH secretion are discussed further in this chapter, with a principal emphasis on the role of *aromatizable androgen* (androgen capable of undergoing biochemical conversion to estrogen by way of the aromatization reaction) and the physiological contexts in which covariations in the somatotrophic and reproductive axes are of central physiological importance. For example, in puberty a concerted activation of the gonadal and somatotrophic axes occurs in boys, with a combined anabolic impact on muscle and skeletal growth, as well as on changes in body composition (1, 29).

In the decade following the maximal growth and anabolic development of late puberty, the healthy aging process is accompanied by progressive reductions in the secretory activity of both the gonadal and somatotrophic axes (1, 17, 30, 31). In both men and women, strongly negative correlations exist between GH release and age that in part are explained by the strong covariation of gonadal sex steroid secretion with age (17, 30, 32). Here, we emphasize the physiological correspondence between effective tissue actions of aromatizable androgen and activation of the GH axis. We also highlight experimental strategies designed to probe specifically the role of androgen acting via the androgen receptor and acting after aromatization to one or more estrogens in regulating the pulsatile secretion of GH.

Strategies for Evaluating the Impact of Androgen on the Somatotrophic Axis in the Human

We and collaborating investigators have combined a range of complementary strategies to evaluate the mechanisms by which androgens control pulsatile GH secretion in men and boys. Briefly, deconvolution analysis (23, 24, 33) has been employed to estimate the number, duration, mass, and amplitude of GH secretory bursts and simultaneously calculate the apparent half-life of endogenous GH in volunteers submitted to 5-, 10-, or 20-min blood sampling over 24 h in the following experimental circumstances: (i) healthy boys evaluated cross-sectionally prior to, during, and immediately following spontaneous puberty (27) ($N = 44$ boys); (ii) healthy men whose ages ranged from 21 to 71 years and who exhibited a broad range of adult serum androgen concentrations (30) ($N = 21$ subjects); (iii) healthy young men treated with a GnRH agonist (leuprolide) or a GnRH antagonist (Nal-Glu-GnRH) with concomitant testosterone or placebo injections so as to create an explicitly defined dose-dependent androgen milieu; (iv) boys with constitutionally delayed puberty treated parenterally with testosterone or the nonaromatizable androgen *5α-dihydrotestosterone* (DHT); and (v) healthy older men with physiologically mild hypoandrogenemia infused with estradiol on DHT for 3.5 days or treated with parenteral testosterone, DHT, or placebo for

3 weeks in a blinded, randomized manner. In addition, we review here responses of the GH axis to administration of specific competitive antagonists of the estrogen or androgen receptor.

Normal Puberty in Boys

Serum androgen concentrations in puberty span a vast spectrum naturally, with prepubertal values (<20 ng/dL) on the one hand and adult androgen concentrations (450–1,000 ng/dL) on the other (1, 27). Accordingly, cross-sectional studies—and preferably, wherever possible, longitudinal studies—of healthy pubertal boys provide a paradigm for investigating the association between physiological blood androgen concentrations and neuroendocrine activity of the somatotrophic axis. Recently, we have applied deconvolution analysis, a technique for estimating endogenous hormone secretion and/or clearance rates simultaneously (23, 24, 33), to 24-h serum GH profiles in 44 boys whose pubertal development spanned the range of Tanner stage I (prepubertal) through fully mature adult. Initial studies used a discrete peak detection methodology (e.g., cluster analysis [25]) and demonstrated that the mean serum GH concentration peak amplitude increased 2- to 3-fold in the mid to later stages of puberty, with no change in the number of detected GH peaks (26).

Further analysis using the multideconvolution technique (23, 24, 33) showed that boys in mid to late puberty exhibited increased mean and pulsatile serum GH concentrations that were mechanistically explained by specific amplification of the mass of GH secretory bursts, rather than by altering GH secretory burst frequency (27). Moreover, the particular mechanism driving an increased mass of GH released per secretory event was a pubertal rise in the maximal rate of GH secretion attained within each secretory burst (augmented secretory burst amplitude) with no alteration in the half-duration (duration of the secretion event at half-maximal amplitude). This increase in GH secretory burst mass could result in principle from *somatostatin* (SRIH) withdrawal, increased GHRH secretion per release episode, decreased negative feedback (IGF-I) signal strength, and/or enhanced sensitivity or secretory capacity of anterior pituitary somatotrope cells (34–40). Further studies utilizing SRIH and/or GHRH antagonists in clinical studies will be required to distinguish among these different mechanisms.

Linear correlation analysis demonstrated that in pubertal boys serum total testosterone concentrations are strongly positively correlated with the calculated daily GH secretion rate ($r = +0.73$, $P < 0.001$) (27). Of considerable interest, the dominant (positive) correlate of serum testosterone concentrations was the mass of GH secreted per burst ($r = +0.50$, $P < 0.001$), as shown in Figure 5.2. Since the mass of GH secreted per burst is proportional to the half-duration and amplitude of

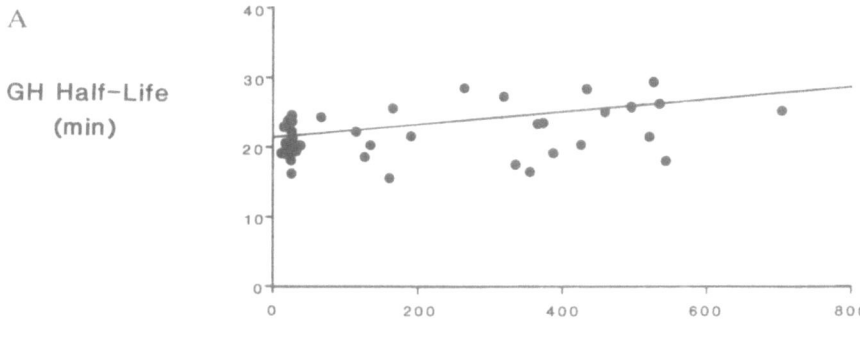

Serum Testosterone Concentration (ug/dL)

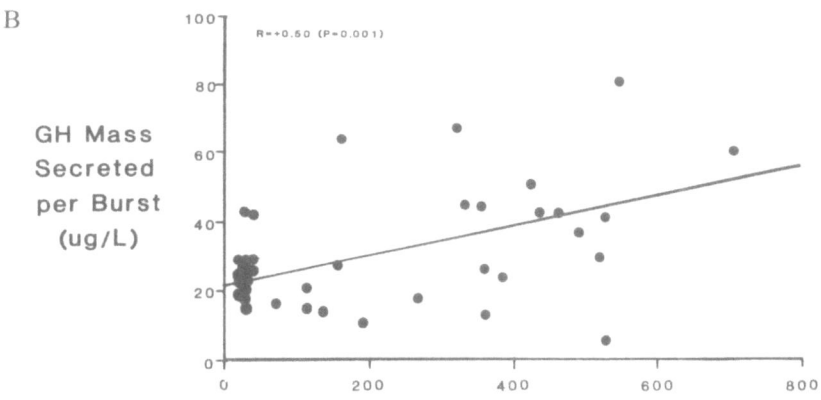

Serum Testosterone Concentration (ug/dL)

FIGURE 5.2. Correlation between specific measures of pulsatile GH secretion or GH half-life as estimated by deconvolution analysis and serum total testosterone concentrations in 44 boys studied at various stages of normal puberty. Volunteers underwent blood sampling as described in the legend of Fig. 5.1. The resulting 24-h serum GH concentration profiles were submitted to deconvolution analysis to compute the number, duration, mass, and amplitude of underlying GH secretory bursts and simultaneously estimate the half-life of endogenous GH, as described in references 23, 24, and 33. The panels show the linear relationships between total serum testosterone concentrations and the calculated half-life of endogenous GH (min) (*A*); the mass of GH secreted per burst (μg of GH secreted per L of distribution volume) (*B*); the daily GH secretory rate (μg/L/ 24 h) (*C*); and as a specificity control, GH secretory burst frequency (pulses/24 h) (*D*). (NS = not significant; $P > 0.05$.) Data are replotted and adapted with permission from Martha, Goorman, Blizzard, Rogol, and Veldhuis (27), © The Endocrine Society, 1992.

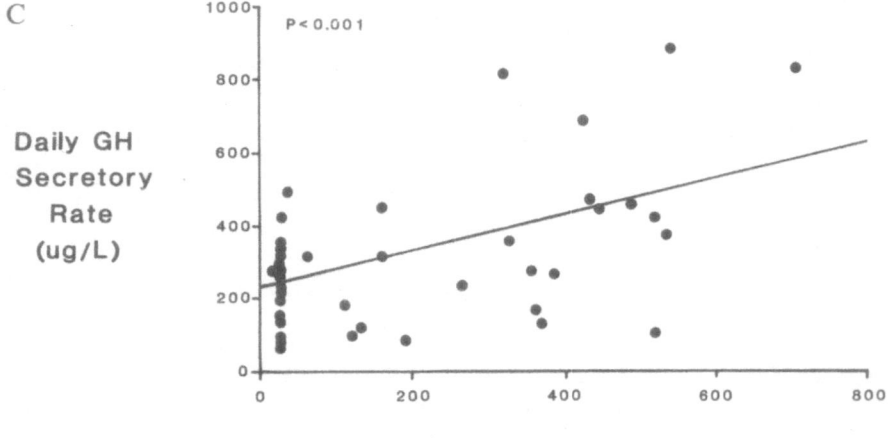

C

Daily GH
Secretory
Rate
(ug/L)

Serum Testosterone Concentration (ug/dL)

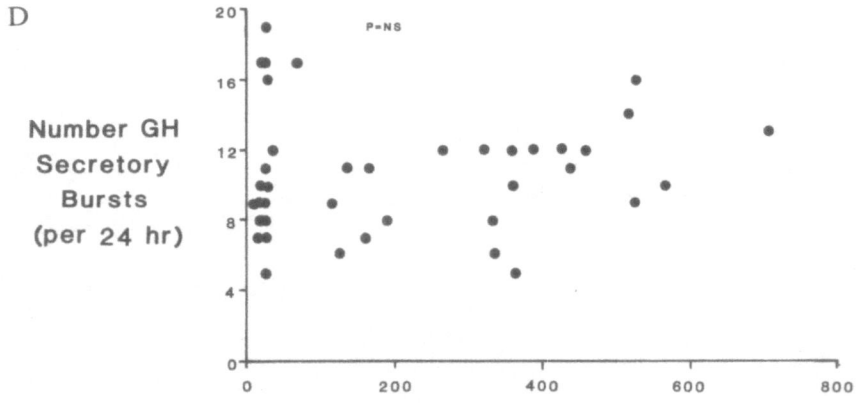

D

Number GH
Secretory
Bursts
(per 24 hr)

Serum Testosterone Concentration (ug/dL)

FIGURE 5.2. *Continued*

the secretory event (24), we evaluated whether serum testosterone concentrations correlated with either of these two measures. This analysis showed that serum testosterone concentrations were positively associated with the amplitude of GH secretory bursts ($r = +0.36$, $P < 0.02$) and their half-duration ($r = +0.37$, $P = 0.02$). The GH half-life and serum testosterone concentrations were also positively correlated ($r = +0.35$, $P = 0.03$) (Fig. 5.2). These relationships between serum total testosterone concentrations and GH secretory burst mass (and amplitude and duration) were specific since testosterone concentrations did not correlate with GH secretory burst frequency (Fig. 5.2) in this cohort of 44 pubertal boys (27).

Further studies in boys with constitutionally delayed puberty treated with testosterone showed that androgen administration increased the mass of GH secreted per burst in a selective manner without altering the GH half-life, secretory burst frequency, or duration (28). Although a qualitatively similar response was observed following administration of *oxandrolone*, a synthetic nonaromatizable androgen, the effect was noted in only 5 boys and was relatively minimal in 4 of the 5 individuals. In addition, the order of baseline versus oxandrolone administration in the 5 boys was not randomized, thus allowing for a possible sequence or intervention effect potentially biasing the second (oxandrolone-treated) evaluation in favor of increased GH secretion. Indeed, other studies suggest that one or more of these explanations may be important since nonaromatizable androgen or androgen acting through the androgen receptor does not appear to stimulate GH secretion; rather, it may be slightly inhibitory (see below).

Healthy Men of Varying Ages and Varying Serum Androgen Concentrations

In a cohort of 21 healthy men, we were able to document a strongly positive association between mean serum GH concentrations, or daily GH secretion rate calculated by deconvolution analysis, and serum total (or free) testosterone concentrations (30) (Fig. 5.3). Of considerable mechanistic import, serum total (and free) testosterone concentrations also correlated positively with GH secretory burst mass or amplitude ($r = +0.533$, $P = 0.02$) and calculated endogenous GH half-life ($r = +0.532$, $P = 0.006$) (Fig. 5.3). The combined tendency of GH secretory burst mass and half-life to increase with increasing serum testosterone concentrations in healthy men explains the strongly positive correlation of serum total testosterone concentrations with the mean serum GH concentration over 24 h since the mean hormone level is controlled jointly by secretion and clearance (41). Indeed, quantitative modeling studies indicate that for a single-pulse generator, an n-fold increase in GH secretory burst mass and an m-fold increase in hormone half-life will result in an m-fold $\times$ n-fold increase in the mean hormone level. Thus, for example, a 2-fold increase in GH secretory burst amplitude combined with a 30% increase in GH half-life would result in a 2.6-fold increase in the mean serum GH concentration.

Since we did not observe a correlation between serum total testosterone concentrations and GH half-life in boys *treated* with testosterone, we infer that its correlation with serum testosterone concentrations in healthy men and boys may reflect the covariation of serum testosterone concentrations with some other correlate of the GH half-life. One plausible covariate of serum testosterone concentrations is the *body mass index* (BMI) or relative adiposity (32). Indeed, in the group of 21 men, we found that BMI was

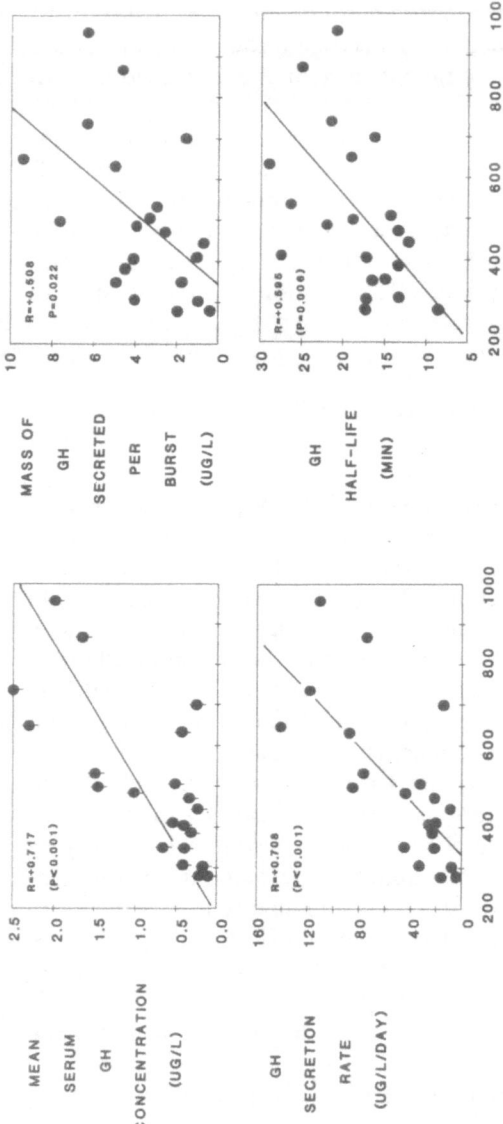

SERUM TOTAL TESTOSTERONE (NG/DL)

FIGURE 5.3. Significant linear correlations between mean (24-h) serum GH concentrations, as well as specific measures of pulsatile GH secretion or clearance, and total serum testosterone concentrations in a cohort of 21 healthy men whose ages range from 21 to 71 years, as described in reference 30. Healthy men underwent blood sampling at 10-min intervals for 24h. Serum GH concentrations were measured by IRMA. The GH series were subjected to multiparameter deconvolution analysis to estimate the number, duration, mass, and amplitude of underlying GH secretory pulses and to calculate the apparent half-life of endogenous GH, as described in references 23 and 24. The individual panels show the relationships between total serum testosterone concentrations and mean 24-h serum GH concentrations (upper-left panel), daily GH secretory rates (lower-left panel), the mass of GH secreted per burst (upper-right panel), and the calculated half-lives of endogenous GH (lower-right panel). Data are replotted and adapted with permission from Iranmanesh, Lizarralde, and Veldhuis (30), © The Endocrine Society, 1992.

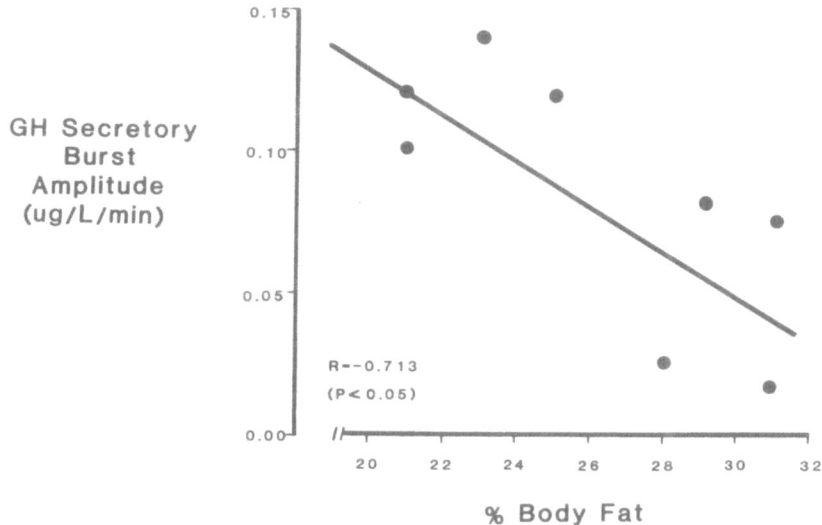

FIGURE 5.4. Negative relationship between percentage body fat and GH secretory burst amplitude in 6 healthy men. Body composition was estimated by underwater weighing. Data used with permission from Iranmanesh, Lizarralde, and Veldhuis (30), © The Endocrine Society, 1992.

correlated negatively with serum total and free testosterone concentrations. However, BMI is also correlated negatively with endogenous GH half-life estimated by deconvolution analysis in men (30, 32) or orchidectomized monkeys (42). Consequently, we hypothesize that the cross-sectional relationship between serum testosterone concentrations and calculated GH half-life in healthy men and boys studied over a wide age span and broad range of BMIs reflects the negative covariation of testosterone with BMI or adiposity. Indeed, we have observed that within a relatively narrow range of age and serum testosterone concentrations, the percentage of body fat as estimated by underwater weighing is a strongly negative correlate of GH secretion (Fig. 5.4) (30).

Steroid Hormone Manipulations

Cross-sectional studies do not allow facile inferences about cause and effect. Accordingly, we have investigated and are studying prospectively the impact of testosterone administration in several models of short-term or long-term testosterone withdrawal. Short-term testosterone withdrawal in healthy young men can be achieved by the administration of the GnRH agonist leuprolide, which results in down-regulation of the pituitary-gonadal axis (1). In this experiment one can search for correlations between GH secretion and serum testosterone concentrations achieved by concomitant parenteral androgen replacement therapy (Fryburg et al., unpublished data).

Moreover, in an alternative experimental model using young men, one can employ the potent and selective GnRH antagonist Nal-Glu-GnRH (43, 44) followed by testosterone replacement. In this paradigm one can ascertain whether testosterone dose dependently supports pulsatile GH secretion in an otherwise acutely androgen-deprived milieu (Pavlou et al., unpublished data). Third, we have administered saline, *estradiol* (E_2), or the nonaromatizable androgen DHT by constant *intravenous* (IV) infusions in healthy older men with mild physiological hypoandrogenemia (45). In this experimental context, we can compare the acute effects of estrogen and nonaromatizable androgen on the somatotrophic axis. We observed that in individuals with long-standing mild androgen deficiency associated with healthy aging, short-term DHT and E_2 infusions over 3.5 days do not differentially regulate GH secretion as estimated in the ultrasensitive GH chemiluminescence assay of 24-h serum pools (Fig. 5.5).

Last, we have replaced testosterone parenterally over 3 weeks in older men and in men with primary testicular failure in order to clarify the impact of testosterone replacement on the activity of the somatotrophic axis. We wish to determine in healthy aging whether androgen replacement can achieve at least partial reconstitution of physiological pulsatile GH secretion with significant (but not necessarily complete) restoration of GH secretory burst mass and amplitude. The GnRH agonist and antagonist strategies may also permit one to delineate the dose-dependent effects of serum concentrations of aromatizable androgen on specific measures of pulsatile GH release. To date, similar stimulatory effects of the nonaromatizable DHT have not been observed in our pilot studies, data that indirectly implicate the pathway of aromatization in mediating androgen action on the GH axis (see below).

An alternative strategy for assessing the specific mechanism of androgen's stimulation of pulsatile GH secretion is the use of selective antagonists of the estrogen or androgen receptor. In this regard, treatment with the antiestrogen tamoxifen reduces serum GH concentrations in men and impairs the stimulatory effect of testosterone (46). Conversely, the antiandrogen flutamide hydrochloride increases pulsatile GH release in healthy men, arguing for an *inhibitory* effect of nonaromatizable androgen acting via the androgen receptor (47). As an additional novel approach, we are evaluating the impact of a specific aromatase inhibitor on pulsatile GH release in healthy older men. We can also suggest investigation of the effects of an inborn error in testosterone's reduction to DHT on the regulation of the GH axis by studying patients with 5α-reductase deficiency.

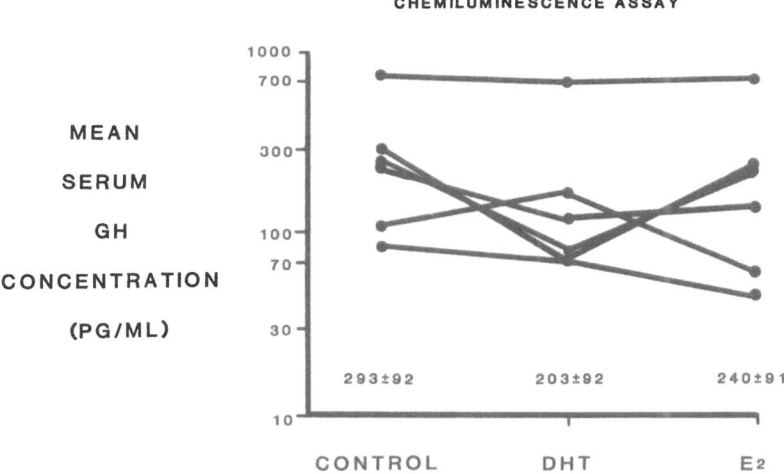

FIGURE 5.5. Lack of impact of acute 3.5-day infusions of E_2, DHT, or saline (control) in healthy older men on the mean 24-h serum GH concentration. Six older men (ages ≥65 years) underwent blood sampling at 10-min intervals for 24 h following 3.5 days of continuous IV infusion of E_2 (48 µg/day), DHT (7.0 mg/ day), or saline (control), as described in reference 48. Pooled aliquots from the 24-h venous sampling were submitted to GH chemiluminescence assay to obtain estimates of mean 24-h serum GH concentrations. This ultrasensitive chemiluminescence GH assay has a sensitivity of 0.005 µg/L, as described in reference 49. Serum E_2 levels rose from approximately 40 to 70 pg/mL during this infusion. Data are unpublished observations from Veldhuis, Iranmanesh, and Urban.

Summary

Given the body of data reviewed above, as developed by multiple experimental strategies, we hypothesize that aromatization to estrogen is important in testosterone's stimulation of the GH axis, whereas its action via the androgen receptor before or after its reduction to DHT is not (Fig. 5.6). The role of the androgen receptor can be tested further by administering testosterone to patients with inborn testosterone receptor defects; for example, complete testicular feminization (1). Considerably more study will be required to elucidate the exact hypothalamic-pituitary mechanisms by which aromatizable androgen stimulates GH secretion in humans.

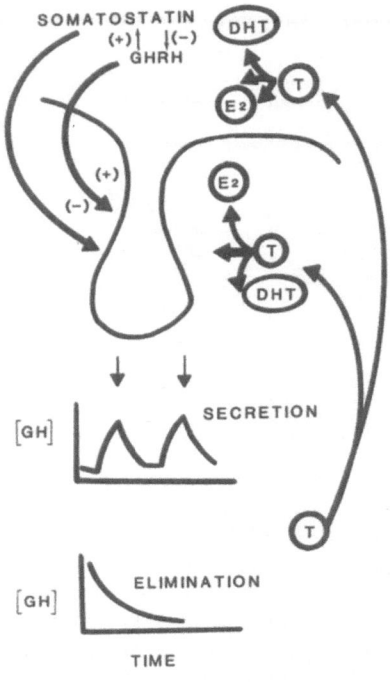

FIGURE 5.6. Proposed schema for the regulatory actions of androgens on pulsatile GH secretion in the human. Effects of androgens can be exerted at the level of the hypothalamic regulatory peptides (GHRH and SRIH [somatostatin]), and/or on the anterior pituitary gland, and/or via modifying IGF-I feedback. GHRH and SRIH show reciprocal interactions also. According to one hypothesis, available aromatizable androgen acts following its conversion via the aromatase reaction to E_2 (when the precursor is testosterone), or estrone (when androstenedione is the precursor) in the pituitary gland or neural tissue, or both. Testosterone acting via the androgen receptor before or after its 5α reduction to DHT either exerts no effect on GH secretion or produces some degree of inhibition. Alternatively, we suggest that apparent inhibition of GH secretion following pharmacological doses of nonaromatizable androgen (e.g., DHT) may reflect suppression of endogenous E_2 production via feedback effects of androgen on the hypothalamic-pituitary gonadal axis, as described in references 1 and 48.

Administration of estrogen to individuals without functional gonads (e.g., prepubertal girls with Turner syndrome) stimulates GH secretion as described in references 21 and 22. Estrogen is also able to suppress hepatic IGF-I production, resulting in withdrawal of IGF-I's negative feedback effect on the somatotrophic axis, as described in references 39, 40, and 61. However, the action of aromatizable androgen in boys and men clearly does not depend on suppression of total plasma IGF-I concentrations since these values increase significantly in normal puberty when serum GH concentrations rise concurrently, as described in reference 27.

The above schema also illustrates that administration of an antiestrogen in healthy men would impede estrogen's stimulatory effect on the GH axis, described in 46, and by increasing testosterone (and, putatively, 5α-reduced androgen in certain target tissues), may partially suppress the GH axis. Conversely, the antiandrogen flutamide can be envisioned as opposing the minimally inhibitory (or null) effect of pure androgen on the somatotrophic axis, as well as promoting an increase in gonadotropin secretion with a consequent rise in gonadal E_2 production, as described in references 47, 62, and 63. The latter may itself evoke an overall stimulation of the GH axis. Thus, the above schema harmonizes available observations following administration of estrogen, aromatizable androgen, nonaromatizable androgen, antiestrogen, and antiandrogen. Not shown are the inhibitory feedback loops by which IGF-I and GH itself can inhibit GH secretion. We speculate that steroid hormones also modify one or both of these feedback control mechanisms.

Acknowledgments. We thank Patsy Craig for her skillful preparation of the manuscript and Paula P. Azimi for the artwork. This work was supported in part by NIH Grant RR-00847 to the Clinical Research Center of the University of Virginia, RCDA 1-KO4-HD-00634 (J.D.V.), VA Merit Review (A.I.), the Diabetes and Endocrinology Research Center NIH Grant DK-38942, the NIH-supported Clinfo Data Reduction Systems, the Pratt Foundation, the University of Virginia Academic Enhancement Fund, and the National Science Foundation Center for Biological Timing (NSF Grant DIR-89-20162).

References

1. Veldhuis JD. Dynamics of the hypothalamic pituitary-testicular axis. In: Yen SSC, Jaffe RB, eds. Reproductive endocrinology. Philadelphia: WB Saunders, 1991:409–59.
2. Massa G, Mulumba N, Ketelslegers JM, Maes M. Initial characterization and sexual dimorphism of serum growth hormone-binding protein in adult rats. Endocrinology 1990;126:1976–80.
3. Painson JC, Tannenbaum GS. Sexual dimorphism of somatostatin and growth hormone-releasing factor signaling in the control of pulsatile growth hormone secretion in the rat. Endocrinology 1991;128:2858–66.
4. Terry L, Saunders A, Aadet J, Willoughby J, Brazeau P, Martin J. Physiological secretion of growth hormone and prolactin in male and female rats. Clin Endocrinol (Oxf) 1977;6(suppl):19s–28s.
5. Clark RG, Jansson JO, Isaksson O, Robinson IC. Intravenous growth hormone: growth responses to patterned infusions in hypophysectomized rats. J Endocrinol 1985;104:53–61.
6. Mode A, Wiersma-Larsson E, Strom A, Zaphiropoulos PG, Gustafsson JA. A dual role of growth hormone as a feminizing and masculinizing factor in the control of sex-specific cytochrome P450 isozymes in rat liver. J Endocrinol 1989;120:311–7.
7. Mode A, Norstedt G, Simic B, Eneroth P, Gustafsson JA. Continuous infusion of growth hormone feminizes hepatic steroid metabolism in the rat. Endocrinology 1981;103:2103–8.
8. Shapiro BH, MacLeod JN, Pampori NA, Morrissey JJ, Lapenson DP, Waxman DJ. Signalling elements in the ultradian rhythm of circulating growth hormone regulating expression of sex-dependent forms of hepatic cytochrome P450. Endocrinology 1989;125:2935–44.
9. Sundseth SS, Alberta JA, Waxman DJ. Sex-specific, growth hormone-regulated transcription of the cytochrome p450 2C11 and 2C12 genes. J Biol Chem 1992;267(6):3907–14.
10. Waxman DJ, Pampori NA, Ram PA, Agrawal AK, Shapiro BH. Interpulse interval in circulating growth hormone patterns regulates sexually dimorphic expression of hepatic cytochrome P450. Proc Natl Acad Sci USA 1991;88:6868–72.
11. Bick T, Hochberg Z, Amit T, Isaksson OGP, Johnolov J. Roles of pulsatility and continuity of growth hormone (GH) administration in the regulation of

hepatic GH-receptors, and circulating GH-binding protein and insulin-like growth factor-I. Endocrinology 1992;131:423–9.

12. Eden S. Age and sex related differences in episodic growth hormone secretion in the rat. Endocrinology 1979;105:555–60.

13. Iranmanesh A, Veldhuis JD. Clinical pathophysiology of the somatotropic (GH) axis in adults. In: Veldhuis JD, ed. Endocrinology and metabolism clinics of North America. Philadelphia: WB Saunders, 1992:783–816.

14. Clark RG, Carlsson LMS, Robinson ICAF. Growth hormone secretory profiles in conscious female rats. J Endocrinol 1987;114:399–407.

15. Jansson J-O, Frohman LA. Differential effects of neonatal and adult androgen exposure on the growth hormone secretory pattern in male rats. Endocrinology 1987;120:1551–7.

16. Evans WS, Faria ACSA, Christiansen E, et al. Impact of intensive venous sampling on characterization of pulsatile GH release. Am J Physiol 1987;252: E549–56.

17. Ho KY, Evans WS, Blizzard RM, et al. Effects of sex and age on the 24-hour profile of growth hormone secretion in man: importance of endogenous estradiol concentrations. J Clin Endocrinol Metab 1987;64:51–8.

18. Hartman ML, Faria ACS, Vance ML, Johnson ML, Thorner MO, Veldhuis JD. Temporal structure of in vivo growth hormone secretory events in man. Am J Physiol 1991;260(1):E101–10.

19. Weltman JY, Veldhuis JD, Weltman A, Kerrigan JR, Evans WS, Rogol AD. Reliability of estimates of pulsatile characteristics of luteinizing hormone (LH) and growth hormone (GH) release in women. J Clin Endocrinol Metab 1990;71(6):1646–52.

20. Faria ACS, Bekenstein LW, Booth RA Jr, et al. Pulsatile growth hormone release in normal women during the menstrual cycle. Clin Endocrinol (Oxf) 1992;36:591–6.

21. Mauras N, Rogol AD, Veldhuis JD. Specific, time-dependent actions of low-dose estradiol administration on the episodic release of GH, FSH and LH in prepubertal girls with Turner's syndrome. J Clin Endocrinol Metab 1989;69: 1053–8.

22. Mauras N, Rogol AD, Veldhuis JD. Increased hGH production rate after low-dose estrogen therapy in prepubertal girls with Turner's syndrome. Pediatr Res 1990;28(6):626–30.

23. Veldhuis JD, Carlson ML, Johnson ML. The pituitary gland secretes in bursts: appraising the nature of glandular secretory impulses by simultaneous multiple-parameter deconvolution of plasma hormone concentrations. Proc Natl Acad Sci USA 1987;84:7686–90.

24. Veldhuis JD, Johnson ML. Deconvolution analysis of hormone data. Methods Enzymol 1992;210:539–75.

25. Veldhuis JD, Johnson ML. Cluster analysis: a simple, versatile and robust algorithm for endocrine pulse detection. Am J Physiol 1986;250:E486–93.

26. Martha PM Jr, Rogol AD, Veldhuis JD, Kerrigan JR, Goodman DW, Blizzard RM. Alterations in the pulsatile properties of circulating growth hormone concentrations during puberty in boys. J Clin Endocrinol Metab 1989;69:563–70.

27. Martha PM Jr, Goorman KM, Blizzard RM, Rogol AD, Veldhuis JD. Endogenous growth hormone secretion and clearance rates in normal boys as

determined by deconvolution analysis: relationship to age, pubertal status and body mass. J Clin Endocrinol Metab 1992;74:336–44.

28. Ulloa-Aguirre A, Blizzard RM, Garcia-Rubi E, et al. Testosterone and oxandrolone, a non-aromatizable androgen, specifically amplify the mass and rate of growth hormone (GH) secreted per burst without altering GH secretory burst duration or frequency or the GH half-life. J Clin Endocrinol Metab 1990;71:846–54.

29. Thompson RG, Rodriguez A, Kowarski AA, Migeon CJ, Blizzard RM. Integrated concentrations of growth hormone correlated with plasma testosterone and bone age in preadolescent and adolescent males. J Clin Endocrinol Metab 1978;62:1341–4.

30. Iranmanesh A, Lizarralde G, Veldhuis JD. Age and relative adiposity are specific negative determinants of the frequency and amplitude of GH secretory bursts and the half-life of endogenous GH in healthy men. J Clin Endocrinol Metab 1991;73:1081–8.

31. Carlson HE, Gillin JC, Gorden P, Synder F. Absence of sleep related growth hormone peaks in aged normal subjects and in acromegaly. J Clin Endocrinol Metab 1972;34:1102–5.

32. Veldhuis JD, Iranmanesh A, Ho KKY, Lizarralde G, Waters MJ, Johnson ML. Dual defects in pulsatile growth hormone secretion and clearance subserve the hyposomatotropism of obesity in man. J Clin Endocrinol Metab 1991;72:51–9.

33. Veldhuis JD, Faria A, Vance ML, Evans WS, Thorner MO, Johnson ML. Contemporary tools for the analysis of episodic growth hormone secretion and clearance in vivo. Acta Paediatr Scand 1988;347:63–82.

34. Tannenbaum GS. Neuroendocrine control of growth hormone secretion. Acta Paediatr Scand Suppl 1991;372:5–16.

35. Smith PN, Howe PRC, Oliver JR, Willoughby JO. Growth hormone releasing factor immunoreactivity in rat hypothalamus. Neuropeptides 1984;4: 109–15.

36. Thomas GB, Cummins JT, Francis H, Sudbury AW, McCloud PI, Clarke IJ. Effect of restricted feeding on the relationship between hypophysial portal concentrations of growth hormone (GH)-releasing factor and somatostatin, and jugular concentrations of GH in ovariectomized ewes. Endocrinology 1991;128:1151–8.

37. Kasting NW, Martin JB, Arnold MA. Pulsatile somatostatin release from the median eminence of the unanesthetized rat and its relationship to plasma growth hormone levels. Endocrinology 1981;109:1739–45.

38. Sato M, Takahara J, Fujioka Y, Niimi M, Irino S. Phyisological role of growth hormone (GH)-releasing factor and somatostatin in the dynamics of GH secretion in adult male rat. Endocrinology 1988;123:1928–33.

39. Berelowitz M, Szabo M, Frohman LA, Firestone S, Chu L. Somatomedin-C mediates growth hormone negative feedback by effects on both the hypothalamus and the pituitary. Science 1981;212:1279–81.

40. Tannenbaum GS, Guyda HJ, Posner BI. Insulin-like growth factors: a role in growth hormone negative feedback and body weight regulation via brain. Science 1983;220:77–80.

41. Veldhuis JD, Lassiter AB, Johnson ML. Operating behavior of dual or multiple endocrine pulse generators. Am J Physiol 1990;259:E351–61.

56   J.D. Veldhuis et al.

42. Dubdey AK, Ahanukoglu A, Hansen BC, Kowarski AA. Metabolic clearance rates of synthetic human growth hormone in lean and obese male rhesus monkeys. J Clin Endocrinol Metab 1988;67:1064–7.
43. Urban RJ, Pavlou SN, Rivier JE, Vale WW, Dufau ML, Veldhuis JD. Suppressive actions of a gonadotropin-releasing hormone (GnRH) antagonist on LH, FSH, and prolactin release in estrogen-deficient postmenopausal women. Am J Obstet Gynecol 1990;162:1255–60.
44. Pavlou SN, Veldhuis JD, Lindner J, et al. Persistence of concordant LH, testosterone and alpha subunit pulses following LHRH antagonist administration in normal men. J Clin Endocrinol Metab 1990;70:1472–8.
45. Veldhuis JD, Urban RJ, Lizarralde G, Johnson ML, Iranmanesh A. Attenuation of luteinizing hormone secretory burst amplitude is a proximate basis for the hypoandrogenism of healthy aging in men. J Clin Endocrinol Metab 1992;75:707–13.
46. Weissberger AJ, Ho KKY. Activation of the somatotropic axis by testosterone in adult males: evidence for the role of aromatization. J Clin Endocrinol Metab 1993;1407:1412.
47. Metzger DL, Kerrigan JR. Androgen receptor blockade with flutamide enhances growth hormone secretion in late pubertal males: evidence for independent actions of estrogen and androgen. J Clin Endocrinol Metab 1993;76:1147–52.
48. Urban RJ, Dahl KD, Padmanabhan V, Beitins IZ, Veldhuis JD. Specific regulatory actions of dihydrotestosterone and estradiol on the dynamics of FSH secretion and clearance in man. J Andrology 1991;12:27–35.
49. Iranmanesh A, Grisso B, Veldhuis JD. Low basal and persistent pulsatile growth hormone secretion are revealed in normal and hyposomatotropic men studied with a new ultrasensitive chemiluminescence assay. J Clin Endocrinol Metab 1994;78:526–35.
50. Clark RG, Carlsson LMS, Rafferty B, Robinson ICAF. The rebound release of growth hormone (GH) following somatostatin infusion in rats involves hypothalamic GH-releasing factor release. J Endocrinol 1988;119:397–404.
51. Harel Z, Tannenbaum GS. Synergistic interaction between insulin-like growth factors-I and -II in central regulation of pulsatile growth hormone secretion. Endocrinology 1992;131(2):758–64.
52. Miki N, Ono M, Miyoshi H, Tsushima T, Shizume K. Hypothalamic growth hormone-releasing factor (GRF) participates in the negative feedback regulation of growth hormone secretion. Life Sci 1981;44:469–76.
53. Ono M, Miki N, Demura H. Effect of antiserum to rat growth hormone (GH)-releasing factor on physiological GH secretion in the female rat. Endocrinology 1991;129:1791–6.
54. Aguila MC, McCann SM. Stimulation of somatostatin release in vitro by synthetic human growth hormone-releasing factor by a nondopaminergic mechanism. Endocrinology 1985;117:762–5.
55. Katakami H, Downs TR, Frohman LA. Inhibitory effect of hypothalamic medial preoptic area somatostatin on growth hormone-releasing factor in the rat. Endocrinology 1988;123:1103–9.
56. Kraicer J, Sheppard MS, Luke J, Lussier B, Moor BC, Cowan JS. Effect of withdrawal of somatostatin and growth hormone (GH)-releasing factor on GH release in vitro. Endocrinology 1988;122:1810–7.

57. Miki N, Ono M, Shizume K. Withdrawal of endogenous somatostatin induces secretion of growth hormone-releasing factor in rats. J Endocrinol 1988;117: 245–52.
58. Plotsky PM, Vale W. Patterns of growth hormone-releasing factor and somatostatin secretion into the hypophysial-portal circulation of the rat. Science 1985;230:461–5.
59. Terry LC, Martin JB. The effects of lateral hypothalamic-medial forebrain stimulation and somatostatin antiserum on pulsatile growth hormone secretion in freely behaving rats: evidence for a dual regulatory mechanism. Endocrinology 1981;109:622–7.
60. Weiss J, Cronin MJ, Thorner MO. Periodic interactions of GH-releasing factor and somatostatin can augment GH release in vitro. Am J Physiol 1987;253:E508–16.
61. Dawson-Hughes B, Stern D, Goldman J, Reichlin S. Regulation of growth hormone and somatomedin-C secretion in postmenopausal women: effect of physiological estrogen replacement. J Clin Endocrinol Metab 1986;63:424–32.
62. Urban RJ, Dahl KD, Lippert MC, Veldhuis JD. Endogenous androgen and estrogen modulate immunoradiometric and bioactive FSH secretion and clearance in young and elderly men. J Androl 1992;13:579–86.

6

Growth Hormone and IGF-I as Anabolic Partitioning Hormones

David R. Clemmons

Growth hormone (GH) is an anabolic hormone that stimulates protein synthesis in normal volunteers. Although it also stimulates protein catabolism, the increase in protein synthesis is significantly greater than the concomitant increase in catabolism, resulting in net protein accretion. Two types of interventions have been used to induce catabolism in normal volunteers. These include caloric restriction and administration of glucocorticoids (1, 2). Normal volunteers who received high doses of prednisone to induce nitrogen loss responded to GH with significant improvement in nitrogen balance and an increase in nonoxidative leucine metabolism, which is an index of stimulation of protein synthesis (3). Similarly, several studies in calorically restricted normal volunteers have shown that GH results in protein conservation (4, 5).

However, the molecular mechanism by which GH stimulates protein synthesis in normal volunteers and in patients with catabolic states is not completely defined. In catabolic patients GH has been shown to increase both protein synthesis and protein breakdown, with a greater increase in synthesis (6, 7). The catabolic patients that have been studied include patients with burns (8, 9), patients recovering from gastrointestinal surgery (6, 7), acute intensive care unit patients (10), patients with chronic renal failure (11), and patients undergoing hyperalimentation therapy (12) in whom significant weight loss has occurred. Likewise, chronic medical conditions, such as *acquired immunodeficiency syndrome* (AIDS) (13) and chronic obstructive lung disease (14), that are associated with increased catabolism have been shown to be responsive to GH. Most of these studies have measured an improvement in either nitrogen balance or protein synthesis as determined by stable isotope infusion techniques.

These findings, taken together, suggest that GH can cause a significant anabolic response in both normal volunteers and patients with catabolism. However, the response to GH is complicated by the fact that there is an increase in protein breakdown as well as synthesis; therefore, the mechanism must involve coordination between these two responses. Why GH stimulates both an increase in catabolism and an increase in protein synthesis is unclear, although various theories have been put forth concerning the need for increased tissue protein turnover during recovery from acute disease states.

The relationship between GH's catabolic actions and anabolic effects may be an important one. Structural analysis of the *GH receptor* (GH-R) has shown that it is a member of the cytokine receptor family (15). Its receptor structure is similar to *interleukin-3* (IL-3) and *interleukin-6* (IL-6), as well as *tumor necrosis factor α* (TNFα) (16). These cytokines are known to cause catabolism of fat and glycogen stores, as well as stimulate protein breakdown. This suggests that the conserved GH-R structure may indicate either that the GH-R has dual functions or that it has a relatively greater role in stimulating the catabolic events.

Catabolism of fat is an important component of the mechanism by which GH stimulates anabolism. Ward et al., in a detailed study in postoperative patients, have shown that the marked increase in fat breakdown that occurs in patients who are receiving hyperalimentation therapy serves as a source of calories that are needed to enhance protein synthesis (7). In that study fat as a source of energy was increased from 47% of calories to 72% of calories after GH. Likewise, Jiang et al., in studying postoperative patients who were being fed a hypocaloric diet, showed that the rate of fat breakdown was markedly accelerated after GH (6).

In GH-deficient and non-GH-deficient patients, GH results in significant partitioning, with a major rise in free fatty acids and glycerol, indicating triglyceride breakdown; this rise is sustained for several hours (17). The increase in glycerol occurs repetitively with each dose of GH and does not attenuate significantly over time, even in the face of weight loss and significant caloric restriction. In GH-deficient patients this ultimately results in a highly significant reduction in fat stores with a concomitant increase in lean tissue (18). This suggests that the GH-R may be intimately involved in mediating fat breakdown and that the release of free fatty acids into the circulation may provide the fuel substrate that is necessary to maintain the high rates of protein synthesis induced by GH while also achieving increased rates of catabolism. Maintaining this accelerated rate of protein turnover may require a high rate of fat breakdown. Whether severely catabolic patients who have less than 2%–4% body fat stores would be able to sustain such rates of fat breakdown over a long period of time in response to GH therapy is an important research question.

Because GH induces fat breakdown and because its receptor is in the cytokine receptor family, the question is raised as to whether stimulation of protein catabolism by GH is also mediated through this receptor. No direct studies have been performed, however, because it has been difficult to show a direct effect of GH on protein anabolism in vitro. It is conceivable that GH is directly stimulating an increase in protein breakdown while stimulating protein synthesis indirectly through *insulin-like growth factor I* (IGF-I). This has been difficult to assess because most tissues that have GH-Rs also synthesize IGF-I. Thus, it is very difficult to study the effect of GH in isolation because there is usually IGF-I present in the system, either as a result of culturing cells in tissues in serum that contains IGF-I or by synthesis of IGF-I by the tissue of relevance.

The ability of GH to stimulate protein synthesis and tissue accretion in normal and catabolic patients is at least partially mediated by IGF-I. GH has been shown to induce IGF-I synthesis in multiple tissues, including bone, cartilage, and connective tissue, as well as muscle (19). Therefore, to the extent that these cell types and tissues possess GH-Rs, they may be able to synthesize IGF-I locally, and locally acting IGF-I can enhance protein synthesis through its receptor. The *IGF-I receptor* (IGF-I-R) is not in the cytokine class of receptors, and infusions of IGF-I into rats or humans do not stimulate an increase in protein breakdown (20). More importantly, if plasma amino acid levels are maintained within the normal range, infusions of IGF-I stimulate protein synthesis (21). This suggests that the effects of GH on protein synthesis may be predominately mediated through IGF-I. In contrast to protein synthesis in muscle, fat cells do not contain IGF-I-Rs; therefore, IGF-I does not counteract the effect of GH on fat breakdown.

We have conducted several studies in normal human volunteers to study the capacity of IGF-I, GH, or a combination of IGF-I and GH to increase protein accretion using caloric restriction of normal volunteers as a model of catabolism. In our first study of IGF-I, 6 volunteers who were within 10% of *ideal body weight* (IBW) were calorically restricted to 50% of a normal caloric intake for a period of 2 weeks. During week 1, we obtained frequent measurements of IGF-I and nitrogen excretion, as well as *blood urea nitrogen* (BUN), glucose, and C-peptide. During week 2, they received either GH, 0.05 mg/day *subcutaneously* (sc), or IGF-I at an infusion rate of 12 µg/kg/h for 16 h per day. The dietary restriction to 20 kcal/kg IBW induced a negative nitrogen balance to -236 ± 45 mmol/day. This degree of catabolism was relatively constant throughout the last 4 days of the restriction period. Following injections of GH, the subjects improved nitrogen balance to -65 ± 40 mmol/day on IGF-I, and this response was maintained throughout the treatment period (22). GH induced a similar degree of improvement in nitrogen retention. Serum urea nitrogen was decreased with both treatments, indicating transport of nitrogen into lean tissue. The degree of decrease was slightly greater with IGF-I than with GH.

Changes in carbohydrate metabolism were quite different, however. Specifically, GH caused blood glucose to increase from 4.75 ± 1.01 to

5.48 ± 1.00 mmol/L, and C-peptide increased by 47%, indicating induction of insulin resistance. In contrast, IGF-I induced significant hypoglycemia, such that the infusions had to be stopped on several occasions, and mean daily plasma glucose was reduced. Concomitantly, C-peptide was suppressed by 66%, suggesting that IGF-I was acting as a primary stimulant of glucose transport in these patients.

The IGF-I levels that were achieved were also different in the two groups. IGF-I rose from 294 ± 52 to 1092 ± 244 μg/L during the last 4 days of infusions. In contrast, GH increased IGF-I to only 501 ± 86 μg/L, or a 1.8-fold increase over baseline values. To determine if the binding profiles of IGF-I in serum were altered and whether this correlated with the change in peak serum IGF-I levels, *IGF binding proteins* (IGFBPs) 1, 2, and 3 were measured. IGFBP-3 rose to approximately 3000 ng/mL, but then levels tapered during infusion concomitantly with the decline in IGF-I. In contrast, subjects who received GH had a significant increase in IGFBP-3 values and no tapering during the last 3 days of treatment (23). IGFBP-2 rose significantly (2.2-fold) during the IGF-I infusions, but was unchanged with GH administration. Therefore, the IGF-I infusion resulted in a significantly different pattern of IGFBPs as compared to GH. The rapidly inducible IGF-I transporters, IGFBP-1 and -2, are induced with IGF-I, and this would be expected to result in a decrease in the half-life of the IGF-I that is bound to those transporters. In contrast, the stable, high-molecular weight complex of IGFBP-3 and *acid-labile subunit* (ALS) increases with GH and decreases with IGF-I, suggesting that the distribution of IGF-I that is bound to transporter would be markedly different under the two treatment regimens.

In summary, these findings suggest that infusions of IGF-I are anabolic, but that this requires plasma concentrations that are in the acromegalic range and may be associated with significant toxicity, such as hypoglycemia. Therefore, this treatment may be useful only in subjects who are receiving high rates of glucose infusions or who have underlying type II diabetes. In contrast, GH clearly is anabolic in such a setting, although the degree of improvement is limited to 30%; if a greater effect is desired, higher doses that are clearly associated with toxicity would be required.

Because of these problems with both drugs, we wondered whether it might be possible to achieve a substantially greater anabolic effect by giving a combination of GH and IGF-I to such patients. The rationale was that combined therapy would utilize the cytokine-like activities of GH to mobilize substrates, such as free fatty acids, and yet not compromise the responsiveness to IGF-I. Likewise, since fat cells do not contain IGF-I-Rs, the fat-mobilizing effect of GH would not be compromised. Furthermore, since GH and IGF-I have opposite effects on glucose metabolism, this regimen might be expected to maintain more near-normal glucose homeostasis as compared to administering either GH or IGF-I alone. Since some induction of insulin appears to be necessary for IGF-I to induce an optimal increase in protein synthesis, it also appeared that suppression of insulin by IGF-I might be counteracting some of its beneficial effects.

To exactly measure the response compared to the previous study, 7 normal-weight volunteers were recruited. The study design was identical to the first study. Specifically, the subjects were calorically restricted (20 kcal/kg IBW) for 7 days and then underwent an additional 7-day treatment period during which the caloric restriction was continued (24). The response variables that were measured were similar to the previously described study. During the treatment interval IGF-I was infused exactly as previously—that is, 12 µg/kg/h for 16 h and this was compared to subjects receiving this infusion plus GH injections. Each subject served as his own control.

The diet once again induced a negative nitrogen balance to −136 ± 41 mmol/day. This was partially reversed with IGF-I infusion, which increased nitrogen retention by 108 ± 29 mmol/day. This degree of change was similar to that induced in the previous study. However, the combination of drugs caused all subjects to enter a positive nitrogen balance, which improved by 262 ± 43 mmol/day, and this was maintained throughout the treatment period. This difference was highly significant. When the effect on urinary nitrogen excretion was compared across the three treatments, GH caused a 26% reduction; IGF-I, a 23% reduction; and the combined therapy, a 56% reduction. Thus, the combined therapy appeared to be at least additive at these doses.

The metabolic effects on carbohydrate metabolism were also quite distinct. The combined therapy allowed maintenance of normal C-peptide concentrations, the levels being similar to control, whereas IGF-I caused suppression. Likewise, the combined therapy allowed maintenance of relatively normal glucose homeostasis, with fasting blood glucose not changing significantly during the study (4.3 ± 1.0 mmol/L), whereas IGF-I again suppressed it to 3.8 ± 0.8 mmol/L. Serum urea nitrogen fell dramatically with combined therapy, and the decrease was significantly greater than that which occurred with IGF-I. There was minimal toxicity. All subjects had parotid gland enlargement, 2 had edema, and 4 had severe episodes of hypoglycemia on IGF-I alone, although none had severe hypoglycemia requiring termination of the IGF-I infusion if GH was also given.

Changes in IGFBPs were also quite distinct. The group receiving combined therapy had a 30% increase in IGFBP-3 and ALS, whereas those receiving IGF-I had a 63% suppression of IGFBP-3 and ALS. Since ALS is GH dependent, this suggests that GH is inhibited by IGF-I infusion; this results secondarily in a decrease in ALS with a concomitant decrease in IGFBP-3. This decrease in IGF-I carrying capacity no doubt shifts the amount of IGF-I that is bound to IGFBP-1 and -2, which are induced by IGF-I therapy, but not by combined therapy. This would result in a substantial reduction of plasma half-life of IGF-I in the subjects who received IGF-I alone, suggesting that the infused IGF-I is cleared much more rapidly in these subjects as compared to those subjects receiving combined therapy. Whether this series of changes is adequate to explain the difference in anabolism that was observed is unclear. Other potential mechanisms that would explain this difference are the maintenance of more normal insulin levels that would allow better stimulation of protein synthesis or some other direct effect of GH on protein anabolism that was not measured in the study.

Interestingly, the catabolic effects of GH on fat metabolism are maintained with this regimen. Unlike insulin, IGF-I does not stimulate lipid synthesis since there are no IGF-I-Rs contained in fat cells. Hussein et al. have shown that combined therapy for 5 days results in a marked increase in plasma free fatty acids, a marked increase in the rate of lipid oxidation, and a decrease in whole-body fat stores (25). If this effect of combined treatment could be maintained for extended periods, it could result in substantial nutrient partitioning. In summary, the combination of GH plus IGF-I appears to induce marked improvement in anabolism and marked decreases in whole-body fat stores. How long this effect can be maintained in patients who are catabolic is unclear. However, it appears that the combined therapy holds great promise for reversing catabolic conditions with many types of treatment regimens.

In a third study we have recently tested the hypothesis that IGF-I plus GH may be anabolic in patients with severe underlying catabolic conditions. Eight patients with AIDS who were between 76% and 88% of IBW were recruited. They were fed a normal diet in our clinical research unit for 2 weeks and then received 1 week of either IGF-I alone, GH alone, or IGF-I plus GH. These subjects all entered positive nitrogen balance on GH plus IGF-I. Specifically, the reduction in urinary urea nitrogen on combined therapy was 48%. All subjects entered and remained in positive nitrogen balance throughout the combined treatment period. Unlike the normal volunteers, parotid tenderness was the only complication. There was no edema, arthralgia, or headaches and no significant hypoglycemia. Changes in IGF-I serum levels and IGFBPs are currently being analyzed. It is clear, then, that catabolic subjects can respond to combined therapy with an improvement in anabolic state, although the degree to which their reduction in whole-body fat stores would limit this response on chronic therapy is unknown. The findings suggest that patients who are acutely catabolic even with limited fat stores may be able to mobilize sufficient energy to respond with increased protein accretion to this regimen. Future studies should address the efficacy of this regimen in a variety of catabolic states.

Acknowledgments. The author wishes to thank Ms. Leigh Elliott for her help in preparing the manuscript. This work was supported by NIH Grants AG-02331 and HD-28081. The studies were performed on the Clinical Research Unit of UNC Hospitals, which is supported by Grant RR-000031.

References

1. Clemmons DR, Snyder DK, Williams R, Underwood L. Treatment with growth hormone conserves lean body mass during dietary restriction in obese subjects. J Clin Endocrinol Metab 1987;64:878–83.
2. Horber FF, Haymond MW. Human growth hormone prevents protein catabolic side effects of prednisone in humans. J Clin Invest 1990;86:265–72.
3. Mauras N, Horber FF, Haymond MW. Low dose recombinant human insulin-like growth factor-I fails to affect protein anabolism but alters islet cell secretion in humans. J Clin Endocrinol Metab 1992:1192–7.
4. Snyder DK, Clemmons DR, Underwood LE. Treatment of obese, diet-restricted subjects with growth hormone for 11 weeks: effects on anabolism, lipolysis and body composition. J Clin Endocrinol Metab 1988;67:54–61.
5. Snyder DK, Clemmons DR, Underwood LE. Dietary carbohydrate content determines responsiveness to growth hormone in humans. J Clin Endocrinol Metab 1989;69:745–52.
6. Jiang ZM, He GZ, Zhang SY. Low dose growth hormone and hypocaloric nutrition attenuate the protein catabolic response after major operation. Ann Surg 1989;210:513–25.
7. Ward HC, Halliday D, Sim AW. Protein and energy metabolism with biosynthetic human growth hormone after gastrointestinal surgery. Ann Surg 1987;206:56–61.
8. Herndon DN, Barrow RE, Kunkel KR, Broemeling L, Rutan RL. Effects of recombinant human growth hormone on donor-site healing in severely burned children. Ann Surg 1990;212:424–31.
9. Sherran SK, Denling RH, LaLoude C, Erikson E, Wilmone RW. Growth hormone enhances re-epithelialization of human split thickness skin graft donor sites. Surg Forum 1993;40:37–9.
10. Zeigler TR, Young LS, Ferrari-Balivera E, Demling B, Wilmore EW. Use of growth hormone combined with nutritional support in the critical care unit. J Parenter Enteral Nutr 1990;14:574–81.
11. Zeigler TR, Lazarus JM, Yough LS, Habeim R, Wilmore DW. Effects of recombinant growth hormone in adults receiving hemodialysis. J Am Soc Nephrol 1991;2:1130–5.
12. Zeigler TR, Bombeau JL, Young CS. Recombinant human growth hormone enhances the metabolic efficacy of parenteral nutrition: a double blind, randomized controlled study. J Clin Endocrinol Metab 1992;74:865–73.
13. Mulligan K, Grunfeld C, Hellerstin MK, Neese RA, Scheinbelan M. Anabolic effects of recombinant human growth hormone in patients with wasting associated with human immunodeficiency virus infection. J Clin Endocrinol Metab 1993;77:956–63.

14. Pape GS, Freidman M, Underwood LE, Clemmons DR. The effect of growth hormone on weight gain and pulmonary function in patients with chronic obstructive lung disease. Chest 1991;99:1495–500.
15. Kelly PA, Djiane J, Postel-Vinay MC, Edery JC. The prolactin/growth hormone receptor family. Endocr Rev 1991;12:235–51.
16. Bazan JF. Structural design and molecular evolution for cytokine receptor super family. Proc Natl Acad Sci USA 1990;87:6934–8.
17. Costin G, Kogut MD, Frasier SD. Effect of low-dose human growth hormone on carbohydrate metabolism in children with hypopituitarism. J Pediatr 1972; 80:796–803.
18. Belcher HJCR, Mercer D, Judkins KC, et al. Biosynthetic human growth hormone in burned patients: a pilot study. Burns 1989;15:99–107.
19. Roberts CT, Brown AO, Graham DE, et al. Growth hormone regulates the abundance of insulin-like growth factor-I RNA in adult rat liver. J Biol Chem 1986;261:10025–8.
20. Turkalj I, Keller N, Ninnis R, Vossner S, Stauffecker W. Effect of increasing doses of recombinant insulin-like growth factor on glucose, lipid and leucine metabolism. J Clin Endocrinol Metab 1993;75:1186–91.
21. Jacob RJ, Nederostek D, Wong L, et al. Low doses of IGF-I stimulate muscle protein synthesis in anabolic rats if hypoaminoacidemia is prevented [Abstract #655]. Annu meet Endocr Soc, Las Vegas, NV, June 14–16, 1993.
22. Clemmons DR, Smith-Banks A, Celniker AC, Underwood LE. Reversal of diet-induced catabolism by infusion of recombinant insulin-like growth factor-I (IGF-I) in humans. J Clin Endocrinol Metab 1992;75:234–8.
23. Young SCJ, Smith-Banks A, Underwood LE, Clemmons DR. Effects of recombinant IGF-I and GH treatment upon serum IGF binding proteins in calorically restricted adults. J Clin Endocrinol Metab 1992;75:603–8.
24. Kupfer SR, Underwood LE, Baxter RC, Clemmons DR. Enhancement of the anabolic effects of growth hormone and insulin-like growth factor-I by the use of both agents simultaneously. J Clin Invest 1993;91:391–7.
25. Hussein MA, Schmitz O, Menge CA, et al. Effects of growth hormone and insulin-like growth factor I on body fuel metabolism and insulin sensitivity in GH deficient humans [Abstract #1595]. 75th annu meet Endocr Soc, Las Vegas, NV, June 14–16, 1993.

Part II

Pubertal, Menstrual, Gestational, and Menopausal Adaptation

7

Growth Hormone Secretory Dynamics During Puberty

ALAN D. ROGOL, PAUL M. MARTHA, JR., MICHAEL JOHNSON,
JOHANNES D. VELDHUIS, AND ROBERT M. BLIZZARD

Growth Hormone: Infant and Child

Growth hormone (GH) is released in an intermittent, pulsatile manner in the fetus and on throughout life. GH is first detectable in the human fetal pituitary by week 9 of gestation (1). Premature infants have higher circulating concentrations of GH compared to term infants. Quantitative analysis of GH secretory profiles (see below) indicates an elevation in production rate, burst amplitude, and mass of GH secreted per burst in premature compared with term infants. The simultaneously derived half-lives of disappearance do not differ between these two groups of newborns; thus, augmented secretion rather than decreased clearance accounts for the differences in circulating GH concentrations. Taken together, the data suggest that increased GH secretory activity in premature infants reflects an increase in hypothalamic *GH releasing hormone* (GHRH) activity and/or reduced *somatostatin* (SRIH) tone (2).

During childhood there are apparently no differences in GH secretion between boys and girls, although several investigators have noted a significant positive correlation between physical stature and circulating levels of GH (3) or between the amount of GH secreted per day and the height of children (4). In addition, Hindmarsh and colleagues (5) reported a relationship between height velocity and mean 24-h GH levels in short prepubertal children. When this issue was investigated in more detail in short boys by Kerrigan and colleagues (6–8), no significant differences in pulsatile GH release were found between normally growing prepubertal boys and the short subjects; however, a subset of short prepubertal boys with significantly delayed bone age had subnormal GH release, as indicated by a low sum of GH pulse areas, burst mass, and sum of GH pulse amplitudes (6). The report of a significant correlation among all subjects

69

between growth velocity and the sum of GH pulse amplitudes lends strong support to the hypothesis that alterations of amplitude-modulated GH release underlie the pathophysiology of suboptimal growth in some short prepubertal children.

Variability in GH Secretion

Prepuberty

There is marked between-subject variability in physiologic GH release in normally growing boys and girls even before the effects of gonadal steroid hormones are noted (9, 10). Attempts to correlate absolute values (e.g., mean 12-h or 24-h GH level) with growth rate are difficult because of the confounding effects of the *GH binding proteins* (GHBPs), IGF-I, the *IGF binding proteins* (IGFBPs), and those derivatives of intermediary metabolism and body composition that regulate them. These factors preclude a simple relationship between circulating mean GH levels and linear growth velocity. Thus, it may be impossible to predict growth velocity— or, for that matter, growth hormone sufficiency—for an individual from mean GH levels or from the pulsatile pattern of circulating GH. The existing data suggest that normality is most appropriately defined individually.

To be better able to study the effects of gonadal steroid hormones on GH production at adolescence, we initially investigated the quantitative aspects of variable GH release in normal prepubertal boys. Nine boys were evaluated. Each had at least 3 consecutive 24-h study periods (venous sampling every 20 min for 24 h at 4-month intervals) before the onset of any pubertal sexual development. Figure 7.1 shows the within-subject (Fig. 7.1A) and between-subject (Fig. 7.1B) variability of 4 representative boys in the 24-h GH concentration versus time pattern. The within-subject variability for the mean range of GH concentration for a group of 11 prepubertal boys evaluated 3–6 times, as well as the entire range of values for mean 24-h GH concentrations of the complete group, is shown in Figure 7.2. Note the much lower variability within individual subjects compared to the group of subjects.

FIGURE 7.1. Individual 24-h GH concentration vs. time profiles. *A:* Individual 24-h GH concentration vs. time profiles for 1 boy evaluated at 4-month intervals during the prepubertal state. (CA = chronologic age; T = serum testosterone concentrations in ng/dL at 0600h; IGF-I = IGF-I concentration in U/ML.) *B:* Individual 24-h GH concentration vs. time profiles for 4 normally growing prepubertal boys. Note difference in vertical scales between *A* and *B*.

A   GROWTH HORMONE

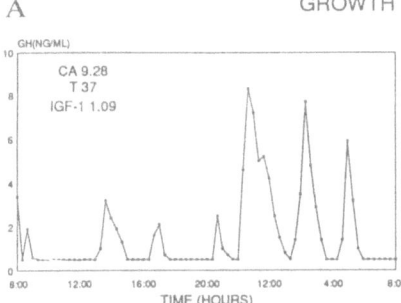

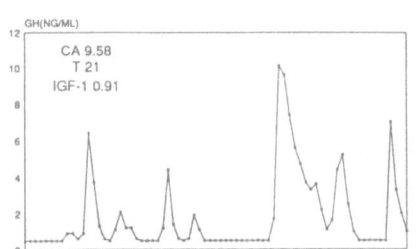

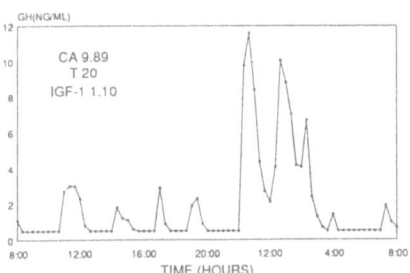

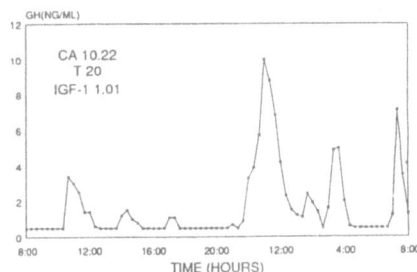

B   **GROWTH HORMONE**

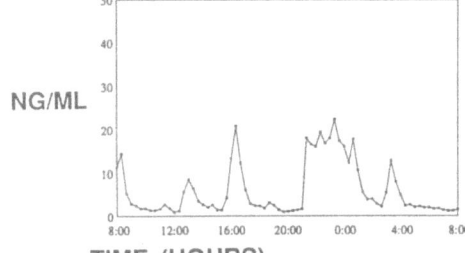

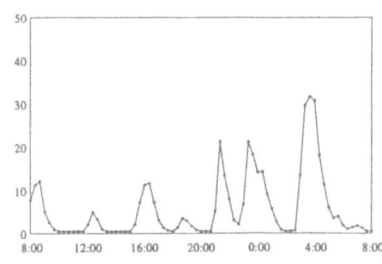

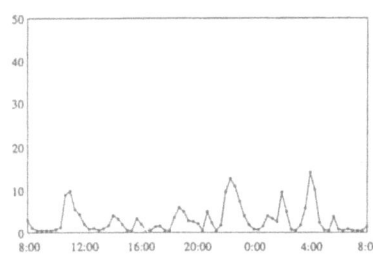

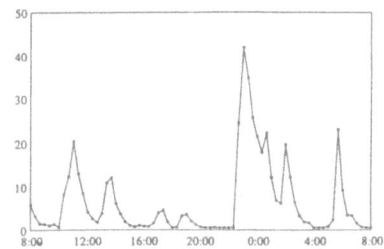

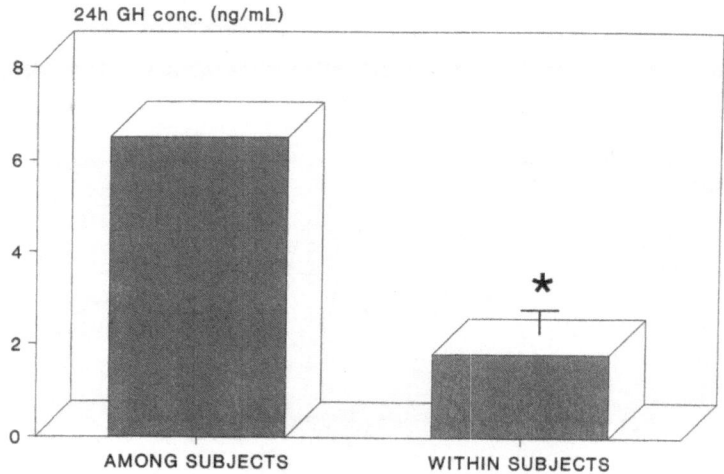

FIGURE 7.2. Variability of mean 24-h GH concentration: comparison of among-subject (left) and within-subject (right) variability ($n = 11$). Vertical axis represents mean 24-h GH concentration (ng/mL).

TABLE 7.1. Variability in the secretory and clearance parameters for GH in late prepubertal boys ($n = 11$).

	Intersubject CV[a] (%)	Intrasubject CV (%)
Daily secretory rate	44	23 ± 4[c]
Burst frequency	18	11 ± 2[d]
Mass per burst	46	24 ± 5[c]
T½[b]	30	23 ± 4[d]

[a] CV = coefficient of variation.
[b] T½ = half-life of disappearance.
[c] $P < 0.001$.
[d] $P < 0.05$.

To gain additional insight into the neuroendocrine events giving rise to the GH concentration profiles, we chose to employ deconvolution analysis (11) and compare daily secretory rate, secretory burst amplitude (maximal secretory rate), burst frequency, GH mass per burst, and half-life of disappearance. Although there were no differences when analyzed at half-year intervals from 9.5 to 11 years, it was clear that there is far greater variability between subjects than within subjects for all parameters, as shown in Table 7.1. The variability found between subjects agrees well with the data obtained from the prepubertal boys in our previous cross-sectional study (12). These results suggest that between-subject differences in these normally growing boys arise more from differences in the amount of GH released during a secretory episode than from the frequency of these events. It is difficult to define precisely the factor(s) responsible for the relatively wide between-subject variability. However, the quantity of GH secreted correlated inversely with the mean intrasubject *body mass index* (BMI) (BMI-SDS, a surrogate for relative fatness; BMI = weight/[height]^2), 24-h production rate ($r = -0.67$, $P = 0.049$), and GH secretory burst amplitude ($r = -0.73$, $P = 0.026$). The 24-h GH burst frequency did not correlate significantly with the BMI-SDS.

Puberty

Puberty is characterized by the onset of development of the secondary sexual characteristics and the impressive acceleration of linear growth. The secondary sexual characteristics are a result of androgen production from the adrenals in both sexes (adrenarche) and testosterone from the testes in the male and estrogens from the ovaries in females (gonadarche). Although the rapid growth spurt had previously been attributed directly to the rising concentrations of gonadal steroid hormones, an indirect effect mediated through GH and the IGFs is now considered extremely important.

During early childhood linear growth velocity steadily decreases from its initially rapid rate to reach a steady state of approximately 5.5 cm/year. As puberty approaches, the rate of growth slows slightly before its sudden acceleration to reach a peak during midadolescence. This peak occurs later in the process of development in boys than in girls. It then diminishes toward zero as epiphyseal fusion approaches. At the zenith of the pubertal growth spurt, the growth velocity often rises to rates greater than at any other time since early infancy. There appears to be little doubt that the neuroendocrine axis subserving the GH secretory system plays a pivotal role, for an adequate pubertal growth spurt cannot occur without sufficient quantities of GH (13). However, GH alone is apparently not sufficient since an important physiological synergism exists between the gonadal and somatotrophic axes coincident with normal pubertal development. Thus, the combined growth-promoting effect of concerted activation of these axes is required for normal pubertal development to proceed.

Cross-Sectional Studies

We initially sought to determine the role of androgens (testosterone) in the pubertal elevation of circulating IGF-I concentrations. Parker and colleagues (14) showed that testosterone administered *intramuscularly*

(I.M.) would stimulate IGF-I production in prepubertal boys who could release GH, but not in those who were GH deficient. These findings were more fully developed by investigating the alterations in the pulsatile release of GH as boys enter and progress through pubertal development. Investigations by Link and coworkers (15), Mauras et al. (16), and Martha and colleagues (9) indicate that concomitant with a rise in IGF-I concentrations, mean circulating GH levels increase during puberty around the time of the midpubertal growth spurt in normal boys and in delayed-pubertal boys administered testosterone therapy. The mode of this increase is through an augmentation in the size (amplitude and pulse area), rather than the number, of detectable GH pulses.

An increase in GH pulse increment over baseline is the primary mechanism producing the increase in mean 24-h GH concentration. Changes in pulse width (duration) contribute little. Shortly after cessation of linear growth, the 24-h pattern of GH secretion returns toward prepubertal levels, with the result that concentration profiles in young men (9) are remarkably similar to those in prepubertal boys, but greater than those in older men despite a continued rise in serum testosterone concentration.

To investigate the mechanisms by which androgens increase mean circulating GH concentrations in pubertal boys, we applied a multiple-parameter deconvolution model of circulating GH concentration versus time profiles from 48 normal boys (12). The 24-h GH profiles were characterized by 10.7 ± 0.5 (mean $\pm$ SE) discrete pituitary secretory bursts with a mean secretion half-duration of 35.6 ± 1.4 min and mass of GH secreted per burst of 25.5 ± 2.4 ng/mL. Values for total 24-h GH production rate according to the subjects' stage of pubertal development are presented in Table 7.2.

Estrogens may also affect pubertal growth and GH secretion. They are considered to have a biphasic effect: first stimulating and then, in larger doses, inhibiting linear growth. Mean GH levels were found to be significantly increased at breast stages 2 through 4 by Rose and colleagues (10). Pulse frequency and percentage of GH values below the detection limit did not change significantly with pubertal stage. However, the per-

TABLE 7.2. Daily GH production rate in boys at different pubertal stages ($n = 54$).

	PR[a,b] (μg/24 h)
Prepubertal	24 ± 2
Early puberty	18 ± 3[c]
Late puberty	31 ± 5
Postpuberty	13 ± 5[c]
Adult	14 ± 2[c]

[a] Production rate (PR) per kilogram body weight ($\pm$SE).
[b] Assuming a distribution volume of 7.9%.
[c] Statistically different from late pubertal, $P < 0.05$.

centage of GH values below the detection limit was significantly lower in
stage 4 girls than in boys, possibly indicating an alteration in SRIH
secretion or action. Mean pulse amplitude was significantly increased at
all advanced stages of puberty as compared with that in the younger,
prepubertal girls. Even in these girls of normal height and weight, mean
nighttime GH level correlated inversely with BMI, indicating a complex
relationship that includes body composition (10). Similar correlation
analysis revealed that the 24-h GH secretory rate in boys varied inversely
with the subject's BMI-SDS ($r = -0.65$, $P < 0.01$) (12).

Levine-Ross and colleagues introduced the concept that low doses of
ethinyl estradiol (100 ng/kg/day) could enhance linear growth without
unduly advancing skeletal maturation and suggested such therapy for
hypogonadal girls with Turner syndrome (17). The physiological actions
of such low doses of estrogen have not been clearly defined. Mauras and
colleagues studied the possible alterations in endogenous GH secretion in
such young girls (18). There were consistent and significant increases in
mean GH concentration and mean GH pulse amplitude without a detectable change in GH pulse frequency in 7 patients with Turner syndrome
following 5 weeks of low-dose (100 ng/kg/day) ethinyl estradiol therapy.
These findings were not accompanied by any significant changes in plasma
IGF-I concentration, serum estradiol concentration, or urinary cytological
maturational indexes. Thus, this hypogonadal model, which presumably
applies to normal prepubertal girls, demonstrates exquisite somatotrope
sensitivity to low-dose estrogen action.

To explore further the mechanism of this action of estrogen, Mauras
and colleagues applied multiple-parameter deconvolution techniques to
GH concentration versus time profiles in girls with Turner syndrome who
were treated with ethinyl estradiol, 100 ng/kg/day (19). The endogenous
GH production rate more than doubled after 5 weeks of ethinyl estradiol
therapy (baseline 194 ± 22 to $412 \pm 66 \mu g/L/12 h$; $P < 0.05$). This change
was predominantly due to an increase in detectable secretory burst frequency (5.3 ± 0.6 to 7.9 ± 0.5 per 12 h, from baseline to 5 weeks of
therapy) without significant change in the mass of GH released in each
secretory burst or in the half-time of disappearance of circulating GH.
We cannot exclude a primary effect of estrogen on GH secretory burst
mass leading to an apparent rise in (detectable) GH secretory burst
frequency. More frequent blood sampling and enhanced assay sensitivity
would be required to adequately explore this consideration. Although the
precise neuroendocrine mechanisms subserving an increase in GH production rate after therapy with low doses of estrogen are not known, the
marked augmentation of the GH secretory rate suggests a significant
enhancement of somatotrope responsiveness to GHRH action and/or
increased GHRH release. Such effects could be secondary to direct facilitative actions of low-dose estrogen on the somatotrope, decreased SRIH
inhibitory tone, and/or amplified GHRH pulse generation.

TABLE 7.3. Linear correlations between growth velocity and hormone concentrations in boys as they progress through puberty ($n = 8$).

	IGF-I	T[a]	24-h mean GH conc.	Sum secretory burst	
				Amplitudes	Areas
r[b]	0.67	0.71	0.46	0.44	0.40
P[c]	<0.001	<0.001	<0.001	0.001	0.003

[a] Testosterone.
[b] Linear correlation coefficient.
[c] Probability (α-value).

Thus, in these cross-sectional studies there is a consistent interaction between gonadal steroid hormones, growth velocity, pubertal progression, and circulating GH levels. Remarkable variability in the quantity and mode of GH release is permitted in normally growing children and adolescents. The complex system in the general circulation to regulate the amount and pattern of GH secretion—GH, GHBPs, IGF-I, IGFBPs, and those derivatives of body composition that regulate them—preclude a simple relationship between circulating mean GH levels and linear growth velocity. Thus, it may be difficult in individual subjects to predict growth velocity or GH sufficiency from mean GH levels or from the circulating GH concentration versus time profiles. Even so, evaluation of the mode of GH release as noted above remains a research tool that is presently without practical therapeutic application to large groups of short, slowly growing children. Thus, we turned to a prospective, longitudinal design (outlined below) to attempt to reduce variability within the data sets to allow a more nearly complete understanding of the interactions between the somatotrophic and gonadal axes.

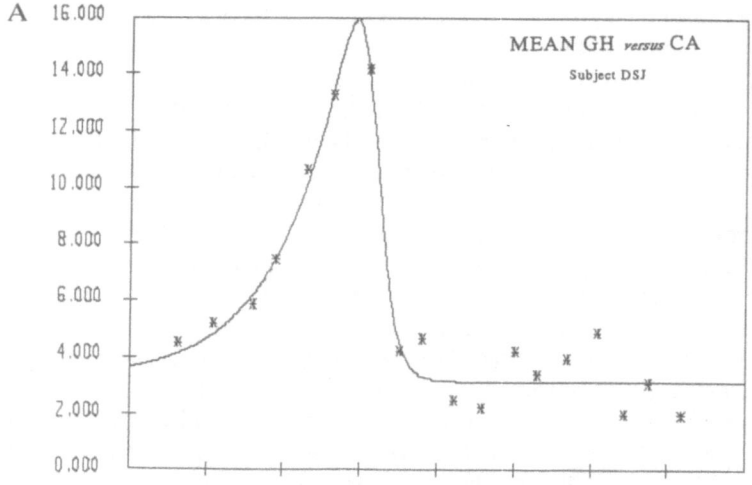

Longitudinal Study

To test further the hypothesis that androgens stimulate GH release and augment GH secretory patterns and to examine the mechanisms of the pubertal growth spurt, we have undertaken a prospective, longitudinal study evaluating hormonal changes in 24 normally growing boys as they

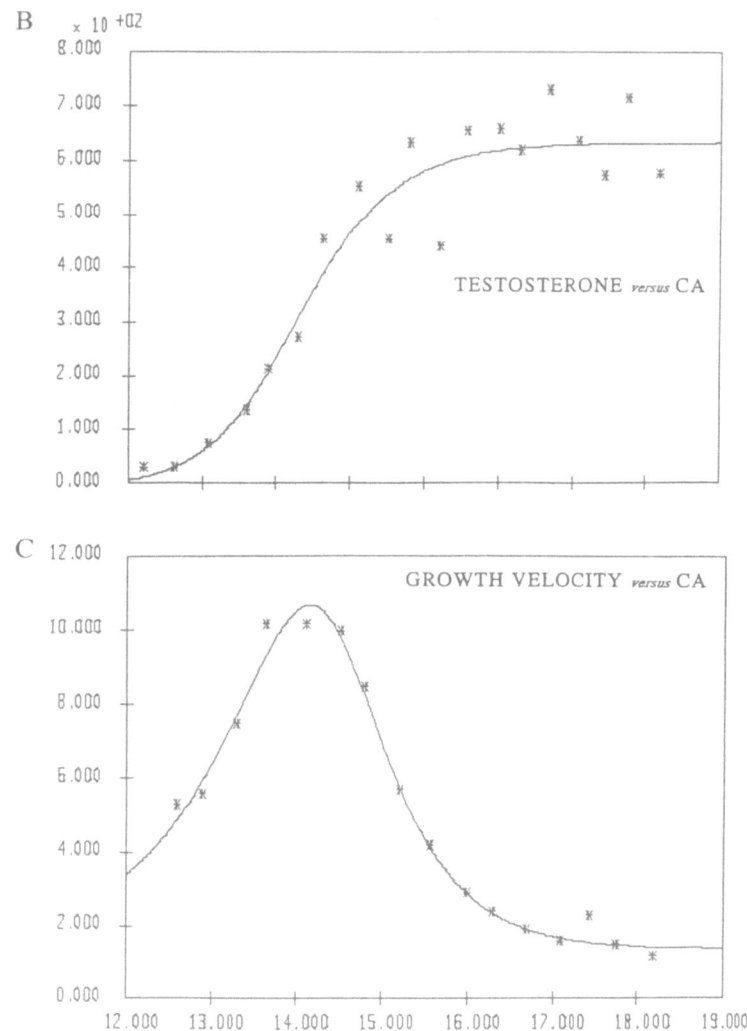

FIGURE 7.3. Composite of mean GH concentration (*A*), 0600-h testosterone concentration (*B*), and growth velocity (*C*) vs. chronological age (CA) in an individual boy as he progresses through puberty.

pass through puberty. Evaluations at approximately 4-month intervals include physical examination and measurement of growth velocity, 24-h GH profiles, and plasma IGF-I and serum testosterone concentrations. This study design with each boy serving as his own control during the variably timed pubertal process should permit better insight into the timing of specific events (e.g., peak growth velocity and maximal GH secretory rate) than have the previous cross-sectionally designed investigations.

Preliminary data from boys ($n = 8$) who have had at least 5 evaluations and who have begun to progress through pubertal development during 55 study periods covering 17 to 36 months are shown in Table 7.3. Not all show the "expected" correlation among mean GH concentration, increasing morning testosterone level, and accelerating growth velocity. If all *prepubertal* studies analyzed to date (11 boys studied 3–6 times, $n = 48$) are considered together, there is no apparent relationship between growth velocity and mean 24-h GH concentration ($r = 0.09$, $P = 0.57$), the sum of the secretory burst amplitudes ($r = 0.08$, $P = 0.59$), or the pulse frequency ($r = 0.06$, $P = 0.30$).

Despite the longitudinal design there appears to be great variability in the amount of GH produced per day in normally growing boys. The relatively constant prepubertal growth velocity can occur with widely disparate pulsatile GH patterns. Although most boys show a marked increase in GH production as the early morning testosterone concentration rises, there are striking exceptions to this pattern.

In summary, rising gonadal steroid hormone concentrations augment GH secretion in both adolescent boys and girls. The increased levels of these hormones stimulate the rapid growth and sexual development, but

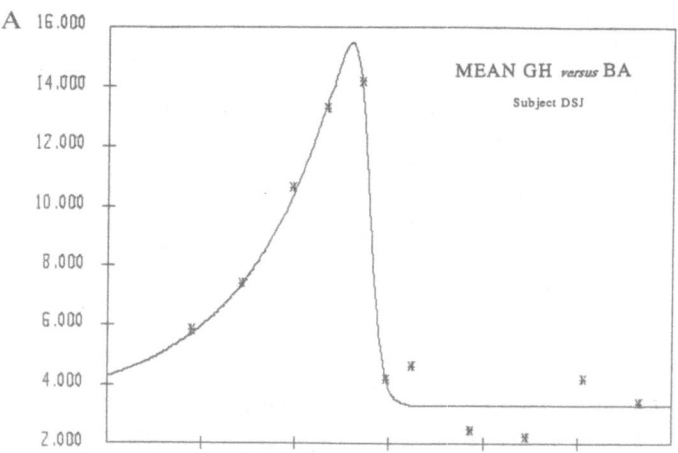

any straightforward relationship is dampened by the intermediate GHBP concentration, the production both locally and systemically of IGF, and by the presence of several IGFBPs for the latter, although the GH-dependent IGFBP-3 appears most relevant.

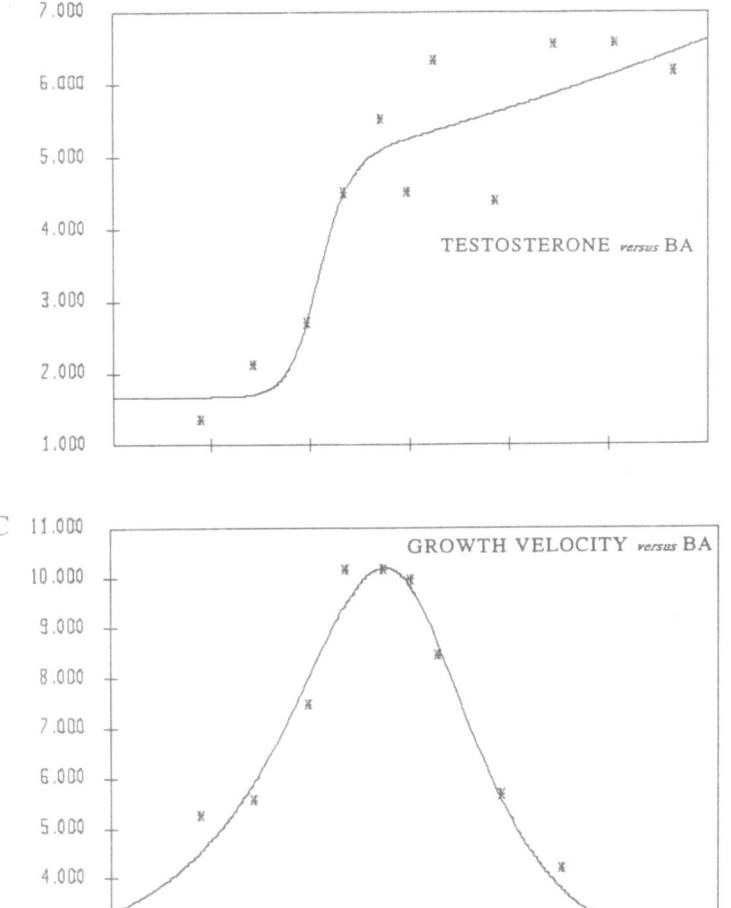

FIGURE 7.4. Composite of mean GH concentration (*A*), 0600-h testosterone concentration (*B*), and growth velocity (*C*) vs. bone age ([BA]; Fels Institute method) in the same individual boy as shown in Fig. 7.3 as he progresses through puberty.

Modeling of Growth and Hormonal Data

The nature of growth, especially as demonstrated by individual growth curves, makes it a suitable candidate for mathematical modeling (20). Many empirical models (functions) have been employed, but parsimonious models with a small number of parameters that can be related to biological events seem most relevant. In addition, modeling may provide further insight into functional relationships or correlations, either within the growth process (of body-segment lengths or breadths) or to biological markers of the growth process, such as hormone levels or other parameters reflecting hormone secretion and clearance.

We have used the robust, well-validated model of Preece and Baines (20) to evaluate growth and hormonal data in boys as they attain and advance through puberty. We have also attempted to employ higher-order polynomial models. A large number of coefficients can be determined, but their biological meaning is obscure. Some of the data can be equally well modeled using a standard 4-parameter logistic model, but there is little validation for the biological definition of the parameters obtained. The biological parameters first evaluated have been the age at *takeoff*, the age at *maximal growth rate*, and the time and the slope from takeoff to maximal growth rate. These parameters were chosen since a priori, these seem related to changes in GH and gonadal steroid hormone levels and also to the tempo of puberty.

Some preliminary data are shown in Figures 7.3 and 7.4, in which we compare the mean 24-h GH level, the testosterone level, and growth velocity versus chronological or bone age. The latter can provide a very good perspective concerning the tempo of puberty. Please note that the biological effect of testosterone is more tightly related to bone age and, thus, to the tempo of puberty, rather than to chronological age. It will be only after a more extensive data set is analyzed that we can more clearly define the hormonal patterns in relation to the chronologic- and bone age-indexed biologically relevant events of takeoff, maximal velocity of growth, and the time between them. These parameters—velocity and tempo—are likely tightly correlated to alterations in the gonadotropic, gonadal, and GH axes.

Acknowledgments. We are indebted to Ms. Sandra Jackson and her staff of nurses at the General Clinical Research Center of the University of Virginia Health Sciences Center for their expert patient care and to our research assistants, Ms. Melanie Harlowe, Ms. Ginger Bauler, and Ms. Katherine Kern, for their technical support. This work is supported in part by Grant RR-00847 to the General Clinical Research Center at the University of Virginia Health Sciences Center, RCDA K94-HD-00634 from NICHD (J.D.V.), the Diabetes and Endocrinology Research

Center (M.L.J.), and the NSF Center for Biological Timing (A.D.R., M.L.J., and J.D.V.).

References

1. Kaplan SL, Grumbach MM, Shepherd TH. The ontogenesis of human fetal hormones, I. Growth hormone and insulin. J Clin Invest 1972;51:3080–93.
2. Wright NM, Northington FJ, Miller JD, Veldhuis JD, Rogol AD. Elevated growth hormone secretory rate in premature infants: deconvolution analysis of pulsatile growth hormone secretion in the neonate. Pediatr Res 1992; 32:286–90.
3. Albertsson-Wikland K, Rosberg S. Analyses of 24h growth hormone profiles in children: relation to growth. J Clin Endocrinol Metab 1988;67:493–500.
4. Albertsson-Wikland K, Rosberg S, Libre E, Lundberg L-O, Groth T. Growth hormone secretory rates in children as estimated by deconvolution analysis of 24-h plasma concentration profiles. Am J Physiol 1989; 257:E809–14.
5. Hindmarsh P, Smith PJ, Brook CGD, Matthews DR. The relationship between height velocity and growth hormone secretion in short prepubertal children. Clin Endocrinol (Oxf) 1987;27:581–91.
6. Kerrigan JR, Martha PM Jr, Blizzard RM, Christie CM, Rogol AD. Variations of pulsatile growth hormone release in healthy short prepubertal boys. Pediatr Res 1990;28:11–4.
7. Kerrigan JR, Martha PM Jr, Veldhuis JD, Blizzard RM, Rogol AD. Altered growth hormone secretory dynamics in prepubertal males with constitutional delay of growth. Pediatr Res 1993;33:278–83.
8. Veldhuis JD, Blizzard RM, Rogol AD, et al. Properties of spontaneous growth hormone secretory bursts and half-life of endogenous growth hormone in boys with idiopathic short stature. J Clin Endocrinol Metab 1992; 74:766–73.
9. Martha PM Jr, Rogol AD, Veldhuis JD, Kerrigan JR, Goodman DW, Blizzard RM. Alterations in the pulsatile properties of circulating growth hormone concentrations during puberty in boys. J Clin Endocrinol Metab 1989;69:563–70.
10. Rose SR, Municchi G, Barnes KM, et al. Spontaneous growth hormone secretion increases during puberty in normal girls and boys. J Clin Endocrinol Metab 1991;73:428–35.
11. Veldhuis JD, Faria A, Vance ML, Evans WS, Thorner MO, Johnson ML. Contemporary tools for the analysis of episodic growth hormone secretion and clearance in vivo. Acta Paediatr Scand 1988;347:63–82.
12. Martha PM Jr, Gorman KM, Blizzard RM, Rogol AD, Veldhuis JD. Endogenous growth hormone secretion and clearance rates in normal boys, as determined by deconvolution analysis: relationship to age, pubertal status, and body mass. J Clin Endocrinol Metab 1992;74:336–44.
13. Aynsley-Green A, Zachmann M, Prader A. Interrelationship of the therapeutic effects of growth hormone and testosterone on growth in hypopituitarism. J Pediatr 1976;89:992–9.

14. Parker MW, Johanson AJ, Rogol AD, Kaiser DL, Blizzard RM. Effect of testosterone on somatomedin-C concentrations in prepubertal boys. J Clin Endocrinol Metab 1984;58:87–90.
15. Link K, Blizzard RM, Evans WS, Kaiser DL, Parker MW, Rogol AD. The effect of androgens on the pulsatile release and the twenty-four hour mean concentration of growth hormone in peripubertal males. J Clin Endocrinol Metab 1986;62:159–64.
16. Mauras N, Blizzard RM, Link K, Johnson ML, Rogol AD, Veldhuis JD. Augmentation of growth hormone secretion during puberty: evidence for a pulse amplitude-modulated phenomenon. J Clin Endocrinol Metab 1987; 64:596–601.
17. Ross JL, Cassorla FG, Carpenter G, et al. The effect of short term treatment with growth hormone and ethinyl estradiol on lower leg growth rate in girls with Turner's syndrome. J Clin Endocrinol Metab 1988;67:515–8.
18. Mauras N, Rogol AD, Veldhuis JD. Specific, time-dependent actions of low-dose ethinyl estradiol administration on the episodic release of growth hormone, follicle-stimulating hormone, and luteinizing hormone in prepubertal girls with Turner's syndrome. J Clin Endocrinol Metab 1989;69:1053–8.
19. Mauras N, Rogol AD, Veldhuis JD. Increased hGH production rate after low-dose estrogen therapy in prepubertal girls with Turner's syndrome. Pediatr Res 1990;28:626–30.
20. Preece MA, Baines MJ. A new family of mathematical models describing the human growth curve. Ann Hum Biol 1978;5:1–24.

8

Impact of Growth Hormone on the Timing and Progression of Puberty: How Are Growth and Sexual Maturation Linked?

PAUL A. BOEPPLE

Perhaps nowhere else are the interactions between the somatotrophic and reproductive axes as dramatically manifest as when sexual maturation and the adolescent growth spurt combine to comprise human pubertal development. Basic and clinical investigations in recent years have added much to our knowledge regarding these complex interactions, including the modulation of neurosecretory patterns of *growth hormone* (GH) release by gonadal steroids and the modulation of gonadal function by *insulin-like growth factor I* (IGF-I) and its binding proteins (IGFBPs) (1–3). The focus of this chapter is the impact of growth and GH on two discrete aspects of sexual maturation: (i) the timing of the onset of central puberty and (ii) the pace or rapidity with which puberty progresses once under way. While they are areas of great clinical and basic research interest, there is still much that is poorly understood about these developmental processes.

A review of the impact of growth and GH on the timing of the onset of puberty requires that discussion focus on the inputs that combine to trigger the transition from childhood to pubertal patterns of *gonadotropin releasing hormone* (GnRH) secretion. Thus, fundamental, but as yet unanswered, questions regarding the maturation of the *central nervous system* (CNS) and neuroendocrine systems become the predominant ones to consider. To address the second focus of this chapter—the impact of GH and/or IGF-I on the pace or rapidity of pubertal maturation—the discussion must broaden to consider modulating influences at all levels of the reproductive axis that may integrate to affect the pattern and level of gonadal sex steroid secretion.

To address these two broad questions, information is reviewed with regard to the association between body size and the onset of puberty, the timing of puberty and function of the reproductive axis in GH-deficient models prior to and following GH replacement, and the impact of supraphysiologic GH levels on sexual maturation in non-GH-deficient models. Data are reviewed from relevant animal models both in vivo and in vitro, but emphasis is placed on clinical studies in the human whenever possible.

GH Deficiency and the Onset of Puberty

The somatotrophic axis is replete with feedback loops that combine to regulate GH secretion. *Growth hormone releasing hormone* (GHRH or GRF) and *somatostatin* (SRIH) modulate each other's secretion, and GH, both directly and indirectly via IGF-I generation, plays an important role in modulating its own secretion (4). While the controls of growth and puberty are clearly linked, it is exceedingly difficult in in vivo models to isolate and examine the impact of GH and/or IGF-I directly on GnRH or pituitary gonadotropin secretion. In an effort to understand these interactions more completely, many studies have sought to determine the impact of GH deficiency on the one hand or GH administration to intact animals or humans on the other.

While most models of GH deficiency are associated with significant delays in the onset of sexual maturation, the precise mechanisms are by no means clear. Many investigational models that have been employed to induce GH deficiency in animals invariably impact on the reproductive axis as well. Certainly, hypophysectomy and cranial irradiation are not valid models in which to examine subtle interactions between the somatotrophic and reproductive axes, but techniques that purport to be more specific in their impact on the GH axis must be viewed with skepticism regarding their selectivity as investigational probes. For instance, the induction of GHRH/GH deficiency in rats by neonatal treatment with monosodium glutamate has been shown to be associated with a wide range of additional hypothalamic effects (5, 6).

Clinical observations in patients with GH deficiency may be subject to the same confounding variables present in these animal models. Most cases of GH deficiency have been shown to result from defects in the hypothalamic regulation of GH secretion, and it is not unreasonable to presume that even patients with what is clinically categorized as isolated GH deficiency may well have hypothalamic dysfunction that extends beyond the somatotrophic axis (7). Thus, observations regarding the timing of puberty in patients with isolated GH deficiency may not reflect only the impact of alterations in GH and IGF-I secretion. For instance, the fact that LH and FSH responses to GnRH are depressed in some patients with isolated GH deficiency may represent partial GnRH deficiency that while not sufficient to impair normal pubertal development clinically, could nevertheless impact on the timing and pace of pubertal maturation (8). With these caveats in mind regarding the uncertainties regarding the pathophysiology involved, it has been a consistent finding that patients with isolated GH deficiency begin their pubertal development later than the general population (9–12).

Given the complex nature of the defects that result in GH deficiency, the generation of animal models or the study of clinical disorders characterized by specific gene deletions at various levels of the somatotrophic axis may well yield more precise information about how GH-deficient states affect the timing of puberty. An alternative immunologic approach has also been employed to respond to the need for a *pure probe* into the GH axis (13). Postnatal male rats treated with anti-GHRH antisera were rendered GH deficient throughout the course of their passive immunization. Like other models examined, this discrete block of GH synthesis and secretion was also associated with a significant delay in sexual maturation (Fig. 8.1) (13).

In such studies the investigational strategies employed to induce GH deficiency are invariably associated with altered body composition and rates of weight gain and linear growth. The same, of course, is true in clinical studies of GH-deficient patients. As such, the direct impacts of GH and/or IGF-I deficiency cannot be examined independent of the indirect effects of these derangements on body size and metabolic status.

One attempt to isolate GH effects per se from those that relate to body size and growth rates is represented in the innovative approach offered by the GH deficiency that is associated with infection by the tapeworm, *Spirometra mansonoides*. Infection with the larval form of *S. mansonoides* results in the production of a factor that appears to (i) suppress endogenous GH secretion, (ii) stimulate IGF-I production, and (iii) thus permit normal rates of growth despite the presence of a GH-deficient state (14, 15). Infection of both male and female immature rats with *S. mansonoides* induced GH deficiency and was associated with a delay in sexual maturation despite the fact that the rate of body weight gain had been no different from controls (16, 17). The authors concluded, therefore, that GH deficiency per se had significantly delayed the timetable of sexual maturation. While an intriguing attempt to dissociate and isolate the effects of GH deficiency from its attendant, usually unavoidable, changes in growth and body size, this model hardly represents a pure probe into the somatotrophic axis. It may well be that more subtle changes in body composition or nutritional and systemic effects of the tapeworm infection itself could have had an impact on the regulation of GnRH secretion and pubertal development.

Much has been written about the interplay between nutritional status and body composition and the activity of the reproductive axis. Large

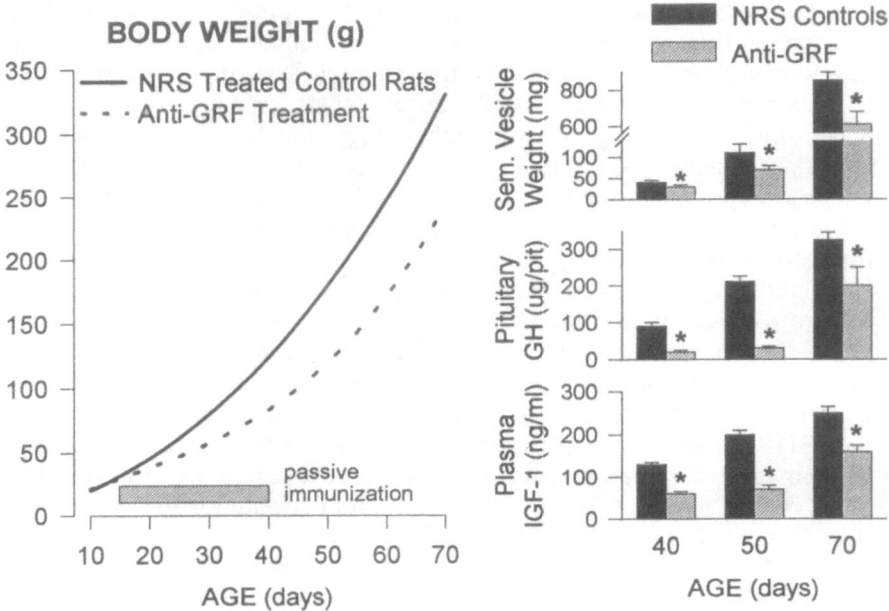

FIGURE 8.1. Induction of GH deficiency in male rats by passive immunization against GHRH (GRF) beginning at postnatal day 15. Growth and sexual maturation, as assessed by body weight and seminal vesicle weight, respectively, were both delayed relative to controls receiving normal rabbit sera (NRS). Redrawn with permission from Arsenijevic, Wehrenberg, Conz, Eshkol, Sizonenko, and Aubert (13), © The Endocrine Society, 1984.

population studies established a correlation between the weight—or, more specifically, the proportion of body fat—and the age at which adolescent girls attain menarche (18–20). Some of these same studies, as well as many others, have established that within rather precise limits, chronic weight loss or loss in body fat results in either primary or secondary amenorrhea stemming from reversible GnRH deficiency (21).

Building on these observations from the chronic, severe undernutrition associated with anorexia nervosa, other investigators have evaluated the impact of more subtle, short-term nutritional manipulations on the function of both the reproductive and somatotrophic axes. Acute fasting for intervals of only 24–48 h results in dramatic alterations in the neurosecretory patterns of both LH and GH release. Studies in both the monkey and the human have demonstrated that short-term fasting suppresses LH secretion, presumably on the basis of decreases in hypothalamic GnRH secretion (22, 23). Whether this suppression of the reproductive axis is causally linked to the fall in IGF-I and increase in GH secretion that also accompanies short-term fasting in the human is not clear (24, 25). It is perhaps more likely that both axes are responding to as yet poorly defined metabolic signals that transduce information about the state of nutrition in the periphery to the CNS.

The complex interplay between nutrition, growth, and sexual maturation was examined in an interesting report by Bourguignon and colleagues that drew on both clinical studies in patients with precocious puberty and an animal model that examined the impact of recovery from nutritional deprivation (26). They began with the observation that several of their cohort of patients with precocious puberty were children who were adopted from underdeveloped countries and went on to hypothesize that the catch-up growth from an undernourished state, if it occurs at a critical point in development, may trigger the early onset of puberty. Supporting evidence was provided by studies of GnRH releases from hypothalamic explants from immature rats that had been nutritionally deprived since cessation of the caloric deprivation, if appropriately timed, was associated with early maturation of GnRH release (26).

In summary, GH-deficient states are associated with delays in sexual maturation in both animal models and in patients with isolated GH deficiency. However, the specific link between GH deficiency and a delay in the central activation of puberty remains to be elucidated. Putative metabolic signals that convey information about nutritional status and body composition may well play critical roles.

GH Administration and the Timing of Puberty

Since most of the GH-deficient patients in whom a delay in the onset of puberty has been documented had received GH therapy during childhood, one may conclude that GH administration in this setting does not result in the normalization of the timetable for sexual maturation (9–12). Thus, this discussion focuses on the impact of GH administration in GH-sufficient animal models and patients.

The changes in GH secretion that accompany sexual maturation display important species differences. For instance, puberty in the sheep is associated with a decline in endogenous GH secretion, while just the opposite is true in the human (1, 27). Thus, for the purposes of this chapter, data are drawn only from studies in subhuman primates and clinical investigations in GH-treated, non-GH-deficient patients. However, given this restriction, there are few relevant reports in the literature.

In a prospective, well-controlled study in female rhesus monkeys, Wilson and colleagues evaluated the impact of GH administration beginning at age 20 months on the subsequent timing of the pubertal maturation (28). Compared to control animals, GH-treated monkeys demonstrated modest but significant increases in linear growth and skeletal maturation

(28, 29). Similarly, the initial rise in LH was documented in GH-treated monkeys at an age that was slightly younger than intact controls (Fig. 8.2). However, given that these studies were undertaken in rhesus monkeys housed outside, seasonal influences on the timing of sexual maturation become potentially confounding in this model. Thus, despite results demonstrating a tendency for GH to advance the timing of the onset of puberty in this model, Wilson and collaborators concluded that their data were not strong enough to support the hypothesis that GH administration leads to an acceleration of the timing of central puberty (28).

A similar conclusion has been arrived at by investigators who have reported their experience with GH therapy in non-GH-deficient short children (11, 12, 30). GH-treated short children have entered puberty somewhat later than the mean for the population despite GH administration that commenced in the prepubertal period (11, 12). No study to my knowledge has employed a prospective, randomized design to address this specific question. Given the broad variance in the timing of pubertal milestones in the population, large numbers of patients would be required to provide sufficient power to such a study. Thus, it is unlikely that any study will be able to do more than compare the timing of pubertal onset in children treated with GH to population-based norms.

In summary, studies in subhuman primates and short children have shown that the impact of GH administration to GH-sufficient subjects results in little, if any, advancement in the onset of pubertal development. The modest acceleration of puberty in the female rhesus monkeys treated with GH, while of questionable clinical significance, once again raises the issue that the GH effects on the maturation of GnRH secretion may well be mediated indirectly through metabolic signals generated in the periphery.

GH and the Pace of Pubertal Maturation

Although the data are scant regarding the tempo of pubertal progression in GH-deficient patients not receiving GH therapy, what little there is suggests that sexual maturation progresses slowly, even when puberty is induced by exogenous sex steroid administration (31). This is not surprising given the large body of data showing that in both animals and humans the biologic actions of sex steroids are significantly diminished in the setting of GH deficiency (32).

In both subhuman primates and in patients, GH administration appears to increase the rate of sexual maturation once pituitary gonadotropin secretion has exhibited pubertal increases (11, 12, 28, 30). In the female rhesus monkeys treated with GH, the milestones of menarche and first ovulation were attained earlier than in control animals (28). While

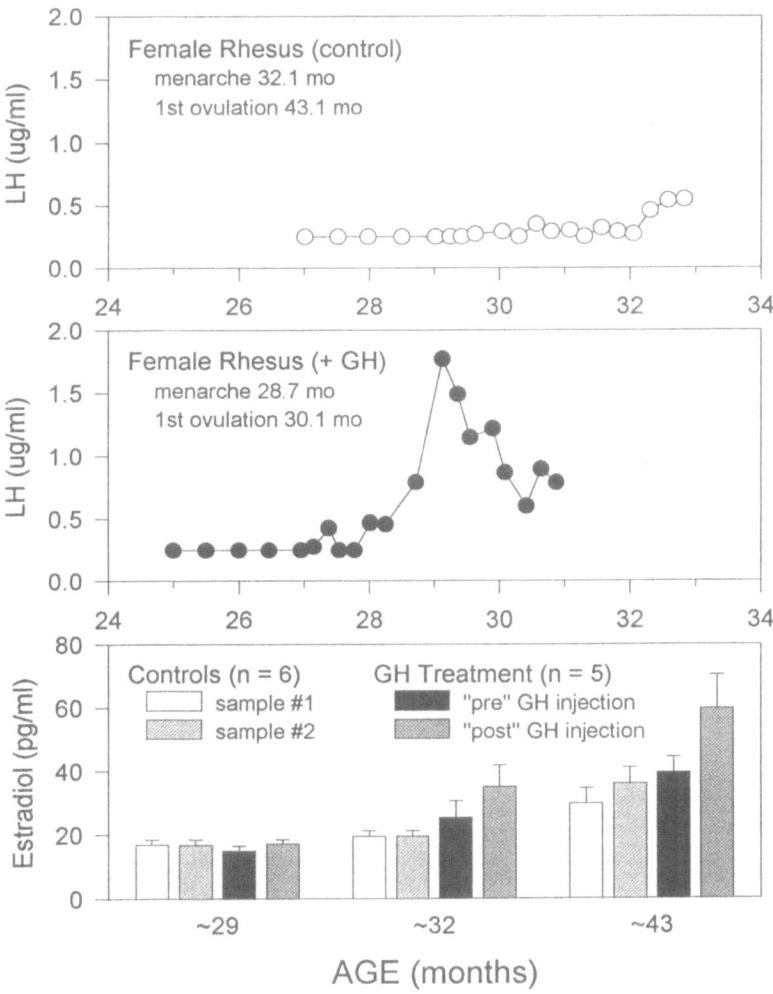

FIGURE 8.2. Study examining the impact of GH administration on the tempo of sexual maturation in female rhesus monkeys. Wilson et al. documented both earlier and more rapidly progressive puberty relative to controls. The upper and middle panels depict age-related increases in LH secretion in 2 representative monkeys, with the GH-treated animal (middle panel) displaying an earlier pubertal LH rise associated with earlier pubertal milestones. The bottom panel indicates that ovarian estradiol secretion matured earlier in the GH-treated monkeys and that GH administration appeared to augment the impact of endogenous gonadotropin on gonadal steroid secretion. Redrawn from Wilson, Gordon, Rudman, and Tanner (28), © The Endocrine Society, 1989.

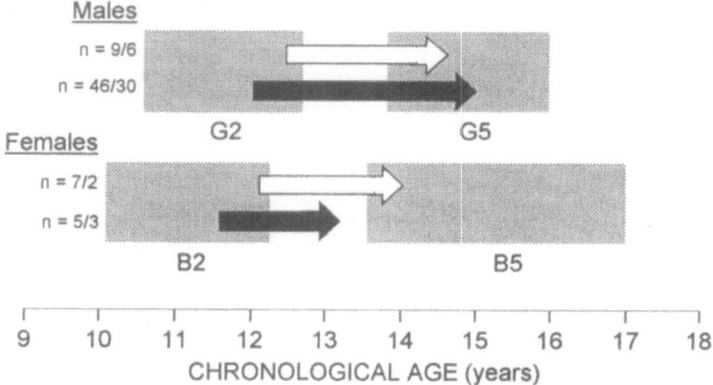

FIGURE 8.3. Progression of puberty (stages 2 to 5) during GH therapy in short, normal children (open arrow) and GH-deficient children (solid arrow). Following the onset of puberty (attainment of genitalia or breast stage 2) at an age that was somewhat later than the general population, Stanhope et al. documented that children treated with GH ($15–30\,IU/m^2$/week) progressed through puberty more rapidly than expected, attaining stage 5 at or prior to the mean age for the general population. The numbers of GH-treated patients who were evaluated at the onset/completion of puberty are indicated. The shaded areas represent the mean $\pm$ 1 SD for attaining breast (B) or genitalia (G) stages 2 and 5 in the general population, as described in references 37 and 38. Redrawn with permission from Stanhope, Albanese, Hindmarsh, and Brook (12), © S. Karger AG, Basel, 1992.

seasonal influences must again be taken into account, the data regarding serum estradiol levels in the two groups of animals were consistent with GH therapy resulting in an augmentation of ovarian steroid secretion (Fig. 8.2). Thus, the increases in estradiol secretion that accompanied GH administration provide a clear basis for an accelerated progression through puberty.

Clinical studies in both GH-deficient and non-GH-deficient patients who received GH therapy during their pubertal years also provide support for the concept that GH administration accelerates the pace of pubertal progression (Fig. 8.3) (11, 12). However, other studies have demonstrated a far less dramatic impact of GH therapy, with boys, but not girls, displaying a modest acceleration in pubertal tempo (30). Nevertheless, on balance, it appears justified to conclude that GH administration does appear to affect the pace of sexual maturation once the "central trigger" of puberty has been "pulled."

Several of the other chapters in this volume have as their focus the likely explanation for these findings; that is, the modulatory role that gonadal IGF-I plays in gonadotropin-stimulated sex steroid production (2, 3). This chapter does not address these issues in any detail. Rather, it closes with a consideration of the therapeutic consequences of GH and IGF-I regulation of gonadal function during puberty.

Clinical Implications

GH therapy in non-GH-deficient short children has resulted in accelerated rates of linear growth, but the increases in predicted height during therapy have been greater in prepubertal children than in those entering puberty while receiving GH (30). During long-term therapy the anticipated gains in adult stature may diminish in both GH-deficient and non-GH-deficient patients if GH administration does in fact result in an accelerated puberty. In addition, previous data have indicated that GH-deficient patients with associated GnRH/gonadotropin deficiency attain greater adult heights than patients with isolated GH deficiency who enter puberty spontaneously (9, 33). Thus, several studies have now adopted the approach that has proven successful for prolonging the growth period in children with central precocious puberty and have reported that the induction of pituitary-gonadal suppression in pubertal patients receiving GH results in increased adult-height predictions (33–36). While such an approach is not one that should be advocated outside of investigational settings at this time, it is quite clear that further elucidation of the physiologic interactions between the somatotrophic and reproductive axes is of direct clinical import to the optimal management of children with growth disorders.

References

1. Kerrigan JR, Rogol AD. The impact of gonadal steroid hormone action on growth hormone secretion during childhood and adolescence. Endocr Rev 1992;13:281–98.
2. Adashi EY, Resnick CE, D'Ercole AJ, Svoboda ME, Van Wyk JJ. Insulin-like growth factors as intraovarian regulators of granulosa cell growth and function. Endocr Rev 1985;6:400–20.
3. Adashi EY. Intraovarian regulation: the proposed role of insulin-like growth factors. Ann NY Acad Sci 1993;687:10–2.
4. Tannenbaum GS. Neuroendocrine control of growth hormone secretion. Acta Paediatr Scand Suppl 1991;372:5–16.
5. Millard WJ, Martin JB Jr, Audet J, Sagar SM, Martin JB. Evidence that reduced growth hormone secretion observed in monosodium glutamate-treated rats is the result of a deficiency in growth hormone-releasing factor. Endocrinology 1982;110:540–50.
6. Meister B, Ceccatelli S, Hokfelt T, Anden NE, Anden M, Theodorsson E. Neurotransmitters, neuropeptides and binding sites in the rat mediobasal hypothalamus: effects of monosodium glutamate (MSG) lesions. Exp Brain Res 1989;76:343–68.

7. Thorner MO, Vance ML, Evans WS, et al. Physiological and clinical studies of GRF and GH. Recent Prog Horm Res 1986;42:589–640.

8. Sauder SE, Corley KP, Hopwood NJ, Kelch RP. Subnormal gonadotropin responses to gonadotropin-releasing hormone persist into puberty in children with isolated growth hormone deficiency. J Clin Endocrinol Metab 1981; 53:1186–92.

9. Burns EC, Tanner JM, Preece MA, Cameron N. Final height and pubertal development in 55 children with idiopathic growth hormone deficiency, treated for between 2 and 15 years with human growth hormone. Eur J Pediatr 1981;137:155–64.

10. Hibi I, Tanaka T. Final height of patients with idiopathic growth hormone deficiency after long-term growth hormone treatment: Committee for Treatment of Growth Hormone Deficient Children, Growth Science Foundation, Japan. Acta Endocrinol (Copenh) 1989;120:409–15.

11. Darendeliler F, Hindmarsh PC, Preece MA, Cox L, Brook CG. Growth hormone increases rate of pubertal maturation. Acta Endocrinol (Copenh) 1990;122:414–6.

12. Stanhope R, Albanese A, Hindmarsh P, Brook CG. The effects of growth hormone therapy on spontaneous sexual development. Horm Res 1992;38 (suppl 1):9–13.

13. Arsenijevic Y, Wehrenberg WB, Conz A, Eshkol A, Sizonenko PC, Aubert ML. Growth hormone (GH) deprivation induced by passive immunization against rat GH-releasing factor delays sexual maturation in the male rat. Endocrinology 1989;124:3050–9.

14. Steeman SL, Glitzer MS, Ostlind DA, Mueller JF. Biological properties of the growth hormonelike factor from the plerocercoid of *Spirometra mansonoides*. Recent Prog Horm Res 1971;27:97–120.

15. Garland JT, Daughaday WH. Feedback inhibition of pituitary growth hormone in rats infected with *Spirometra mansonoides*. Proc Soc Exp Biol Med 1972;139:497–9.

16. Ramaley JA, Phares CK. Delay of puberty onset in females due to suppression of growth hormone. Endocrinology 1980;106:1989–93.

17. Ramaley JA, Phares CK. Delay of puberty onset in males due to suppression of growth hormone. Neuroendocrinology 1983;36:321–9.

18. Frisch RE, Revelle R. Height and weight at menarche and a hypothesis of critical body weights and adolescent events. Science 1970;169:397–9.

19. Frisch RE, McArthur JW. Menstrual cycles: fatness as a determinant of minimum weight for height necessary for their maintenance or onset. Science 1974;185:949–51.

20. Frisch RE. Pubertal adipose tissue: is it necessary for normal sexual maturation? Evidence from the rat and human female. Fed Proc 1980;39:2395–400.

21. Boyar RM. Endocrine changes in anorexia nervosa. Med Clin North Am 1978;62:297–303.

22. Cameron JL, Weltzin TE, McConaha C, Helmreich DL, Kaye WH. Slowing of pulsatile luteinizing hormone secretion in men after forty-eight hours of fasting. J Clin Endocrinol Metab 1991;73:35–41.

23. Schreihofer DA, Parfitt DB, Cameron JL. Suppression of luteinizing hormone secretion during short-term fasting in male rhesus monkeys: the role

of metabolic versus stress signals [see comments]. Endocrinology 1993;132: 1881–9.

24. Cemmons DR, Klibanski A, Underwood LE, et al. Reduction of plasma immunoreactive somatomedin C during fasting in humans. J Clin Endocrinol Metab 1981;53:1245–7.

25. Ho KY, Veldhuis JD, Johnson ML, et al. Fasting enhances growth hormone secretion and amplifies the complex rhythms of growth hormone secretion in man. J Clin Invest 1988;81:968–75.

26. Bourguignon JP, Gerard A, Alvarez Gonzalez ML, Fawe L, Franchimont P. Effects of changes in nutritional conditions on timing of puberty: clinical evidence from adopted children and experimental studies in the male rat. Horm Res 1992;38(suppl 1):97–105.

27. Suttie JM, Kostyo JL, Ebling FJ, et al. Metabolic interfaces between growth and reproduction, IV. Chronic pulsatile administration of growth hormone and the timing of puberty in the female sheep. Endocrinology 1991;129:2024–32.

28. Wilson ME, Gordon TP, Rudman CG, Tanner JM. Effects of growth hormone on the tempo of sexual maturation in female rhesus monkeys. J Clin Endocrinol Metab 1989;68:29–38.

29. Wilson ME, Tanner JM. Long-term effects of recombinant human growth hormone treatment on skeletal maturation and growth in female rhesus monkeys with normal pituitary function. J Endocrinol 1991;130:435–41.

30. Hopwood NJ, Hintz RL, Gertner JM, et al. Growth response of children with non-growth-hormone deficiency and marked short stature during three years of growth hormone therapy. J Pediatr 1993;123:215–22.

31. van der Werff ten Bosch JJ, Bot A. Growth of males with idiopathic hypopituitarism without growth hormone treatment. Clin Endocrinol (Oxf) 1990;32:707–17.

32. Zachmann M. Interrelations between growth hormone and sex hormones: physiologic and therapeutic consequences. Horm Res 1992;38(suppl 1):1–18.

33. Hibi I, Tanaka T, Tanae A, et al. The influence of gonadal function and the effect of gonadal suppression treatment on final height in growth hormone (GH)-treated GH-deficient children. J Clin Endocrinol Metab 1989;69:221–6.

34. Boepple PA, Mansfield MJ, Crawford JD, Crigler JF Jr, Blizzard RM, Crowley WF Jr. Analysis of growth data in children with central precocious puberty: the impact of long-term GnRH agonist therapy. In: Grave GD, Cutler GB Jr, eds. Sexual precocity: etiology, diagnosis, and management. New York: Raven Press, 1993:69–83.

35. Stanhope R, Brook CG. The effect of gonadotrophin releasing hormone analogue on height prognosis in growth hormone deficiency and normal puberty. Eur J Pediatr 1988;148:200–2.

36. Toublanc JE, Couprie C, Garnier P, Job JC. The effects of treatment combining an agonist of gonadotropin-releasing hormone with growth hormone in pubertal patients with isolated growth hormone deficiency. Acta Endocrinol (Copenh) 1989;120:795–9.

37. Marshall WA, Tanner JM. Variations in pattern of pubertal changes in girls. Arch Dis Child 1969;44:291–303.

38. Marshall WA, Tanner JM. Variations in the pattern of pubertal changes in boys. Arch Dis Child 1970;45:13–23.

9

Role of Growth Hormone in the Promotion of Linear Skeletal Growth

CLAES OHLSSON, JÖRGEN ISGAARD, ANDERS LINDAHL, AND OLLE G.P. ISAKSSON

The *epiphyseal growth plate* is a cartilaginous structure located between the bony epiphysis and the metaphyseal bone. These structures are developmentally regulated and cannot be identified during fetal life. In most species the epiphysis is first seen as a bone nucleus at the time of birth. The growth plate and articular cartilage are then gradually separated by accumulating bone tissue during subsequent growth. Longitudinal bone growth is a result of endochondral ossification in the long bones. During this process a cartilaginous structure is produced by chondrocyte proliferation in the epiphyseal growth plate, and this cartilaginous template is then degenerated and replaced by bone tissue in the direction of the metaphysis (Fig. 9.1).

Several hormones are needed for normal bone growth, but *growth hormone* (GH) is the only hormone that dose dependently stimulates longitudinal bone growth (1). The mechanism(s) of action for GH in regulating longitudinal bone growth is reviewed in this chapter. Furthermore, the interaction between GH and three other growth-promoting factors—thyroid hormones, sex steroids, and IGF-I—is discussed.

In Vivo Effects of GH and IGF-I

It has been well established for several decades that GH stimulates bone growth in vivo. Hypophysectomy has marked effects on longitudinal bone growth and the histological features of the growth plate, and these effects can be reversed by GH substitution therapy (2). Although early studies did not address the question of site of action of GH, it was generally

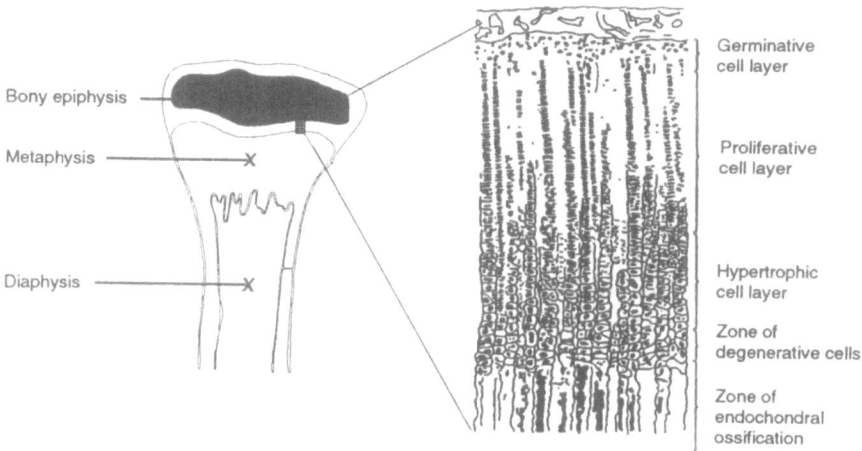

Germinative
cell layer

Proliferative
cell layer

Hypertrophic
cell layer

Zone of
degenerative cells

Zone of
endochondral
ossification

Bony epiphysis

Metaphysis

Diaphysis

FIGURE 9.1. The proximal tibial epiphyseal growth plate and its cellular organization in the rat. Reprinted with permission from Lindahl, Isgaard, and Isaksson (74).

assumed that GH acted directly on cells in the growth plate. However, difficulties in demonstrating the in vitro effects of GH on growth-related parameters, such as sulphate uptake in costal cartilage of hypophysectomized rats, and the finding that the addition of serum from normal but not hypophysectomized rats stimulated sulphate uptake formed the basis for the hypothesis that GH stimulates growth indirectly by stimulating the production of serum factors (somatomedins) that mediate the growth-promoting effects of GH (3).

Approximately 10 years ago this hypothesis was challenged when GH injected directly into the tibial growth plate of hypophysectomized rats resulted in unilateral bone growth, while the same dose given systemically was ineffective (4). This finding was later confirmed in several studies (5–7); it is now well established that GH can interact directly with peripheral tissues. In a recent study it was demonstrated that GH transgenic mice with specific expression of GH in osteoblasts displayed significantly more bone growth than nontransgenic controls without significant systemic effects (8).

Difficulties in obtaining sufficient quantities of purified IGF-I delayed in vivo studies on the effect of IGF-I on body growth, although IGF-I purified from human serum was found to have a small but significant stimulatory effect on total body growth, as well as on tibial epiphyseal width, of hypophysectomized rats when administered continuously via *subcutaneously* (sc) implanted osmotic minipumps (9). Similar results were obtained by Skottner et al. (10), who found that recombinant methionyl *human IGF-I* (hIGF-I) given by sc infusion, by injections twice daily, or by continuous IV infusion had a small stimulatory effect on longitudinal bone growth. Moreover, Guler et al. (11) demonstrated that infusion of recombinant hIGF-I to hypophysectomized rats stimulated body growth as well as accumulated bone growth, although the magnitude of the stimulatory effect was comparatively small. Using the GH-deficient dwarf rat, it was found that *human GH* (hGH) and hIGF-I increased body weight to approximately the same extent (12). However, the effect of hGH on accumulated bone growth was more pronounced compared to the hIGF-I-infused group, suggesting that GH particularly stimulated epiphyseal cartilage growth.

Studies of transgenic mice expressing IGF-I have demonstrated a 1.3-fold increase in body weight without a concomitant increase in skeletal growth (13). However, more recently, it was demonstrated that mice generated from a cross of transgenic mice expressing IGF-I and mice lacking GH-expressing cells had increased body weight and linear growth compared to GH-deficient transgenic littermates (14). However, they did not grow more than their nontransgenic siblings, as has been previously shown for GH transgenic mice that grow to approximately twice the size of their normal littermates (15). In a recent study by Heinrichs and coworkers (16), 2 siblings with Laron syndrome (GH insensitivity syndrome) were treated for 17 months with recombinant IGF-I. After an initial growth-promoting effect of IGF-I, growth response could no longer be observed after 1 year of therapy.

Local administration of IGF-I into the growth plate has been shown to stimulate epiphyseal width as well as accumulated bone growth (5–6), showing that IGF-I has direct growth-promoting effects on cartilage in vivo. Stimulation of longitudinal bone growth by locally administered GH is completely abolished if antibodies to IGF-I are co-infused with GH (7). This finding supports the hypothesis that the effect of GH on accumulated bone growth, at least in part, is dependent on IGF-I, either circulating or locally produced. IGF-I may thus have an important mediator role for the expression of the growth-promoting effect of GH at the tissue level.

Using histochemistry, it was demonstrated that an IGF-I-like immuno-reactivity was present in proliferative chondrocytes of rat epiphyseal growth plate (17). Moreover, it was shown that hypophysectomy decreases the number of fluorescent cells and that GH treatment partially restored the immunoreactivity. To address the question of whether IGF-I is actually produced in the growth plate, in situ hybridization technique was used to demonstrate the presence of IGF-I mRNA in the proliferative and hyper-trophic chondrocytes of rat tibial growth plates (18). Hypophysectomy of the rats resulted in both decreased hybridization signal and reduction of cell number compared to intact controls. Replacement therapy with GH partially normalized the expression of IGF-I mRNA, confirming earlier results showing a GH-dependent expression of IGF-I mRNA in rib growth plate of hypophysectomized rats (19).

Results of early studies on hypophysectomized rats given GH replacement therapy according to different protocols indicate that a pulsatile GH plasma pattern is more favorable for body growth than an intermediate and rather constant plasma GH level (20). In alignment with this finding, it has been demonstrated that a pulsatile GH treatment is more effective than continuous GH infusion in increasing IGF-I mRNA in rib growth plate (21).

In Vitro Effects of GH and IGF-I

It is evident that both GH and IGF-I stimulate longitudinal bone growth in vivo, but there seems to be a discrepancy between the potency of effects demonstrated in vivo versus the effect in vitro for the two peptides. GH appears to be more potent than IGF-I in promoting growth in vivo, while IGF-I appears to be a more potent mitogen in vitro (12, 22). Hypothetically, the stimulatory effect of GH or IGF-I could be explained by an increased rate of differentiation of precursor cells, or an increased rate of cell multiplication of differentiated cells, or both. To distinguish these two processes at the tissue level, however, is not possible since stimulation of either process results in tissue growth.

In vitro effects of GH have been difficult to demonstrate, and this fact was one of the main reasons for the postulation of the somatomedin hypothesis. It took more than 20 years before the hypothesis was challenged by investigators demonstrating that somatomedin/IGF-I was produced in tissues other than the liver. Thus, fibroblasts derived from human skin and lung produced IGF-I, and GH in vitro stimulated this production (23, 24). Furthermore, intact fetal tibia maintained in organ cultures released IGF-I into the culture medium in response to GH (25). The cellular interactions between GH and IGF-I have been extensively studied in 3T3 preadipocytes, a suitable model for studies of adipose cell differentiation/proliferation (26–28).

Confluent 3T3 preadipose cells that are cultured in the presence of a low concentration of serum are able to convert to adipocytes. During the process of differentiation, there are marked alterations in the expression of various cellular enzymes (29–31). The enzyme glycerophosphate dehydrogenase is expressed early in the process of differentiation and is dependent on GH (32, 33). Furthermore, GH exerted no effects on exponentially growing cells, but was active only when cells were in a resting state. Subsequent experiments demonstrated that GH responsiveness is restricted to a transient period during differentiation and that the responsiveness is then reduced in the following 10 days. After that time the cells become refractory to GH stimulation. IGF-I is unable to substitute for GH in adipose differentiation of 3T3 cells (34). We have recently been able to confirm these results in primary isolated epiphyseal chondrocytes (35). The reason for this phenomenon is poorly understood, but could depend on a postconfluent growth inhibition.

From the above-related experiments, it became evident that the somatomedin hypothesis for GH action was invalid for the proliferation/differentiation of fibroblasts and adipose tissues. In subsequent years it also became evident that primary isolated chondrocytes were sensitive to both GH and IGF-I (22, 35–37) and expressed binding sites for both IGF-I and GH (38–41). However, it was difficult to demonstrate whether GH or IGF-I acts on cells at the same stage or at different stages of differentiation.

Primary isolated epiphyseal chondrocytes cultured in a semisolid culture system stabilized with agarose are able to form colonies of various sizes (42). If cells are isolated from the different layers of the growth plate, cells form clones that relate to their previous location in the growth plate. Thus, cells isolated from the proximal part of the growth plate—that is, from the germinative and early proliferative cell layer—form large clones, while cells isolated from the proliferative cell layer form smaller clones. Terminally differentiated cells in the hypertrophic cell layer are unable to form any colonies (22). These results indicate that the capacity of clonal growth is increased in the proximal direction. Furthermore, the clonal capacity is also dependent on the age of the animal since both clonal size and cloning efficiency are decreased with age (42).

Both GH and IGF-I had direct effects in vitro and potentiated chondrocyte colony formation. However, the stimulatory effect of the two hormones differed between different clones. GH preferentially stimulated the formation of large undifferentiated colonies, suggesting that GH interacts with undifferentiated progenitor cells (prechondrocytes or early differentiated proliferative cells) with an inborn high capacity for clonal growth (43). The small-sized and intermediate-sized colonies were stimulated by IGF-I. When the epiphyseal growth plate was microdissected, the stimulatory effect of GH was confined to the cells isolated from the proximal part of the growth plate (44). IGF-I stimulated the formation of small- and middle-sized colonies where the progenitor cells were isolated from the intermediate or distal part of the growth plate. These results further support the notion that IGF-I stimulates the clonal growth of chondrocytes in the proliferative cell layer of the growth plate.

Binding data further support the fact that GH and IGF-I have different target populations in the growth plate; the GH binding/GHR immunoreactivity is heterogenous, and GH and IGF-I bind to different populations of cultured epiphyseal chondrocytes (45). Furthermore, the stimulatory effect of GH and IGF-I on primary isolated epiphyseal chondrocytes in monolayer cultures differs; a sustained effect of GH, but not IGF-I, on ^{3}H-thymidine incorporation has been demonstrated (35). These findings suggest that only a short exposure to GH is needed to prime a prechondrocyte to initiate a subsequent clonal expansion, while the IGF-I effect reflects a stimulatory effect on already differentiated proliferative chondrocytes. The effect could also be compared with a commitment versus a progression factor influencing the cell cycle.

Recently, we were able to confirm in vivo the notion that GH and IGF-I indeed have different target populations. Using long-term infusion of ^{3}H-thymidine with a concomitant local GH/or IGF-I infusion to the growth plate, we were able to demonstrate that GH, but not IGF-I, was able to induce a mitogenic effect in the germinative cell layer (46).

Effects of Thyroid Hormones in Relation to GH

Both clinical studies in humans and experimental animal studies have shown that thyroid hormones are crucial for optimal bone growth (47–49). It is well known that thyroid hormones stimulate GH secretion both in vivo (50, 51) and in vitro (52, 53), indicating that thyroid hormones can exert effects on longitudinal bone growth indirectly via regulation of GH secretion. However, mechanisms that are GH independent have been suggested. This is supported by the finding that thyroid hormones stimulate longitudinal bone growth in hypophysectomized rats (54) and that they are required for the formation of hypertrophic cells in the epiphyseal plate, while GH is required for a normal proliferative cell layer in both thyroidectomized and hypophysectomized rats (48, 55). Furthermore, clinical studies have demonstrated that thyroid hormones are important for normal bone maturation, and GH cannot replace thyroid hormones in this maturation process (56). These results indicate that GH and thyroid hormones exert different effects on the epiphyseal growth plate. Recently, it was shown that thyroid hormones can stimulate IGF-I production from rat liver in a GH-independent manner (57, 58). However, the physiological importance of this IGF-I stimulation remains to be elucidated.

In tissue culture thyroid hormones increase the number of hypertrophic chondrocytes and recruit new cells into the maturational zone of growth plate cartilage from fetal pig scapulae (59). Furthermore, it was recently shown that epiphyseal chondrocytes in monolayer culture have receptors for thyroid hormones and that thyroid hormones increase alkaline phosphatase activity in those cells (60, 61). Cell proliferation was decreased concomitantly with the thyroid hormone-induced alkaline phosphatase activity (61). It is thus conceivable that thyroid hormones inhibit the proliferation of cells undergoing a clonal expansion and stimulate the transition to a nondividing state associated with terminal differentiation (Fig. 9.2).

Three possible mechanisms for the stimulatory effect of thyroid hormones on longitudinal bone growth are summarized in Figure 9.2: (i) modulation of GH secretion, (ii) direct stimulation of IGF-I production

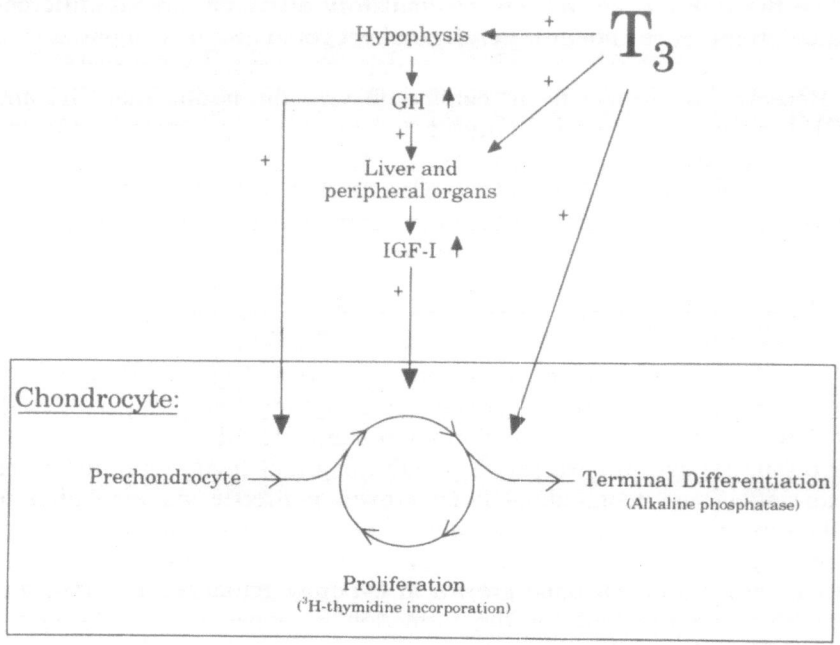

FIGURE 9.2. Hypothetical model for the stimulatory effect of thyroid hormone on longitudinal bone growth. Reprinted with permission from Ohlsson, Isgaard, Tornell, Nilsson, Isaksson, and Lindahl (75).

in the liver, and (iii) direct interaction with the epiphyseal cartilage. The relative importance of these different pathways is unknown.

Sex Steroid Effects in Relation to GH

Sex steroids are crucial for increased longitudinal bone growth during puberty, as illustrated in central precocious puberty where the final height is reduced despite an initial marked increase in growth velocity (62). It is well established that sex steroids regulate GH secretion (63, 64) and thereby indirectly stimulate bone growth. However, there are conflicting results concerning the ability of sex steroids to raise circulating IGF-I levels without changing the GH secretion. No changes were found in IGF-I levels after sex steroid administration to hypopituitary patients under a constant regimen of GH therapy, while sex steroids increased IGF-I levels with unchanged GH secretion in gonadal dysgenesis patients (62).

Another hypothetical mechanism for sex steroids to stimulate longitudinal bone growth is a regulation of *GH binding protein* (GHBP) and/or *IGF binding proteins* (IGFBPs), thereby modulating the effect of circulating GH and/or IGF-I. A recent study in girls with Turner syndrome that demonstrated that estrogen increases while oxandrolone decreases serum levels of the GHBP (65) is in favor of such a mechanism.

Several clinical and in vivo studies indicate that sex steroids can interact with the epiphyseal cartilage: (i) Children with GH insensitivity syndrome demonstrate a growth spurt at sexual maturation (66); (ii) in hypophysectomized prepubertal lambs testosterone stimulates growth (67); and (iii) direct local administration of testosterone into rat tibial epiphyseal plate results in an increased epiphyseal width (68). Furthermore, in a GH transgenic animal model with sex steroid-independent GH secretion, we have shown that androgens can regulate growth without changing systemic GH and/or IGF-I concentrations. Gonadectomy in male GH transgenic animals resulted in disproportional bone growth, indicating that the spine, but not the long bones, are stimulated by androgens (69). These results together strongly support the hypothesis that there are GH-independent mechanisms for sex steroids to regulate bone growth.

In vitro both androgen and estrogen receptors have been demonstrated on cartilage cells (70, 71). Cartilage was able to convert testosterone into its more active metabolite, dihydrotestosterone, suggesting that cartilage is a target organ for androgens (72). Sex steroids stimulated proteoglycan synthesis by epiphyseal chondrocytes, but the effect was dependent on the age of the animal donors. Cartilage cells from rabbit fetuses and from animals at the age of gonadal maturation were most responsive to sex steroids (73). In conclusion, some of the effects of sex steroids on longitudinal bone growth are probably via regulation of GH secretion, but both clinical and experimental studies demonstrate that sex steroids also have effects that are independent of GH secretion.

Concluding Remarks

The regulation of longitudinal bone growth is complex, and several factors, such as nutritional, endocrine, paracrine, and autocrine factors, are necessary for optimal longitudinal bone growth. In this chapter the mechanism behind the growth-promoting effect of GH has been discussed. Both GH and IGF-I can promote bone growth, but IGF-I cannot replace all the effects of GH. Furthermore, data are presented indicating that both sex steroids and thyroid hormones, besides their regulation of GH secretion, interact directly with epiphyseal growth plates. The results indicate that GH interacts with a prechondrocyte, followed by an IGF-I-regulated clonal expansion that is terminated by a thyroid hormone-induced terminal differentiation. However, the cellular mechanism of action for sex steroids on the epiphyseal growth plate remains to be studied.

Acknowledgments. This work was supported by grants from the Swedish Medical Research Council (14x-04250 and K90-13F-09248-01A); KabiPharmacia, Stockholm; The Lundberg Foundation; the Göteborg Medical Society; and the Faculty of Medicine, University of Göteborg. We should also like to thank M. Walser and K. Lövstedt for excellent technical assistance.

References

1. Cheek D, Hill DE. Effect of growth hormone on cell and somatic growth. In: Knobil E, Sawyer W, eds. Handbook of physiology; vol IV. Washington, DC: American Physiological Society, 1974:159–85.
2. Sissons HA. The growth of bone. In: Bourne GH, ed. The biochemistry and physiology of bone. New York: Academic Press, 1971:145–80.
3. Salmon WD Jr, Daughaday WH. A hormonally controlled serum factor which stimulates sulphate uptake in vitro. J Lab Clin Med 1957;49:825–36.
4. Isaksson OGP, Jansson J-O, Gause IAM. Growth hormone stimulates longitudinal bone growth directly. Science 1982;216:1237–9.
5. Russell SM, Spencer EM. Local injections of human or rat growth hormone or of purified human somatomedin-C stimulate unilateral tibial epiphyseal growth in hypophysectomized rats. Endocrinology 1985;116:2563–7.
6. Isgaard J, Nilsson A, Lindahl A, Jansson J-O, Isaksson OGP. Effects of local administration of GH and IGF-I on longitudinal bone growth in rats. Am J Physiol 1986;250:E367–72.
7. Schlecter NL, Russell SM, Spencer EM, Nicoll CS. Evidence suggesting that the direct growth promoting effect of growth hormone on cartilage in vivo is mediated by local production of somatomedin. Proc Natl Acad Sci USA 1986;83:7932–4.
8. Baker AS, Hollingshead PG, Pitts-Meek S, Hansen S, Taylor R, Stewart TA. Osteoblast-specific expression of growth hormone stimulates bone growth in transgenic mice. Mol Cell Biol 1992;12:5541–7.
9. Schoenle E, Zapf J, Humbel RE, Froesch ER. Insulin-like growth factor I stimulates growth in hypophysectomized rats. Nature 1982;296:252–3.
10. Skottner A, Clark RG, Robinson ICAF, Fryklund L. Recombinant human insulin-like growth factor: testing the somatomedin hypothesis in hypophysectomized rats. J Endocrinol 1987;112:123–32.
11. Guler HP, Zapf J, Schweiwiller E, Froesch ER. Recombinant human insulin-like growth factor I stimulates growth and has distinct effects on organ size in hypophysectomized rats. Proc Natl Acad Sci USA 1988;85:4889–93.
12. Skottner A, Clark RG, Fryklund L, Robinson ICAF. Growth responses in mutant dwarf rat to human growth hormone and recombinant human insulin-like growth factor I. Endocrinology 1989;124:2519–26.
13. Mathews LS, Hammer RE, Behringer RR, et al. Growth enhancement of transgenic mice expressing human insulin-like growth factor I. Endocrinology 1988;123:2827–33.

14. Behringer RR, Lewin TM, Quaife CJ, Palmiter RD, Brinster RL, D'Ercole AJ. Expression of insulin-like growth factor I stimulates normal somatic growth in growth hormone deficient transgenic mice. Endocrinology 1990; 127:1033–40.
15. Palmiter RD, Brinster RL, Hammer RE. Dramatic growth of mice that develop from eggs microinjected with metallothionein-growth hormone fusion genes. Nature 1982;300:611–5.
16. Heinrichs C, Vis HL, Bergmann P, Wilton P, Bourguignon JP. Effects of 17 months of treatment using recombinant insulin-like growth factor I in two children with growth hormone insensitivity (Laron) syndrome. Clinical Endocrinol (Oxf) 1993;33:647–51.
17. Nilsson A, Isgaard J, Lindahl A, Dahlström A, Skottner A, Isaksson OGP. Regulation by growth hormone of number of chondrocytes containing IGF-I in rat growth plate. Science 1986;233:571–4.
18. Nilsson A, Carlsson B, Isgaard J, Isaksson O, Rymo L. Regulation by GH of insulin-like growth factor I mRNA expression in epiphyseal growth plate as studied by in situ hybridization. J Endocrinol 1990;125:67–74.
19. Isgaard J, Möller C, Isaksson OGP, Nilsson A, Mathews LS, Norstedt G. Regulation of insulin-like growth factor messenger ribonucleic acid in rat growth plate by growth hormone. Endocrinology 1988;122:1515–20.
20. Jansson J-O, Edén S, Isaksson OGP. Sexual dimorphism in the control of growth hormone secretion. Endocr Rev 1985;6:128–50.
21. Isgaard J, Carlsson L, Isaksson OGP, Jansson J-O. Pulsatile intravenous growth hormone infusion to hypophysectomized rats increases insulin-like growth factor messenger ribonucleic acid in skeletal tissues more effectively than continuous GH infusion. Endocrinology 1989;123:2605–10.
22. Lindahl A, Isgaard J, Isaksson O. Growth hormone in vivo potentiates the stimulatory effect of insulin-like growth factor-I in vitro on colony formation of epiphyseal chondrocytes isolated from hypophysectomized rats. Endocrinology 1987;121:1070–5.
23. Atkinson PR, Weidman ER, Bhaumick B, Bala RM. Release of somatomedin-like activity by cultured WI-38 human fibroblasts. Endocrinology 1980;106: 2006–12.
24. Clemmons DR, Underwood LE, Van Wyk JJ. Hormonal control of immuno-reactive somatomedin production by cultured human fibroblasts. J Clin Invest 1981;67:10–9.
25. Stracke H, Schultz A, Moeller D, Rossol S, Schatz H. Effects of growth hormone on osteoblasts and demonstration of sometomedin-C/IGF-I in bone organ culture. Acta Endocrinol (Copenh) 1984;107:16–24.
26. Green H, Kehinde O. Spontaneous heritable changes leading to increased adipose conversion in 3T3 cells. Cell 1976;7:105–13.
27. Négrel R, Grimaldi P, Ailhaud G. Establishment of preadipocyte clonal line from epididymal fat pad of ob/ob mouse that responds to insulin and to lipolytic hormones. Proc Natl Acad Sci USA 1978;75:6054–8.
28. Vannier C, Gaillard D, Grimaldi P, et al. Adipose conversion of ob17 cells and hormone-related events. Int J Obes 1985;9(suppl 1):41–53.
29. Green H. Adipose conversion: a program of differentiation: obesity—cellular and molecular aspects. INSERM 1979;87:15–24.
30. Wise LS, Green H. Studies of lipoprotein lipase during the adipose con-version of 3T3 cells. Cell 1978;13:233–42.

31. Kawamura M, Jensen DF, Wancewicz EV, Joy LL, Khoo JC, Steinberg D. Hormone-sensitive lipase in differentiated 3T3–L1 cells and its activation by cyclic AMP-dependent protein kinase. Proc Natl Acad Sci USA 1981;78:732–6.

32. Pairault J, Green H. A study of the adipose conversion of suspended 3T3 cells by using glycerophosphate dehydrogenase as differentiation marker. Proc Natl Acad Sci USA 1979;76:5138–42.

33. Morikawa M, Nixon T, Green H. Growth hormone and the adipose conversion of 3T3 cells. Cell 1982;29:783–9.

34. Morikawa M. Sensitivity of preadipose 3T3 cells to growth hormone. J Cell Physiol 1986;128:293–8.

35. Ohlsson C, Nilsson A, Isaksson OGP, Lindahl A. Effects of growth hormone and insulin-like growth factor-I on DNA synthesis and matrix production in rat epiphyseal chondrocytes in monolayer culture. J Endocrinol 1992;133:291–300.

36. Madsen K, Friberg U, Roos P, Eden S, Isaksson O. Growth hormone stimulates the proliferation of cultured chondrocytes from rabbit ear and rat rib growth cartilage. Nature 1983;304:545–7.

37. Trippel SB, Corvol MT, Dumontier MF, Rappaport R, Hung HH, Mankin HJ. Effects of somatomedin-C/insulin-like growth factor I and growth hormone on cultured growth plate and articular chondrocytes. Pediatr Res 1989;25:76–82.

38. Eden S, Isaksson OGP, Madsen K, Friberg U. Specific binding of growth hormone to isolated chondrocytes from rabbit ear and epiphyseal plate. Endocrinology 1983;112:1127–9.

39. Trippel S, Van-Wyk J, Foster M, Svoboda M. Characterization of a specific somatomedin-C receptor on isolated bovine growth plate chondrocytes. Endocrinology 1983;112:2128–36.

40. Trippel S, Van Wyk J, Mankin H. Localization of somatomedin-C binding to bovine growth plate chondrocytes in situ. J Bone Joint Surg [Am] 1986;68:897–903.

41. Nilsson A, Lindahl A, Eden S, Isaksson OGP. Demonstration of growth hormone receptors in cultured rat epiphyseal chondrocytes by specific binding of growth hormone and immunohistochemistry. J Endocrinol 1989;122:69–77.

42. Lindahl A, Isgaard J, Carlsson L, Isaksson O. Differential effects of growth hormone and insulin-like growth factor I on colony formation of epiphyseal chondrocytes in suspension culture in rats of different ages. Endocrinology 1987;121:1061–9.

43. Ohlsson C, Isaksson OGP, Lindahl A. Clonal analysis of rat growth plate chondrocytes in suspension culture—differential effects of growth hormone and insulin-like growth factor-I. Growth Reg 1994;4:1–7.

44. Lindahl A, Nilsson A, Isaksson OGP. Effects of growth hormone and insulin-like growth factor-I on colony formation of rabbit epiphyseal chondrocytes at different stages of maturation. J Endocrinol 1987;115:263–71.

45. Bentham J, Ohlsson C, Lindahl A, Isaksson O, Nilsson A. A double-staining technique for detection of growth hormone and insulin-like growth factor-I binding to rat tibial epiphyseal chondrocytes. J Endocrinol 1993:361–7.

46. Ohlsson C, Nilsson A, Isaksson O, Lindahl A. Growth hormone induces multiplication of the slowly cycling germinal cells of the rat tibial growth plate. Proc Natl Acad Sci USA 1992;89:9826–30.
47. Rochiccioli P. Thyroid dysgenesis. In: Delange F, Fisher D, eds. Pediatric and adolescent endocrinology. Basel: Karger, 1985:154–73.
48. Ray R, Asling C, Walker D, Simpson M, Li C, Evans H. Growth and differentiation of the skeleton in thyroidectomized-hypophysectomized rats treated with thyroxin, growth hormone and the combination. J Bone Joint Surg [Am] 1954;36A:94–103.
49. Thorngren K, Hansson L. Effect of thyroxine and growth hormone on longitudinal bone growth in the hypophysectomized rat. Acta Endocrinol (Copenh) 1973;74:24–40.
50. Hervas F, Escorbar G, Escorbar Del Ray F. Rapid effects of single small doses of L-thyroxin and triiodothyronine on growth hormone, as studied in the rat by radioimmunoassay. Endocrinology 1975;97:91–101.
51. Coiro V, Braverman L, Christianson D, Fang S, Goodman M. Effect of hypothyroidism and thyroxine replacement on growth hormone in the rat. Endocrinology 1979;105:641–6.
52. Samuels H, Stanley F, Shapiro L. Dose dependent depletion of nuclear receptors by L-triiodothyronine: evidence for a role in induction of growth hormone synthesis in cultured GH1 cells. Proc Natl Acad Sci USA 1976;73:3877–81.
53. Vale W, Vaughan J, Yamamoto G, Spiess J, Rivier J. Effect of synthetic human pancreatic (tumor) GH releasing factor and somatostatin, triiodothyronine and dexamethasone on GH secretion in vitro. Endocrinology 1983;112:1553–5.
54. Asling W, Simpson M, Li C, Evans H. The effect of chronic administration of thyroxin to hypophysectomized rats on their skeletal growth, maturation and response to growth hormone. Anat Res 1954;119:101–17.
55. Lewinson D, Harel Z, Shenzer P, Silbermann M, Hochberg Z. Effect of thyroid hormone and growth hormone on recovery from hypothyroidism of epiphyseal growth plate cartilage and its adjacent bone. Endocrinology 1989;124:937–45.
56. Virtanen M, Perheentupa J. Bone age at birth; method and effect of hypothyroidism. Acta Pediatr Scand 1989;78:412–8.
57. Ikeda T, Fujiyama K, Takeuchi T, et al. Effect of thyroid hormone on somatomedin-C release from perfused rat liver. Experientia 1989;45:170–1.
58. Wolf M, Ingbar S, Mose A. Thyroid hormone and growth hormone interact to regulate insulin-like growth factor-I messenger ribonucleic acid and circulating levels in the rat. Endocrinology 1989;125:2905–14.
59. Burch WM, Lebovitz HE. Triiodothyronine stimulates maturation of porcine growth plate cartilage in vitro. J Clin Invest 1982;70:496–504.
60. Carrascosa A, Ferrandez M, Audi L, Ballabriga A. Effects of triiodothyronine (T3) and identification of specific nuclear T3-binding sites in cultured human fetal epiphyseal chondrocytes. J Clin Endocrinol Metab 1992;75:140–4.
61. Ohlsson C, Nilsson A, Isaksson O, Lindahl A. Effects of tri-iodothyronine and insulin-like growth factor-I (IGF-I) on alkaline phosphatase activity, 3H-thymidine incorporation and IGF-I receptor mRNA in cultured rat epiphyseal chondrocytes. J Endocrinol 1992;135:115–23.

62. Bourguinon J. Linear growth as a function of age at onset of puberty and sex steroid dosage: therapeutical implications. Endocr Rev 1988;9:467–88.
63. Kerrigan JR, Rogol AD. The impact of gonadal steroid hormone action on growth hormone secretion during childhood and adolescence. Endocr Rev 1992;13:281–98.
64. Wehrenberg WB, Giustina A. Basic counterpoint: mechanism and pathways of gonadal steroid modulation. Endocr Rev 1992;13:299–308.
65. Carlsson LMS, Albertsson-Wikland K. Growth hormone-binding protein concentration in girls with Turner syndrome: effects of GH, oxandrolone and ethinyl estradiol. In: Hibi I, Takano K, eds. Basic and clinical approach to Turner syndrome. Elsevier Science, 1993:225–59.
66. Laron Z, Sarel R, Pertzelan A. Puberty in Laron type dwarfism. Eur J Pediatr 1980;134:79–83.
67. Young I, Mesiano S, Hintz R. Growth hormone and testosterone can independently stimulate the growth of hypophysectomized prepubertal lambs without any alteration in circulating concentrations of insulin-like growth factors. J Endocrinol 1989;121:563–70.
68. Ren S, Malozovski S, Sanches P, Sweet D, Loriaux D, Cassorla F. Direct administration of testosterone increases rat tibial epiphyseal growth plate width. Acta Endocrinol (Copenh) 1989;121:401–5.
69. Sandstedt J, Norjavaara E, Ohlsson C, Törnell J. Gonadectomy results in a disproportional bone growth and reduced weight gain in growth hormone transgenic and normal male mice [Abstract]. 75th annu meet Endocr Soc, 1993:1132.
70. Rosner I, Manni A, Malemud C, Boja B, Moskowitz R. Estradiol receptors in articular chondrocytes. Biochem Biophys Res Commun 1982;106:1378–82.
71. Carrascosa A, Audi L, Ferrandez A, Ballabriga A. Biological effects of androgens and identifications of specific dihydrotestosterone-binding sites in cultured human fetal epiphyseal chondrocytes. J Clin Endocrinol Metab 1990;70:136–40.
72. Takahashi Y, Corvol M, Tsagris L, Carrascosa A, Bok S, Rappaport R. Testosterone metabolism in prepubertal rabbit cartilage. Mol Cell Endocrinol 1984;35:15–24.
73. Corvol M, Carrascosa A, Tsagris L, Blanchard O, Rappaport R. Evidence for a direct in vitro action of sex steroids on rabbit cartilage cells during skeletal growth: influence of age and sex. Endocrinology 1987;120:1422–9.
74. Lindahl A, Isgaard J, Isaksson O. Growth and differentiation. In: Robertson DM, Herington AC, eds. Clinical endocrinology and metabolism: growth factors in endocrinology; vol 5. London: Ballieres Tindall, 1991:671–87.
75. Ohlsson C, Isgaard J, Tornell J, Nilsson A, Isaksson OGP, Lindahl A. Endocrine regulation of bone growth. Acta Paediatr Scand Suppl 1993;391:33–40.

10

Growth Hormone Economy in Normally Cycling Women

W.S. Evans, R.A. Booth, Jr., K.K.Y. Ho, A.C.S. Faria, C.M. Asplin, J.D. Veldhuis, and M.O. Thorner

Although it has long been recognized that the secretion of *growth hormone* (GH) differs between adult men and women, the mechanisms involved remain to be fully elucidated. A number of observations have suggested that the gonadal hormones in general and estrogen in particular may be primary factors subserving gender-associated differences in GH release. In premenopausal women with ostensibly intact ovarian function, both serum concentrations of GH (1) and GH secretory rates (2) are higher than those documented in men. Serum GH is reported to be relatively elevated during the periovulatory (3) and midluteal (1) phases of the human menstrual cycle, times when endogenous estrogen concentrations are increased in comparison to those measured during the early follicular phase of the cycle. Moreover, the exogenous administration of estrogen appears to be associated with alterations in serum GH concentrations. Thus, such concentrations are greater in premenopausal girls (4), premenopausal women (2), postmenopausal women (5), and men (1) treated with estrogen compared to the estrogen-deficient state.

Within the past decade significant investigative interest has focused on potential processes through which gonadal hormones may modulate GH secretion. Clearly, gonadal products could be exerting an effect on the hypothalamic regulators of GH release or directly on somatotrophs within the anterior pituitary or at both loci. In addition, steroids might modify IGF-I feedback regulation of GH secretion. Within this context, the application of recently developed hormone pulse analysis techniques for the characterization of endogenously driven GH release and the availability of *GH releasing hormone* (GHRH) itself have allowed for a markedly improved understanding of the mechanisms governing the secretion of GH.

Evaluation of Hypothalamic Factors Regulating GH Release

It is generally accepted that the pulsatile nature of GH secretion reflects an interplay between two primary hypothalamic factors—GHRH and *somatostatin release inhibiting factor* (SRIF)—on somatotrophs, the cells within the anterior pituitary gland responsible for the secretion of GH (6, 7). Presumably, a pulse of GH represents sufficiently diminished SRIF tone in the setting of adequate GHRH stimulation. Given the fact that hypothalamic activity cannot be monitored directly in the human, the identification and characterization of GH pulses has emerged as a valuable tool with which to assess, albeit indirectly, the hypothalamic factors responsible for GH secretion. Thus, the frequency with which pulses of GH are released may reflect intervals during which hypothalamic neurons curtail secretion of SRIF, but continue to secrete GHRH, or secrete relatively more GHRH than SRIF.

Moreover, certain characteristics of GH pulses, including amplitude and width, may be a consequence, at least in part, of the amount of GHRH impinging on somatotrophs and the duration of this signal. However, caution must be exercised in interpreting parameters such as amplitude and width in that these pulse attributes are likely also to reflect the prevailing response characteristics of the somatotrophs themselves. Stated succinctly, whereas the frequency with which GH pulses are released primarily represents the hypothalamic signal, GH pulse amplitude and width may be determined by a combination of hypothalamic input and somatotroph responsivity.

Evaluation of Pituitary Factors Regulating GH Release

The goal of documenting somatotroph responsivity became closer to reality with the isolation, characterization, and subsequent availability of GHRH for human studies. Numerous studies have now demonstrated that GHRH does stimulate the release of GH and that GHRH-stimulated GH release does vary under differing physiological and pathophysiological circumstances. However, concerns have been raised regarding the interpretation of the GHRH challenge, given that responses may vary significantly as a function of somatostatinergic input to the somatotroph at the time of testing. To address this potential confounding factor, a number of recent studies have begun to utilize agents thought to curtail SRIF secretion such that relatively pure somatotroph responsivity can be assessed.

Evaluation of GH Release: Use of the Premenopausal Woman as a Model

Using a combination of these and other experimental techniques, we and others have sought to define GH release in premenopausal women. Whereas earlier studies compared results obtained in relatively young women during the early follicular phase of the menstrual cycle to results obtained in young men and older men and women, our more recent studies have focused on GH economy during several phases of the menstrual cycle. The latter investigations have been of particular interest in that the effects of three combinations of gonadal hormone input to the hypothalamic-somatotrophic axis can be appraised: (i) relatively low estrogen and progesterone (the early follicular phase), (ii) relatively high estrogen but low progesterone (the late follicular phase), and (iii) relatively high estrogen and high progesterone (the midluteal phase) (8).

GH Release in Men and Women: Effects of Age and Sex

As noted above, certain studies reported over two decades ago suggested that GH secretion is higher in women than in men (1, 2). Other studies, however, failed to confirm these observations (9, 10). Moreover, the issue of whether GH secretion diminishes with age has also been debated by some authors (11–15).

The possibility that somatotroph responsivity might play a role in age- and sex-associated changes in GH release was appraised by GH challenge shortly after the releasing hormone became available for clinical use. Unfortunately, the conclusions drawn from these studies were anything but consistent. Thus, GHRH-stimulated GH release was reported to be markedly decreased in older versus younger men (16) and also to be unaffected by the aging process (17). Similarly, whereas differences in GHRH-stimulated GH release in young men versus young women were reported (18), others found no discernible difference in stimulated release (19–21). The discrepancies in these studies may well reflect differences in experimental design (e.g., sub- vs. supramaximal doses of GHRH administered) in addition to the confounding issues of a physiological nature (e.g., degree of ambient SRIF tone) alluded to above.

In an earlier study we raised the possibility that in addition to the concerns just mentioned, the effects of gender (i.e., mediated by the gonadal hormones) and age on GH secretion might be interrelated, thus preventing proper assessment of one factor without adequately controlling for the other. To address this issue, Ho and colleagues characterized both integrated concentrations of GH and GH pulse patterns using 24-h serum

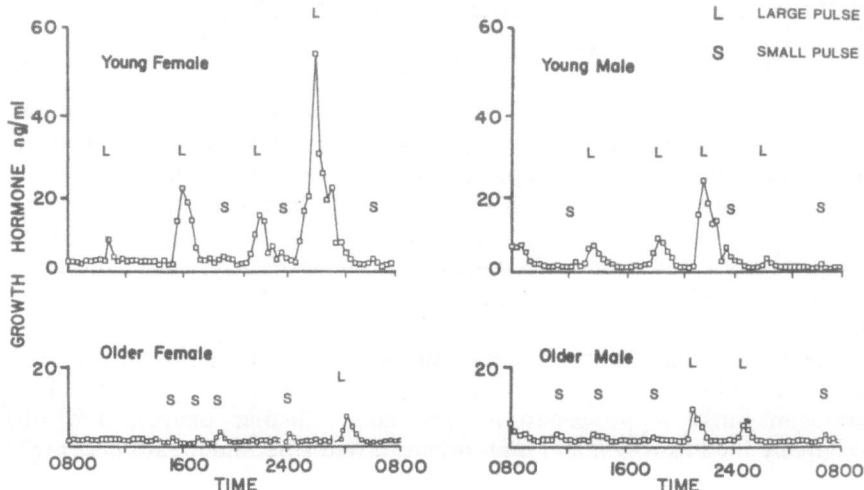

FIGURE 10.1. Serum GH profiles from a young woman, young man, older woman, and older man sampled every 20 min for 24 h. Pulses were categorized as large (L) or small (S) depending on whether the rise was greater or less than 3 times the threshold criterion for a pulse. Reprinted with permission from Ho, Evans, Blizzard, et al. (22), © The Endocrine Society, 1987.

GH concentration-time series obtained in young men, young early follicular phase women, older men, and postmenopausal women (22). It should be noted that this study utilized a relatively infrequent sampling paradigm (blood samples obtained every 20 min for 24 h) and a first-generation pulse detection algorithm. Nevertheless, the results confirmed the suspicion that sex and age do have both independent and interrelated effects on the release of GH.

Figure 10.1 shows the representative GH profiles from younger and older men and women. Visual inspection of these profiles suggested that younger individuals secrete more GH than do older subjects and that young women secrete more GH than do young men. Assessment of 24-h *integrated GH concentrations* (IGHC) (Fig. 10.2) supported this impression; that is, the mean IGHC (μg/min/mL) was higher in younger (5.05 ± 0.32) versus older (3.44 ± 0.3) subjects ($P < 0.003$) and higher in women (4.77 ± 0.34) versus men (3.72 ± 0.34) ($P < 0.025$). Of particular interest, however, was the observation that the sex difference in IGHC could essentially be accounted for by measurements in the young early follicular phase women studied. With regard to the characteristics of pulsatile GH release, no differences in GH pulse frequencies were noted, although there was a trend towards a relatively high frequency in younger women compared to the other groups. In addition, younger subjects

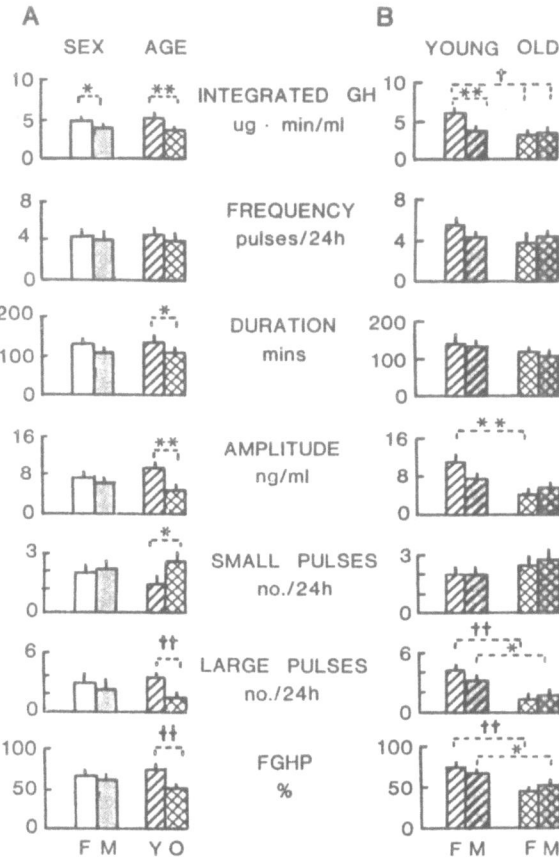

FIGURE 10.2. Comparison of mean (±SEM) serum IGHC, pulse frequency, duration, amplitude, number of small or large pulses, and fraction of GH secreted in pulses (FGHP) in study subjects grouped according to sex or age (*A*) or according to study subgroups of young women, young men, older women, and older men (*B*). Women are indicated by nonshaded bars, men by shaded bars, young subjects by diagonally hatched bars, and older subjects by diamond-hatched bars. (*$P < 0.05$; **$P < 0.01$; †$P < 0.001$; ††$P < 0.00001$). Reprinted with permission from Ho, Evans, Blizzard, et al. (22), © The Endocrine Society, 1987.

exhibited a greater pulse amplitude than did older subjects, but, again, this observation related primarily to the large GH pulse amplitudes seen in the younger individuals.

Given the obvious possibility that the estrogen-sufficient status of the early follicular phase women might account for the above observations, correlations were sought between concentrations of free and total

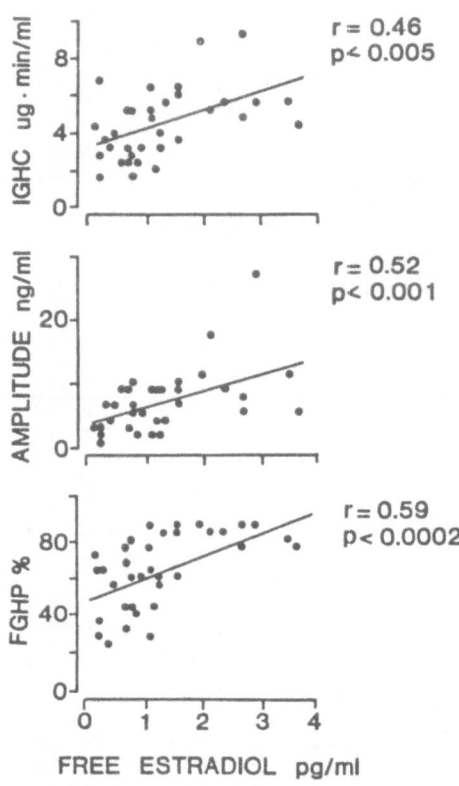

FIGURE 10.3. Serum IGHC (top panel), mean GH pulse amplitude (middle panel), and fraction of GH secreted in pulses (FGHP) (bottom panel) in relation to free estradiol concentrations. Reprinted with permission from Ho, Evans, Blizzard, et al. (22), © The Endocrine Society, 1987.

estradiol and several parameters of GH release. Both free and total estradiol concentrations were found to correlate significantly and positively with IGHC, GH pulse amplitude, and the percentage of GH released within pulses. The actual correlations between these parameters of GH release and free estradiol concentrations are shown in Figure 10.3. It should be noted that neither free nor total testosterone concentrations correlated with any parameter of GH release.

Because of the strong relationship between estradiol concentration and GH release, the effects of sex and age were reassessed after statistical removal of the effects of free estradiol. Under these circumstances, the mean IGHC was not affected by age and/or sex, and mean pulse amplitude was no longer affected by age. The overall conclusion of this study was that estradiol accounts significantly for the observed age and sex-related differences in GH release in the human. Moreover, given that there were no significant differences in GH pulse frequency in young early follicular phase women versus the other groups, these data suggested that increased GH release in young women might primarily reflect only an amplitude-modulated event. However, concerns as to whether the sampling paradigm employed allowed for the recovery of the majority of GH pulses prompted further studies examining these specific issues.

Effects of Sampling Paradigm on Identification of GH Pulses

To address the potential effects of blood sampling intensity on the identification of GH pulses within concentration-time data series, Evans and colleagues obtained samples in normal men at 5-min intervals and subjected the data to analysis using the pulse detection algorithm *cluster* (23). Subsets of the 5-min data series were also analyzed so as to mimic sampling intensities of every 10, 15, 20, and 30 min. This exercise revealed that whereas sampling at 20-min intervals is sufficient to capture the larger GH episodes, sampling at 5- or 10-min intervals divulges additional pulses. More specifically, many, if not all, of the major GH episodes appear to comprise high-frequency pulsatile activity. Although it is not certain as to the exact nature of the cellular mechanisms responsible for this high-frequency activity, the fact that such pulses fail to be enumerated with less intensive sampling techniques provided the impetus for reexamining GH pulsatility in young men and young early follicular phase women using improved sampling and analytical techniques.

GH Release in Young Men Versus Young Early Follicular Phase Women

To reexamine pulsatile GH release in young men and women using more rigorous sampling and analytical methods, Asplin and colleagues obtained blood samples at 10-min intervals for 24 h from 14 young men (mean age: 27.5 ± 1.4 years) and 14 young early follicular phase women (mean age: 28.8 ± 1.3 years) (24). Serum GH was measured by *immunoradiometric assay* (IRMA). IGHC were calculated, and GH pulses were identified and characterized using cluster analysis. Representative GH concentrations from a male and early follicular phase female are shown in Figure 10.4, and the GH pulse characteristics are summarized in Table 10.1.

As can be seen, the mean IGHC calculated for early follicular phase women were significantly higher ($P < 0.05$) than those in young men, thus confirming results obtained in the earlier studies. However, in con-

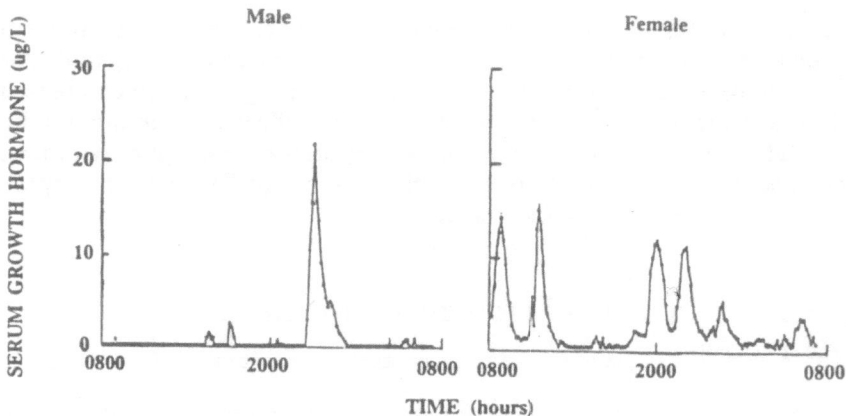

FIGURE 10.4. Profiles of pulsatile serum GH concentrations obtained by sampling at 10-min intervals for 24 h in a normal male and normal female. Adapted with permission from Asplin, Faria, Carlsen, et al. (24), © The Endocrine Society, 1989.

TABLE 10.1. IGHC and GH pulse characteristics in healthy young male and female subjects.

	Men	Women
IGHC (μg/L/min)	3574 ± 617	5287 ± 602[a]
Pulse frequency (/24 h)	4.6 ± 0.7	7.4 ± 0.8[b]
Maximal pulse amplitude (μg/L)	11.1 ± 1.9	10.1 ± 1.0
Incremental pulse amplitude (μg/L)	15.1 ± 5.1	8.4 ± 0.7
Pulse width (min)	90.1 ± 8.6	99.6 ± 7.7
Pulse area (μg/L/min)	560 ± 111	455 ± 64
Interpulse valley mean (μg/L)	1.0 ± 0.1	1.9 ± 0.4[a]

Values are the mean ± SEM.
[a] $P < 0.05$, vs. men.
[b] $P < 0.025$, vs. men.
Source: Adapted with permission from Asplin, Faria, Carlsen, et al. (24), © The Endocrine Society, 1989.

trast to the observations reported by Ho and colleagues based on every 20-min blood sampling (22), GH pulse frequency was found to be greater ($P < 0.025$) in the women than in the men; that is, with 10-min sampling early follicular phase women were found to have approximately 60% more GH pulses than did their male counterparts. In addition to the higher frequency of GH pulses observed in the early follicular phase women, the mean interpulse GH concentration was also found to be greater than in the men; that is, the early follicular phase women had more GH pulses per unit time superimposed on a higher baseline.

However, none of the GH pulse characteristics appraised by cluster analysis (maximal and incremental amplitude, width, and area) differed in the men versus early follicular phase women. These data suggested that if a more aggressive approach to pulse detection is utilized, the higher IGHC in men versus women primarily reflect a frequency-modulated process (thus implying differences in hypothalamic mechanisms), rather than a process involving amplitude modulation. Even so, this study left a number of issues unresolved. For example, would estimates of GH pulse frequency be greater yet if the sampling paradigm had been even more rigorous? What is the physiological significance of the higher mean inter-pulse GH concentration detected in the early follicular phase women? Might the male-female difference in frequency of GH pulses disappear or be reduced if a more sensitive GH assay were utilized? These and numerous related questions remain to be addressed in future studies.

GH Release in Young Women: Studies During the Menstrual Cycle

Taken as a group, the above studies suggested that young women secrete more GH than do young men, older men, and older women and that estrogen may play a predominant role in this phenomenon. Given that young women manifest significantly different gonadal hormone milieus at various times during the menstrual cycle, this group provides an excellent model with which to investigate the impact of endogenously secreted estrogen (and estrogen/progesterone) on GH release. As noted before, alterations in GH release during the menstrual cycle were described some years ago (1, 3, 25). However, other studies, including some reported quite recently, have failed to document menstrual cycle phase-associated changes in serum GH concentrations (10, 26, 27).

The possibility that somatotroph responsivity to GHRH may differ during the several phases of the menstrual cycle has been tested, but the results have not been straightforward. Gelato and colleagues found no difference in the dose of GHRH required to obtain half-maximal GH release in midfollicular versus midluteal phase women (19). Similarly, Evans and colleagues found that a supramaximal dose of GHRH resulted in the same amount of GH release in women during the early follicular, late follicular, and midluteal phases of the cycle (21). However, Lang and colleagues were able to document a strong correlation between GHRH-stimulated GH release and serum estradiol concentrations in women studied during the follicular and midluteal phases of the menstrual cycle (28). Therefore, it remains quite possible that somatotroph responsivity is affected directly or indirectly by estrogen, but that other factors (e.g., somatostatinergic input) need to be controlled in order to demonstrate

TABLE 10.2. Mean (±SEM) serum concentrations of 17β-estradiol and progesterone during the 3 phases of the menstrual cycle.

	Early follicular	Late follicular	Midluteal
17β-estradiol (pmol/L)	86.8 ± 12.6[a]	538 ± 63.6[b]	421 ± 58.5[b]
Progesterone (nmol/L)	1.45 ± 0.21[a]	1.56 ± 0.30[a]	35.4 ± 7.9[b]

[a,b] For each hormone, values identified with different superscripts differ significantly ($P <$ 0.05).
Source: Adapted with permission from Faria, Waranch Bekenstein, Booth, et al. (8), © Blackwell Scientific Publications, 1992.

unequivocally gonadal hormone milieu-associated changes in GH release in response to exogenously administered GHRH.

Quite recently, we applied our improved sampling paradigm and pulse detection methodology to assess endogenously driven GH release in normal women during the menstrual cycle (8). Fifteen young women at each of 3 phases of the cycle were studied. Blood was obtained at 10-min intervals for 24 h, GH was measured by IRMA, and GH pulses were detected and characterized by cluster analysis. Serum concentrations of estradiol and progesterone are shown in Table 10.2. As predicted, concentrations of estradiol and progesterone were minimal during the early follicular phase. Estradiol, but not progesterone, was elevated in the late follicular phase, and both estradiol and progesterone were elevated in the midluteal phase of the menstrual cycle. Representative 24-h GH profiles are shown at each phase of the cycle in Figure 10.5. Integrated serum GH concentrations are depicted in Figure 10.6. Serum IGHC were higher during the late follicular phase compared to the early follicular phase ($P = 0.032$). Concentrations during the midluteal phase were intermediate; that is, between, but not statistically distinguishable from, those calculated for the early and late follicular phases.

The results of identification and characterization of GH pulses with cluster analysis are shown in Table 10.3. No differences in GH pulse number were found during the 3 phases of the menstrual cycle. However, maximal GH pulse amplitude was increased during the late follicular phase compared to the early follicular ($P = 0.008$) and midluteal ($P = 0.008$) phase values. In addition, GH pulse amplitude, when expressed as increment from baseline, revealed a similar relationship; that is, late follicular phase GH incremental amplitudes were higher than those observed during the early follicular ($P = 0.005$) and midluteal ($P = 0.002$) phases of the cycle. This latter observation suggested that the differences in maximal amplitude reflected a change from baseline, rather than pulses of uniform amplitude superimposed on a higher baseline. That this interpretation may be correct is suggested by the mean serum GH interpulse valley concentrations that did not vary during the cycle.

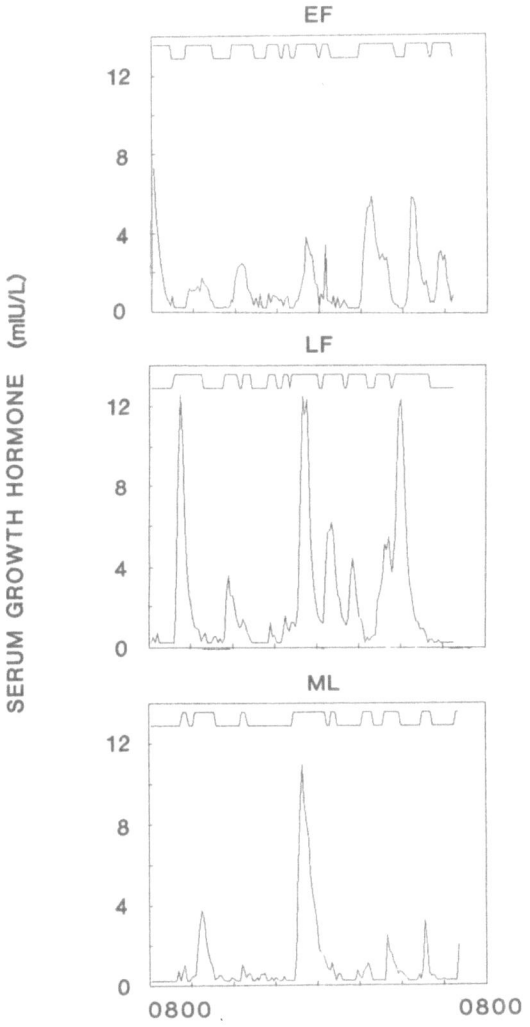

FIGURE 10.5. Illustrative profiles of serum GH concentration-time series obtained during the early follicular (top panel), late follicular (middle panel), and midluteal (lower panel) phases of the menstrual cycle. The deflections at the top of each panel indicate significant GH pulses as detected by cluster analysis. Reprinted with permission from Faria, Waranch Bekenstein, Booth, et al. (8), © Blackwell Scientific Publications, 1992.

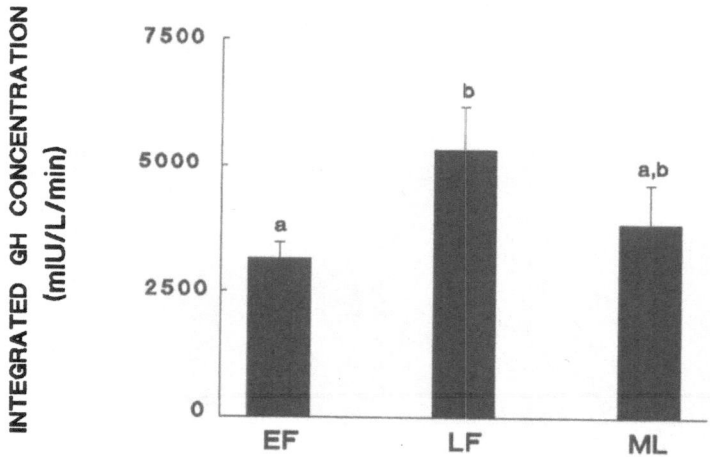

FIGURE 10.6. Mean (±SEM) serum IGHC (mU/L/min) estimated for women during the early follicular (EF) ($n = 15$), late follicular (LF) ($n = 15$), and midluteal (ML) ($n = 15$) phases of the menstrual cycle. Values identified with different superscripts differ significantly ($P < 0.05$). Reprinted with permission from Faria, Waranch Bekenstein, Booth, et al. (8), © Blackwell Scientific Publications, 1992.

TABLE 10.3. Mean (±SEM) GH pulse frequency and pulse characteristics in women at 3 phases of the menstrual cycle.

	Early follicular ($n = 15$)	Late follicular ($n = 15$)	Midluteal ($n = 15$)
Peak number (/24 h)	8.3 ± 0.55	7.9 ± 0.91	8.5 ± 0.66
Maximal peak height (mU/L)	5.7 ± 0.67[a]	8.9 ± 1.00[b]	5.8 ± 0.74[a]
Incremental peak height (mU/L)	4.7 ± 0.58[a]	7.3 ± 0.83[b]	4.4 ± 0.39[a]
Peak width (min)	100 ± 3.84	116 ± 10.56	97.7 ± 5.55
Interpeak valley mean (mU/L)	1.1 ± 0.15	1.7 ± 0.43	1.5 ± 0.46

[a,b] For each characteristic, parameters identified by different superscripts differ significantly ($P < 0.05$).
Source: Reproduced with permission from Faria, Waranch Bekenstein, Booth, et al. (8), © Blackwell Scientific Publications, 1992.

Multiple regression analysis was utilized to confirm or refute the possibility that concentrations of the gonadal hormones (estradiol, progesterone, and testosterone) correlated with parameters of GH release. Estradiol was found to correlate significantly ($P = 0.045$) and positively with GH maximal pulse amplitude. Progesterone correlated significantly ($P = 0.043$) but negatively with the same GH pulse characteristic. No other attribute of pulsatile GH release correlated with estradiol or progesterone, and no GH pulse characteristic correlated with total serum testosterone concentrations in these women.

Considered together, these results suggest that GH concentrations are higher during the estrogen-sufficient but progesterone-deficient late follicular phase of the menstrual cycle; that enhanced concentrations may reflect, at least in part, amplitude modulation rather than frequency modulation of the GH signal; that serum concentrations of estrogen do correlate significantly with enhanced GH release; and that the facilitative effects of estrogen may be blunted by inhibitory effects of progesterone during the midluteal phase of the cycle, thus resulting in IGHC greater than early follicular phase values, but less than those in the late follicular phase.

Future Studies

Improved sampling and analytical techniques have allowed for studies in premenopausal women that have confirmed earlier studies demonstrating differences in GH secretion between men and women. Moreover, our most recent investigations have documented menstrual cycle-dependent alterations in GH release and suggest that these differences may reflect the amount of GHRH reaching the somatotrophs and/or the prevailing state of somatotroph responsivity. However, limitations in analytical techniques and experimental design may still be preventing a complete understanding of the factors regulating GH secretion in women.

The issue of just how frequently blood samples must be obtained in order to capture the majority of GH pulses remains unclear. As noted above, sampling at 5- and 10-min intervals discloses high-frequency GH pulses within the major release episodes (23). Analysis of GH concentration-time series obtained in normal men but subjected to deconvolution procedures (to define secretory episodes, see below) has resulted in similar conclusions (29). Moreover, when blood was sampled overnight at 30-sec intervals, an even higher frequency of GH secretory *volleys* could be defined (30). Therefore, the blood-sampling paradigms typically employed in the available studies in premenopausal women cited above (e.g., 10- to 20-min intervals for 24 h) may not be adequate to allow detection of all high-frequency GH secretory activity.

There is also significant concern that the IRMAs currently used to measure GH do not have adequate sensitivity to describe physiological GH secretion fully. Thus, it has been hypothesized that with the ultrasensitive GH assays that are just now becoming available, the apparent male versus female differences in the frequency with which GH pulses occur may disappear; that is, it is entirely conceivable that men have a component of low-amplitude GH pulsatility that has previously been unresolved because of assay limitations. Indeed, recent application of an ultrasensitive chemiluminescent GH assay discloses persistent, low basal rates of interpulse GH secretion in young men, as well as in hyposomato-trophic states associated with aging, obesity, and hypothyroidism (31).

The methods previously used for the identification and characterization of GH pulses have recently been markedly improved. Pulse detection algorithms, including cluster analysis, appraise fluctuations in hormone concentrations, but do not provide information about exactly how these alterations occur. That is, pulse analysis per se does not allow for inferences regarding the secretory impulse itself or about clearance mechanisms, the two factors that result in perturbations in the hormone concentration-time series. Newer analytical techniques utilizing so-called deconvolution procedures do allow estimates of both secretory activity and clearance factors to be made from serum hormone concentrations. The application of such procedures to data obtained from young women during the menstrual cycle may, for example, elucidate whether the high-amplitude GH pulses documented during the late follicular phase reflect enhanced secretory activity—that is, a greater secretory rate and/or duration of the secretory episode—or a decrease in peripheral clearance of the hormone.

Although improved sampling, assay, and analytical techniques will undoubtedly result in more refined observations concerning GH secretion, certain confounding physiologic issues also demand advances in experimental design. One obvious concern related to GH secretion is the fact that at least two hypothalamically derived factors (GHRH and SRIF) influence GH secretion in opposite directions. As discussed above, questions about somatotroph responsivity to GHRH of both endogenous and exogenous origin have been difficult to address because of uncertainties concerning the degree to which SRIF is affecting GH release.

Quite recently, O'Keane and Dinan administered pyridostigmine, an agent thought to diminish hypothalamic SRIF secretion, to women during the menstrual cycle (32). The GH response to pyridostigmine was greater during the late follicular and midluteal phases of the cycle when compared to the early follicular phase. Moreover, serum estradiol concentrations correlated strongly with pyridostigmine-associated GH release. These results are consistent with the hypothesis that SRIF inhibitory tone may be influenced by endogenously secreted estrogen, which, overall, has a positive effect on GH secretion. However, the question of whether estrogens modulate somatotroph responsivity remains, given that diminished hypothalamic SRIF tone could theoretically have an effect on GHRH secretion. Nevertheless, the concept of controlling for at least one variable (in this case SRIF) in studies related to GH secretion in the human is exceedingly attractive and will, without doubt, be exploited in future studies.

References

1. Frantz AG, Rabkin MT. Effects of estrogen and sex difference on secretion of human growth hormone. J Clin Endocrinol Metab 1965;25:1470–80.
2. Thompson RG, Rodriguez A, Kowarski A, Blizzard RM. Growth hormone: metabolic clearance rates, integrated concentrations, and production rates in normal adults and the effects of prednisone. J Clin Invest 1972;51:3193–9.
3. Genazzani AR, Lemaarchand-Beraud TH, Aubert ML, Felber JP. Pattern of plasma ACTH, hGH, and cortisol during menstrual cycle. J Clin Endocrinol Metab 1975;41:431–7.
4. Mauras N, Rogol AD, Veldhuis JD. Specific, time-dependent actions of low-dose ethinyl estradiol administration on the episodic release of growth hormone, follicle-stimulating hormone, and luteinizing hormone in prepubertal girls with Turner's syndrome. J Clin Endocrinol Metab 1989;69:1053–8.
5. Dawson-Hughes B, Stern D, Goldman J, Reichlin S. Regulation of growth hormone and somatomedin-C secretion in postmenopausal women: effect of physiological estrogen replacement. J Clin Endocrinol Metab 1986;63:424–32.
6. Tannenbaum GS, Ling N. The interrelationship of growth hormone (GH)-releasing factor and somatostatin in generation of the ultradian rhythm of GH secretion. Endocrinology 1984;115:1952–7.
7. Plotsky PM, Vale W. Patterns of growth hormone-releasing factor and somatostatin: secretion into the hypophyseal-portal circulation of the rat. Science 1985;230:461–3.
8. Faria ACS, Waranch Bekenstein L, Booth RA Jr, et al. Pulsatile growth hormone release in normal women during the menstrual cycle. Clin Endocrinol (Oxf) 1992;36:591–6.
9. Taylor AL, Finster JL, Mintz DH. Metabolic clearance and production rates of human growth hormone. J Clin Invest 1969;48:2349–58.
10. Zadik Z, Chalew SA, McCarter RJ Jr, Meistas M, Kowarski AA. The influence of age on the 24-hour integrated concentration of growth hormone in normal individuals. J Clin Endocrinol Metab 1985;60:513–6.
11. Finklestein JW, Roffwarg HP, Boyar RM, Kream J, Hellman L. Age-related change in twenty-four hour spontaneous secretion of growth hormone. J Clin Endocrinol Metab 1972;35:665–70.
12. Carlsen HE, Gillin JC, Gorden P, Synder F. Absence of sleep-related growth hormone peaks in aged normal subjects and in acromegaly. J Clin Endocrinol Metab 1972;34:1102–7.
13. Rudman D, Kutner MH, Rogers M, Lubin MF, Fleming GA, Bain RP. Impaired growth hormone secretion in the adult population: relation to age and adiposity. J Clin Invest 1981;67:1361–9.
14. Prinz PN, Weitzman ED, Cunningham GR, Karacan I. Plasma growth hormone during sleep in young and aged men. J Gerontol 1983;38:519–24.
15. Dudl RJ, Ensinck JW, Palmer HE, Williams RH. Effect of age on growth hormone secretion in man. J Clin Endocrinol Metab 1973;37:11–6.
16. Shibaski T, Shizume K, Nakahara M, et al. Age-related changes in plasma growth hormone response to growth hormone-releasing factor in man. J Clin Endocrinol Metab 1984;58:212–4.
17. Pavlov EP, Merriam GR, Gelato MC, Harman SM, Blackman MR. Responses of growth hormone and somatomedin-C to growth hormone-

releasing factor I-44 NH_2 (GRF-44) in healthy aging men. Clin Res 1984; 32:689A.

18. Smals AEM, Pieters GFFM, Smals AGH, Benraad TJ, van Laarhoven J, Kloppenborg PWC. Sex difference in human growth hormone (GH) response to intravenous human pancreatic GH-releasing hormone administration in young adults. J Clin Endocrinol Metab 1986;62:336–41.

19. Gelato MC, Pescovitz OH, Cassorla F, Loriaux DL, Merrian GR. Dose-response relationships for the effects of growth hormone-releasing factor-(I-44)-NH_2 in young adult men and women. J Clin Endocrinol Metab 1984; 59:197–201.

20. Vance ML, Borges JLC, Kaiser DL, et al. Human pancreatic growth hormone-releasing factor (hpGRF-40): dose-response relationships in normal man. J Clin Endocrinol Metab 1984;58:838–44.

21. Evans WS, Borges JLC, et al. Effects of human pancreatic growth hormone-releasing factor-40 on serum growth hormone, prolactin, luteinizing hormone, follicle-stimulating hormone, and somatomedin-C concentrations in normal women throughout the menstrual cycle. J Clin Endocrinol Metab 1984;59: 1006–10.

22. Ho KY, Evans WS, Blizzard RM, et al. Effects of sex and age on the 24-hour profile of growth hormone secretion in man: importance of endogenous estradiol concentrations. J Clin Endocrinol Metab 1987;64:51–8.

23. Evans WS, Faria ACS, Christiansen E, et al. Impact of intensive venous sampling on characterization of pulsatile GH release. Am J Physiol 1987; 252:E549–56.

24. Asplin CM, Faria ACS, Carlsen EC, et al. Alterations in the pulsatile mode of growth hormone release in men and women with insulin-dependent diabetes mellitus. J Clin Endocrinol Metab 1989;69:239–45.

25. Yen SSC, Vela P, Rankin J, Littell AS. Hormonal relationships during the menstrual cycle. JAMA 1970;211:1513–7.

26. Holst N, Jenssen PG, Burhol E, Haug E, Forsdahl F. Plasma gastrointestinal hormones during spontaneous and induced menstrual cycles. J Clin Endocrinol Metab 1989;68:1160–6.

27. Stone BA, Marrs RP. Growth hormone in serum of women during the menstrual cycle and during controlled ovarian hyperstimulation. Fertil Steril 1991;56:52–8.

28. Lang I, Schernthaner G, Pietschmann P, Kurz R, Stephenson JM, Templ H. Effects of sex and age on growth hormone response to growth hormone-releasing hormone in healthy individuals. J Clin Endocrinol Metab 1987; 65:535–40.

29. Hartman ML, Faria ACS, Vance ML, Johnson ML, Thorner MO, Veldhuis JD. Temporal structure of in vivo growth hormone secretory events in humans. Am J Physiol 1991;260:E101–10.

30. Holl RW, Hartman ML, Veldhuis JD, Taylor WM, Thorner MO. Thirty-second sampling of plasma growth hormone in man: correlation with sleep stages. J Clin Endocrinol Metab 1991;72:854–61.

31. Iranmanesh A, Grisso B, Veldhuis JD. Low basal and persistent pulsatile GH secretion are revealed in normal and hyposomatotropic men studied with a new ultrasensitive chemiluminescent assay. J Clin Endocrinol Metab 1994;78: 526–35.

32. O'Keane V, Dinan TG. Sex steroid priming effects on growth hormone response to pyridostigmine throughout the menstrual cycle. J Clin Endocrinol Metab 1992;75:11–4.

11

Gestational Physiology of the Growth Hormone Gene Family

NANCY E. COOKE, BEVERLY K. JONES, ALAN SALZMAN,
J. ERIC RUSSELL, ANITA MISRA-PRESS, MARGRIT URBANEK,
AND STEPHEN A. LIEBHABER

The *human growth hormone* (hGH) gene family, located in a cluster spanning 48 kb on chromosome 17q22-24 (1), contains 5 genes: the pituitary GH gene (hGH-N) and 4 placentally expressed genes, *chorionic somatomammotropin-like* (hCS-L), hCS-A, *hGH-variant* (hGH-V), and hCS-B (2). These genes evolved by a series of duplication events, the most ancient one giving rise to the distantly related and unlinked (3) single *prolactin* (PRL) gene (4, 5). The linked GH and CS genes, presumed to be generated by 3 recent duplication events, now share >90% nucleotide sequence identity. These genes are expressed in a highly tissue-specific manner; hGH-N is expressed solely in pituitary somatotropes/somatolactotropes, and the hCS and hGH-V genes are expressed exclusively in the syncytiotrophoblastic layer of the placenta (6, 7). In contrast, hPRL is expressed in both pituitary lactotropes/somatolactotropes and placental decidua (8).

The evolution of the GH gene families in rodents is distinct from humans. In rats and mice there are multiple PRL-related genes, one expressed in pituitary and several expressed in the placenta, but only one GH gene (9). It remains unclear whether the molecular mechanisms of GH/PRL gene regulation have diverged in parallel with this divergence in gene organization among different species.

The PRL- and GH-related genes all share a common subset of regulatory elements that includes multiple binding sites for the pituitary-restricted, POU-homeodomain transcription factor pit-1/GHF-1 (10, 11). Expression of pit-1 in the developing pituitary shortly before the expression of GH (12) is critically important for eventual GH gene expression, as well as for normal pituitary development. Rodents and humans homozygous for null mutations of pit-1 fail to develop

124

somatotropes, lactotropes, and thyrotropes in their pituitaries and have a dwarf phenotype (13, 14). Transcription of the hGH-N gene in transfected rat pituitary cell lines requires 2 functional binding sites for pit-1 in its proximal promoter. However, the presence of these cis elements does not fully account for the pituitary specificity of hGH-N gene expression. The hGH-N gene with its proximal promoter region is expressed (inappropriately) when transfected into L cells (15). The hCS-A and hCS-B gene promoters contain functional pit-1 binding sites and are expressed nearly as well as the hGH-N gene promoter in transfected pituitary cells (16), but in vivo are expressed exclusively in the placenta. Therefore, it appears that the presence of pit-1, although necessary, is not fully sufficient to establish the tissue-specific patterns of hGH-N gene expression in transfected, cultured cells.

The extent of cis elements necessary to target GH gene expression has been studied in a number of species. In the case of the rat GH gene, 188 bp of 5' flanking region are sufficient to target expression of marker genes to the anterior pituitary of transgenic mice (17), implicating the 2 pit-1 sites in this function. However, levels of pituitary expression are extremely low. Expression is increased 100-fold by the inclusion of an additional 1 kb of 5' sequences (18). In marked contrast, the hGH-N gene is not effectively expressed in the pituitary of transgenic mice, even when 4.6 kb of 5' and 30 kb of 3' flanking sequences are retained in the injected fragment (19). This suggests that regulatory information in addition to the pit-1 sites and associated proximal promoter elements are needed to establish the tissue-specific expression of the genes of the human GH cluster.

The expression of a gene at reproducible, tissue-specific, and copy number-dependent levels in transgenic mice is the most rigorous test for the adequacy of transcriptional control elements. In most cases, transgene expression is low and inconsistent even when a large extent of homologous promoter sequences is retained. This reflects overriding *position effects* at the site of transgene integration into the host chromatin (20). Since eukaryotic DNA is highly condensed and packaged in chromatin, the transgene must either fortuitously integrate into an open, noncondensed region of chromatin or be able to establish that environment by bringing along its own chromatin-modulating elements in order to achieve high levels of expression (21).

Several genetic elements have been described that are capable of altering chromatin packing and establishing such open domains. These elements have been described in only a limited number of systems and have been referred to as *locus control regions* (LCRs) (22), boundary elements (23), and specialized chromatin structures (24). These elements may colocalize with nuclear scaffold attachment regions (25). For certain linked gene families, these determinants act as dominant signals to reorganize adjacent chromatin into active domains, or else they set the boundaries of such domains. These genetic elements are considered to be activated by trans-acting factors. Nucleosome-free regions of chromatin, such as those necessary for access of trans-regulatory factors, can be detected experimentally by the pronounced sensitivity of the DNA at those sites to nuclease cleavage. Typically, such sites are greater than 2 orders of magnitude more sensitive than bulk chromatin to *DNaseI* digestion. Such *DNaseI hypersensitive sites* (HS) can be divided into two major subtypes: constitutive and inducible. The inducible HS have been further subdivided into tissue-specific and developmentally regulated HS (26).

The β-globin gene cluster has served as a model system for studying the role of chromatin activation in the determination of gene expression. Four tissue-specific HS have been discovered 6–18 kb upstream from the human β-globin gene cluster (22). These were initially suspected to represent an LCR when naturally occurring deletions of these HS resulted in global inactivation of the 5 intact genes in the 50-kb β-globin gene cluster and the inheritance of thalassemia in the family members carrying such deletions.

When the human β-globin gene under the transcriptional control of its promoter is introduced into transgenic mice, its expression is low and nonreproducible. In contrast, when the β-globin gene linked to the 4 tissue-specific HS is introduced into transgenic mice, transgene expression is reproducible, tissue-specific, copy-number dependent, and equivalent to endogenous mouse β-globin gene expression (22). The human HS are reformed in the transgene chromatin, and they induce an increased sensitivity to *DNaseI* over the entire adjacent multigene cluster, resulting in an open chromatin domain specific to erythroid cells (27). It is postulated that this open domain facilitates the access of trans-acting transcriptional factors to their binding sites adjacent to each globin gene, as well as resulting in the early S-phase replication of the globin locus in erythroid cells. Since the human LCR was functional in transgenic mice, it was postulated that critical LCR sequences must be conserved between human and mouse. We have begun investigating the possibility that the human GH cluster of genes is regulated by an LCR.

Regulation of the GH Cluster of Genes by a Locus Control Element

Since the hGH-N gene is not expressed in the pituitaries of transgenic mice when it is under the regulation of its own proximal promoter (19 and unpublished data, see below), we sought to identify additional critical regulatory elements. Genomic clones spanning the 48-kb cluster of 5 GH-related genes were assembled (2), and additional, overlapping cosmid clones were isolated and characterized, constituting a total

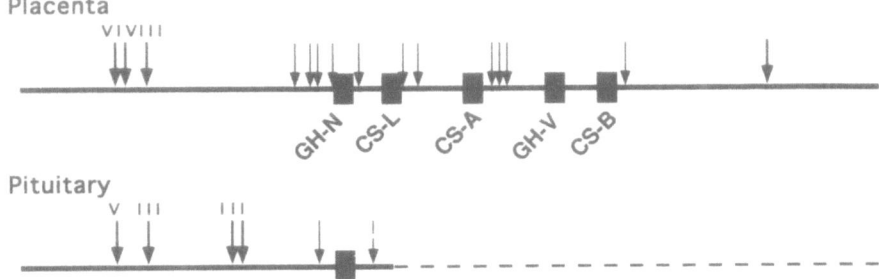

FIGURE 11.1. Map of *DNase*I hypersensitive sites within and surrounding the hGH gene cluster as detected in the nuclear chromatin of placental syncytiotrophoblast and pituitary GH adenoma cells. Each of the 5 structural genes is indicated by a solid rectangle. In the pituitary cells mapping within the cluster was limited to the hGH-N gene; the unmapped region of the cluster is indicated by the dashed line.

contiguous span of 166 kb. A detailed restriction map was generated. Since the hCS-L/hCS-A/hGH-V/hCS-B genes are expressed in the syncytiotrophoblastic epithelium of the placenta, we initially mapped placental *DNase*I HS in chromatin from syncytiotrophoblast nuclei with the assumption that these sites would point to important regulatory regions (26). To carry this out, a technique was developed to isolate syncytiotrophoblast nuclei from fresh human term placentas.

This technique is dependent on the high levels of carbonic anhydrase in syncytiotrophoblasts. Osmotic lysis is induced by the generation of CO_2 in the incubation media. Cell lysis is monitored by observing the release of the multiple nuclei of the syncytiotrophoblasts by light microscopy. The released nuclei, isolated by differential centrifugation, are subjected to digestion with *DNase*I for increasing periods of time. Fragments from a surgically removed, human GH-secreting pituitary adenoma were obtained, and the nuclei were isolated and studied in parallel. Using a set of unique probes derived from the cloned 166-kb GH locus, the positions of HS within and flanking the GH cluster were mapped relative to established restriction sites by indirect end-labeling on Southern blots (Fig. 11.1).

In addition to known HS in close proximity to each of the genes, we detected an additional set of HS in the remote 5' and 3' flanking regions. In syncytiotrophoblast cells, 3 such remote HS sites are located between −32 and −27.5 kb upstream from the hGH-N promoter (HS V, IV, and III), and an additional 3' HS is located 24 kb 3' to the hCS-B. In the pituitary adenoma, 2 HS at −32 and −27.5 colocalized with the placental sites (HS V and III; HS IV was specifically absent), and 2 additional sites unique to the pituitary were identified at −16 and −14 kb (HS II and I).

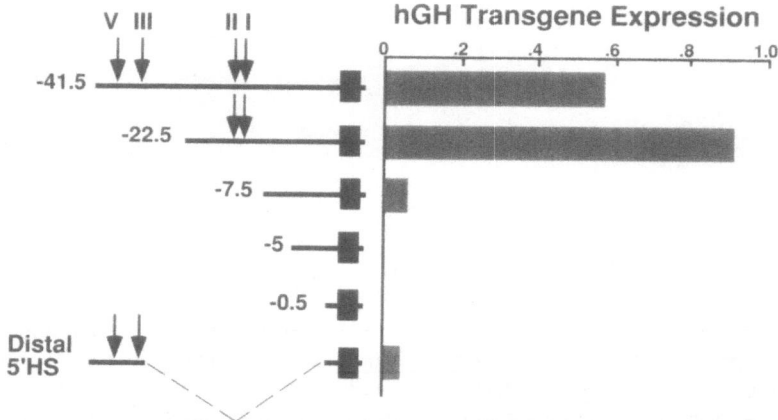

FIGURE 11.2. Expression of the hGH gene in transgenic mice. Transgene constructs introduced into the mouse germline are indicated to the left of the histogram. The hGH structural gene is shown as a solid rectangle, the contiguous 5' flanking region as a line, and the positions of *DNase*I hypersensitive sites by vertical arrows. The extent of 5' contiguous, 5' flanking region relative to the site of transcription initiation is noted by the negative numbers (in kb) to the left of each construct. In the case of the final construct, distal 5' HS, the segment of the 5' flanking region that has been deleted is indicated by the dotted line. The level of hGH mRNA measured in the pituitaries of the transgenic mice is divided by the level of endogenous mGH mRNA (defined as 1.0). Each bar represents the average of at least 3 separate lines (except for the −5, a single line).

The HS analysis in chromatin of the IM-9 B lymphoblast cell line, the K562 erythroleukemia line, the JEG-3 choriocarcinoma line, and the primary pituitary adenoma cells demonstrated that the presence of HS I through V correlates directly with expression of GH or CS genes. The 3 placental HS are faintly detected in JEG-3 cells. These cells express low levels of hCS, while none is present in the K562 cells in which the hGH gene cluster is inactive. The presence of HS in IM-9 cells, one of which may comigrate with HS IV, is intriguing in light of a recent report that suggests that immunoreactive GH may be secreted from IM-9 cells (28 and unpublished data). Confirmation of this will require further study.

To study the function of the upstream HS in vivo, we generated transgenic mouse lines carrying the insert from cosmid clone K2B. This cosmid insert contains the hGH-N gene as well as 41.5 kb of contiguous 5' flanking sequences, thus including all the defined 5' HS (Fig. 11.2). RT/ PCR and RNase protection analyses of hGH-N gene expression in pituitaries of F1 transgenic mice indicate that in 5 of 5 lines, the hGH-N gene is expressed solely in the pituitary, and in all these lines expression is at high levels. Consistent with this finding, hGH-N was detected in the serum of mice in all 5 of these lines. The hGH-N transgene was not expressed in placentas of these transgenic mice. None of these lines showed evidence of abnormal growth or gigantism, suggesting that the human transgene is under physiologic regulation.

To characterize further the function of the HS, several additional transgenic lines were established. These lines carry derivatives of the −41.5hGH-N construct with progressive 5′ deletions (Fig. 11.2). With −41.5hGH-N, the level of pituitary expression normalized to mGH mRNA (to control for in vivo hormonal regulatory effects) increased proportional to transgene copy number. Copy number in the −41.5hGH-N lines varied from 2 to 38 copies. Four out of 4 lines of mice with the −22.5hGH-N insert also expressed hGH-N at high levels and in a copy number-dependent fashion, although this construct lacks HS V and III. Three of 6 lines with less than 8 kb of 5′ flanking DNA failed to express hGH-N at all, 2 expressed at very low levels, and 1 expressed at low levels. These experiments point to the pituitary-specific HS between −21 and −14 as critical to pituitary expression of hGH-N.

We then asked whether the most distal set of HS had a function as well. This was done by directly linking a fragment encompassing HS V through III to the proximal promoter of hGH-N that, as in each construct, includes its 2 promoter proximal pit-1 binding sites. Although the levels of expression in these lines were low, all 3 showed a consistent copy number dependence. In addition, the mice carrying this construct expressed high and unregulated levels of hGH-N in the kidney, moderate levels in brain, and in 1 line high levels in the liver. The pituitary-specific HS (I and II) therefore appear necessary for pituitary-exclusive expression, although the more distal HS do appear to function as chromatin control elements. We conclude that the in vivo establishment of transcriptional competence of the hGH-N gene during development and differentiation of the pituitary requires the participation of upstream regulatory elements in addition to proximal promoter sequences and that these elements have properties of an LCR (29 and unpublished data).

Developmental Expression of the GH Genes in the Placenta

By studying the expression of the 4 GH-related genes in the GH cluster in placental RNA samples obtained at various stages of pregnancy, we determined that all 4 of these genes are expressed in the placenta and are coactivated between 8 and 10 weeks of gestation. The hCS-A and hCS-B genes encode identical protein products. Based on a minor difference in the 3′ nontranslated regions of their mRNAs, they could be distinguished by an RT/PCR assay. From 8 weeks to term, hCS-A mRNA levels increase 30-fold, while hCS-B, as well as hCS-L and hGH-V, mRNA levels increase only 5- to 10-fold. Therefore, the hCS-A:hCS-B ratio at 8 weeks is 1.5, gradually shifting to 5.0 by term (30). The molecular basis for this differential regulation is presently unknown. Since the relative increases in mRNA concentrations in the placenta are significantly less than the corresponding increases in circulating levels of hCS and hGH-V hormones (31, 32), we conclude that the developmental profiles or hormone levels detected in serum reflect both increases in placental mass and increases in transcriptional activation.

Expression Patterns of GH-Related Genes in the Placenta

The hCS-L gene, located just 3' to hGH-N in the cluster, was initially considered to be a pseudogene inactivated by the loss of the intron 2 splice-donor site by a point mutation (33). We have shown that hCS-L transcripts are present in the human placenta and that their levels are induced in parallel with hCS-A, hCS-B, and hGH-V. The missing splice-donor site of intron 2 is functionally replaced by a cryptic donor site 19 bases within intron 2. This novel donor site splices to 2 competing splice-acceptor sites within exon 3 (sites L and L').

To facilitate further study of this gene, we established stably transformed mouse C127 fibroblast cell lines expressing the hCS-L gene under transcriptional control of the *mouse metallothioneine* (mMT) promoter. This was done to circumvent assay interference by the overwhelming levels of structurally related hCS-A and hCS-B transcripts present in placental RNA samples. RNA samples from these lines were analyzed for hCS-L gene expression by RT/PCR mapping, cloning, sequencing, Northern analyses using targeted *RNase*H digestions, and quantitative RNase protection mapping. These studies demonstrate 5 alternatively spliced hCS-L transcripts in these cell lines, with a parallel profile confirmed in placental RNA. Three of these transcripts retain an extended open reading frame. Of particular interest, 1 of these 3 alternative mRNAs, hCS-L(L), retains exon 2 and encodes an intact signal peptide. Thus, it has the potential to encode a hormone of 197 amino acids, sharing roughly 91% amino acid identity with hCS-A and hCS-B. In vitro studies directly confirm efficient and accurate cleavage of the hCS-L(L) signal peptide.

To facilitate further analysis of these proteins, we have raised high-titer polyclonal rabbit antisera to a synthetic peptide specific to the region encoded by the unique 19-base extension of exon 2. The specificity of this antisera was confirmed by its recognition of in vitro translated hCS-L(L), but not the closely related hCS-A. These studies strongly support the potential of hCS-L(L) mRNA to encode a novel placental hormone (34 and unpublished data) and establish the necessary reagents for its further study.

We have documented that hGH-V is not a pseudogene as originally predicted, but instead is a functional gene expressed in a developmentally regulated, tissue-specific fashion in the placenta. We cloned hGH-V mRNAs from placenta and found that the hGH-V gene encodes 2 alternatively spliced mRNAs, hGH-V and hGH-V2 (35). These mRNAs differ by the alternative splicing of intron 4 that is retained in hGH-V2. This alternative splice contrasts with the alternative splicing of the greater than 92% identical hGH-N gene that constitutively utilizes an alternative splice-acceptor within exon 3 at a frequency of 10% (36, 37).

The hGH-V gene is expressed in the villous layer of the placenta and was sublocalized by in situ histohybridization to syncytiotrophoblast cells (38). To facilitate detailed analysis of hGH-V expression and studies of hormone function, we established stably transfected C127 fibroblast cell lines expressing the hGH-V gene under the transcriptional control of the mMT promoter. These cells were shown to express and secrete a 22-kd hGH-V as well as 2 minor hGH-V-related proteins of 24 and 26 kd. The latter 2 species were shown to be N-linked glycosylation products. Both were reduced to 22 kd in size when the cell line was grown in tunicamycin, a specific inhibitor of N-linked glycosylation. A consensus sequence for N-linked glycosylation exists in the hGH-N sequence at Asn140. These results indicate that the hGH-V gene encodes a major 22-kd hGH-V protein that can exist in N-linked glycosylated forms (39).

The retention of intron 4 in the hGH-V2 mRNA by alternative splicing is regulated during placental development, resulting in a 3-fold increase of hGH-V2 mRNA between 10 and 38 weeks. This regulation of hGH-V intron 4 splicing suggests that synthesis of hGH-V2 might be important for placental function (35). The open reading frame of hGH-V2 mRNA extends through intron 4 and into exon 5, predicting the expression of a 26-kd protein from hGH-V2 mRNA. Three percent of hCS-A transcripts also retains intron 4. Intron 4 of hCS-A was sequenced from a number of independent alleles, and, contrary to previous reports, an open reading frame was found throughout intron 4, completing the analogy between "hCS-A2" and hGH-V2 (30).

Although we have demonstrated that hGH-V2 mRNA translates to the predicted 26-kd protein product in vitro, it does not appear to be secreted into the media of the hGH-V-transfected C127 cell line. Since the predicted carboxyterminus of hGH-V2 contains a hydrophobic domain not present in hGH-V (35), we speculate that hGH-V2 may be membrane associated. We tested this hypothesis by injecting hGH-V2 mRNA into *Xenopus* oocytes in parallel with mRNAs encoding hGH-N and hGH-V. The injected oocytes were metabolically labeled, and the conditioned media were assayed for secreted proteins by immunoprecipitation. In a preliminary experiment radiolabeled hGH-N and hGH-V proteins, but not hGH-V2 proteins, were immunoprecipitated. In contrast, a 26-kd hGH-V2 protein was immunoprecipitated from the *Xenopus* oocyte lysate. Pulse-chase labeling confirmed that the hGH-V2 protein, but not hGH-N or hGH-V proteins, is retained in the oocyte. Therefore, we predict that hGH-V2 may represent a unique GH isoform that is cell membrane associated (40 and unpublished data). These studies demonstrate that the hGH-V gene may express 2 gestational GH isoforms.

To study the function of the 22-kd secreted hGH-V, we analyzed the ability of this putative hormone to bind to each of the two classes of GH/PRL receptors, somatogen and lactogen. The sources of ligands in these experiments were conditioned media from the hGH-V and hGH-N transfected cell lines. Conditioned media from these cell lines were tested for displacement of hGH-N from rabbit liver microsomes (somatogen receptor), as well as from the serum *hGH binding protein* (hGHBP), and displacement of oPRL from rat liver microsomes (lactogen receptor). The results of this series of studies demonstrate that hGH-V binds with high affinity to all 3 receptors tested. Remarkably, when the affinities of hGH-V and hGH-N were directly compared, hGH-V displayed a marked preference for the somatogen receptor. The IC_{50} somatogen:IC_{50} lactogen ratio for hGH-V is 7.4-fold higher than that for hGH-N. This was confirmed by noting a markedly decreased affinity of hGH-V for the *PRL receptor* (PRL-R) isoform expressed in the Nb2 lymphoma cell line (41, 42).

On the basis of these data, we predicted that the expression of hGH-V during mid to late gestation, coupled with its preference for the somatogen receptor, may be responsible for the acromegaloid facial features detected in some women during pregnancy (42). To test the prediction that hGH-V is a biologically active somatogen, we studied the ability of hGH-V-conditioned media to stimulate growth in hypophysectomized rats. To study lactogen bioactivity, we assayed hGH-V's ability to stimulate mitogenesis of Nb2 lymphoma cells, an established model of PRL-R activation. The hGH-V was as effective as hGH-N in stimulating body weight gain in the hypophysectomized rat, while its mitogenic bioactivity was significantly less than hGH-N. Thus, these bioactivities paralleled their respective receptor binding profiles (43).

These studies demonstrate that hGH-V is a biologically active hormone possessing both lactogen and somatogen bioactivity. The finding that hGH-V is approximately 7-fold more purely somatogenic than pituitary hGH-N makes it more similar to the purely somatogenic subprimate GHs. Present in maternal circulation during gestation (32), hGH-V may act to promote mammary gland development, a known role for the somatogenic GHs of other species (44, 45). *GH receptors* (GH-Rs) have been detected in mammary tissue (46), a *GH binding protein* (GHBP) has been detected in rabbit milk (47), and transgenic mice overexpressing GH in mammary tissue develop precocious mammary differentiation and lactation (48). Alternatively, hGH-V might act locally on the fetus or in the placenta.

Placental GH-R

The finding that hGH-V circulates in maternal serum (32, 49), as do hCS-A and hCS-B, suggested the possibility that an additional *human GH-R* (hGH-R) isoform(s) might exist and might be present locally in the placenta. To search for hGH-R isoforms in the placenta, a human placental cDNA library was screened using the liver hGH-R cDNA (50). Two distinct clones were isolated and characterized that span the entire hGH-R cDNA, establishing that the receptor is expressed in placenta. Clone pGH-R-P2 corresponds to nucleotides 1502–4371 of the previously defined liver GH-R, while clone pGH-R-P1 encompasses 418 bp of 5′ nontranslated region, the entire coding region, and 857 bp of 3′ nontranslated region. The first 45 bp of 5′ nontranslated region are identical to the published sequence, but the remainder are unique. The coding region is identical to the previously reported hepatic hGH-R (51) with one exception; 66 bases corresponding to positions 72–137 of the liver cDNA are not present in the placental hGH-R cDNA. This represents a precise deletion of exon 3 (hGH-Rd3). Its absence does not shift the reading frame of this mRNA, but the new splice junction does alter the amino acid encoded by the first partial codon of exon 4 (Asn to Asp), eliminating a potential N-linked glycosylation site in the hGH-Rd3 form. We conclude that the hGH-R gene is expressed in human term placenta and that exon 3 is deleted by an alternative splicing mechanism (52). This finding has been confirmed by others (53).

To determine if the hGH-Rd3 mRNA detected in term placental RNA by cDNA cloning was representative of the hGH-R mRNAs in this tissue, and to determine if there were any additional alterations in the placental GH-R population, the GH-R coding region was scanned by RT/PCR using 4 sets of amplification primers collectively spanning the entire mRNA (Fig. 11.3). RNA from the IM-9 lymphocyte line, a known source of GH-R mRNA, was used as a positive control. The cDNA fragments generated from IM-9 and placental RNA were identical and corresponded to the sizes predicted for the liver GH-R with one exception. Only the fragment predicted by the exon 3 deletion was visualized in placenta, but in IM-9 cells the exon 3-containing fragment was present as well as the exon 3-deleted fragment. Full-length GH-R mRNA was expected, but the GH-Rd3 form was unanticipated since IM-9 cells had been considered to possess only a single, high-affinity GH-R. A survey of all 4 placental layers, as well as a spectrum of tissue culture cell lines and tissues, demonstrated widespread expression of the

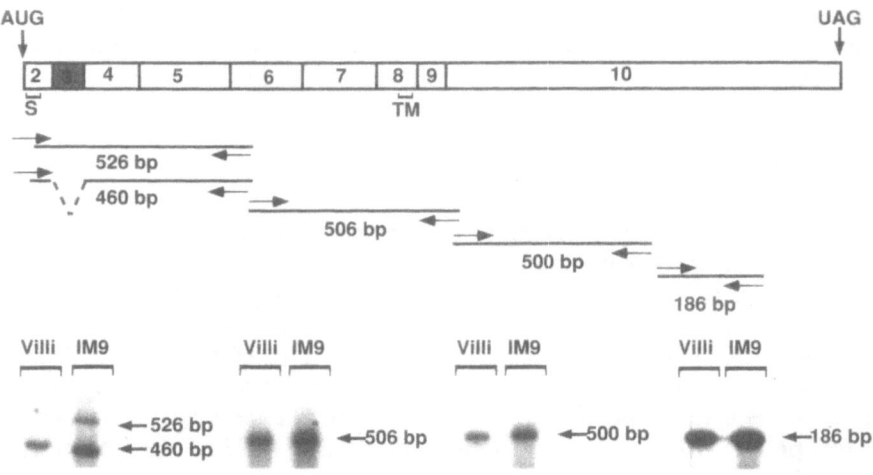

FIGURE 11.3. Structural analysis of the hGH-R mRNA by scanning RT/PCR. The structure of hGH-R mRNA is shown on the top line. Each of the boxes represents the location of exonic regions in the mRNA. (Exon 1 is not shown and consists of 5′ nontranslated sequences.) The positions of the AUG initiation codon, AUG termination codon, signal sequence (S), and transmembrane domain (TM) are shown. Four sets of primers used to reverse-transcribe and amplify regions of the hGH-R mRNA are indicated by arrows. The expected size of each amplified fragment is shown. In the case of the most 5′ set of primers, the amplification of the exon 3-deleted and exon 3-containing forms of the mRNA are shown. In each amplification set the most 3′ primer is labeled with ^{32}P to allow direct visualization of the cDNA fragment. At the bottom of the figure are autoradiographs of the analyses carried out on mRNA isolated from either term placental villi (villi) or from IM-9 lymphocytes (IM9). The size of each amplified fragment following gel electrophoresis is indicated.

hGH-R gene and tissue-specific patterns of GH-R exon 3 splicing. Examples include the choriocarcinoma cell line JEG-3 that expresses only hGH-R and the hepatoma-derived line Hep3B line that expresses only hGH-Rd3. We conclude that GH-Rd3 is a major form of GH-R mRNA and is expressed in tissues other than the placenta as well (52).

To determine whether the hGH-Rd3 was biologically active, we compared the function of the hGH-R and hGH-Rd3 receptors (Fig. 11.4). We set up a *Xenopus* oocyte microinjection system in which high levels of hGH-R and hGH-Rd3 can be expressed on the surface of *Xenopus* oocytes that have been microinjected with synthetic hGH-R and hGH-Rd3 mRNA transcripts. Using this approach, we compared the relative affinities of the hGH-R and hGH-Rd3 for a series of potential ligands, including hGH-N, 20-kd hGH-N, hGH-V, hCS, and ovine PRL. In each case the binding of the ligand to the 2 receptor isoforms was

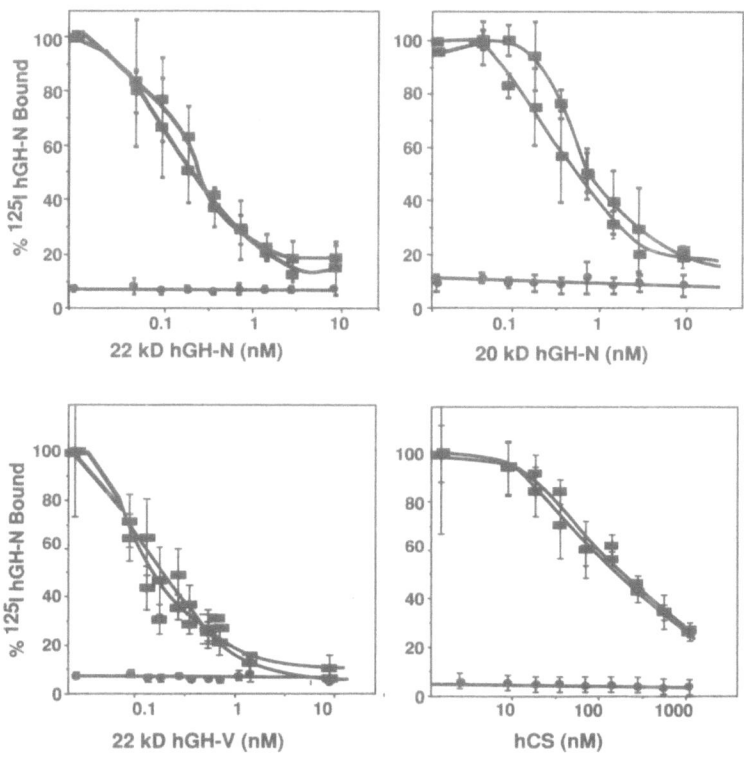

FIGURE 11.4. Radioreceptor analysis of hGH-R and hGH-Rd3 binding to 22-kd hGH-N, 20-kd hGH-N, 22-kd hGH-V, and hCS ligands. *Xenopus* oocytes were injected with synthetic mRNA encoding hGH-R, hGH-Rd3, or injection buffer alone (shaded boxes, open boxes, or shaded circles, respectively), incubated overnight to allow receptor synthesis, and then incubated with [125I]hGH along with defined concentration of cold competitor ligands. Bound [125I] was then determined in each case. The level of cold competitor is shown on the abscissa, and the level of [125I]hGH binding is shown on the ordinate. The level of binding in the absence of competitor is defined as 100%.

identical, indicating that the hGH-Rd3 is likely to be a functional receptor. The rates of internalization of the 2 receptor isoforms were also shown to be identical (Fig. 11.5). The expression and/or potential expression of hormones from hCS-A, hCS-B, hCS-L, and hGH-V genes in the syncytiotrophoblastic cells of the placenta in parallel with the hGH-Rd3 form of the GH-R suggest the possibility that an autocrine or paracrine loop exists in the placenta to mediate the bioactivity of the GH-related hormones locally (55).

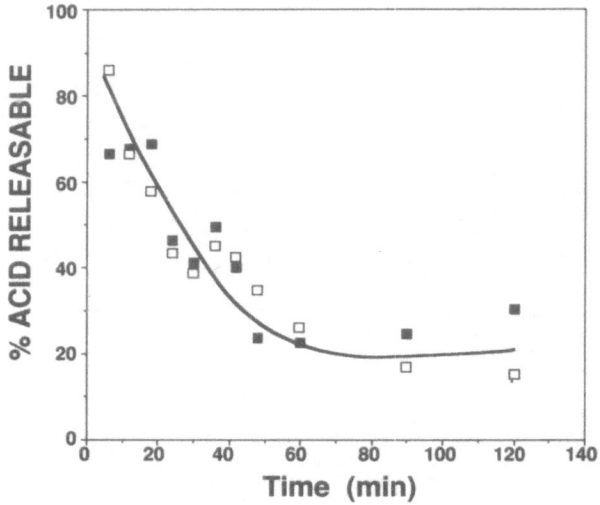

FIGURE 11.5. Rates of internalization of the 2 hGH-R isoforms, hGH-R and hGH-Rd3. Synthetic mRNAs encoding each of the 2 hGH-R isoforms, hGH-R and hGH-Rd3, were injected into separate sets of *Xenopus laevis* oocytes. After incubation to allow synthesis of the receptor proteins, the cells were incubated with [^{125}I]hGH at 4°C to allow ligand binding. They were then shifted to room temperature to allow receptor internalization. The rate of receptor internalization was measured by the loss of acid-releasable [^{125}I] from the surface of the intact oocytes. The time course of the experiment is shown on the abscissa. The open and closed boxes represent data from hGH-R- and hGH-Rd3-expressing oocytes, respectively.

Summary

The human GH gene cluster is expressed in the somatotrope cells of the anterior pituitary and the syncytiotrophoblastic epithelial layer of the placenta (Fig. 11.6). This expression is dependent on activation of the chromatin domain encompassing the cluster. We have identified cis-acting elements that are essential in this process of transcriptional activation. Details of this process appear to differ between these two tissues and may serve to underlie the mutually exclusive pituitary versus placental pattern of expression of the encoded genes. Two genes in the cluster, hGH-V and hCS-L, previously considered to be pseudogenes, are in fact expressed in a developmentally controlled manner during gestation. The hGH-V is biologically active and is the most purely somatogenic hormone thus defined in humans. The impact of this potent somatogen on the course of pregnancy can now be studied. The complexity of the hormones produced by the hGH gene cluster is increased by alternative splicing pathways that generate a number of novel isohormones. In parallel, a novel form of the

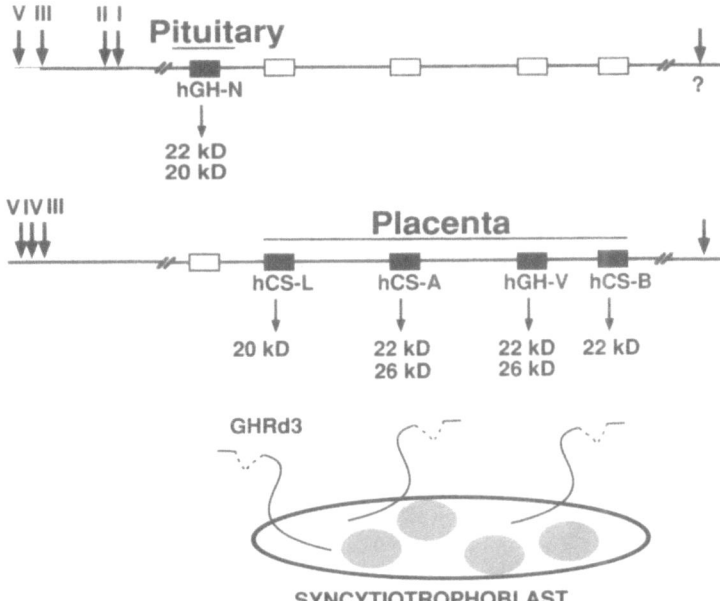

FIGURE 11.6. Summary. The hGH gene cluster and the pattern of gene expression from this cluster in the pituitary and placenta are shown on the 1st and 2nd lines, respectively. The genes that are actively expressed are shown as solid rectangles, and the position of the *DNase*I hypersensitive sites mapped in the chromatin of the 2 tissues are shown by the vertical arrows. The existence of the HS 3' to the cluster in the pituitary chromatin has not been tested. Encoded protein sizes are shown below each of the respective genes. At the bottom is a representation of a syncytiotrophoblast of the placental villus expressing both hGH-Rd3 and its potential ligands encoded by the hGH cluster (shown just above). The dotted region of the hGH-R represents the excised exon 3 of the hGH-Rd3 placental isoform. An autocrine/paracrine loop is suggested by the proximity of ligands and receptor in the syncytiotrophoblast.

hGH-R is generated in the placenta by the selective exclusion of exon 3 from the mRNAs processed in the nuclei of the syncytial layer of the villi. This receptor isoform is functionally active and may serve to mediate the effects of the locally expressed hGH-V, hCS, and possibly hCS-L genes on placental growth and function.

References

1. George DL, Phillips J, Francke U, Seeburg PH. The genes for growth hormone and chorionic somatomammotropin are on the long arm of human chromosome 17 in region q21 to pter. Hum Genet 1981;57:138–41.

2. Chen Y, Liao Y-C, Smith DH, Barrera-Saldaña HA, Gelinas RE, Seeburg PH. The human growth hormone locus: nucleotide sequence, biology, and evolution. Genomics 1989;4:479–97.

3. Owerbach D, Rutter WJ, Cooke NE, Martial JA, Shows TB. The prolactin gene is located on chromosome 6 in humans. Science 1981;212:815–6.

4. Cooke NE, Baxter JD. Structural analysis of the prolactin gene suggests a separate origin for its 5' end. Nature 1982;297:603–6.

5. Cooke NE. Mechanisms for evolutionary divergence within the prolactin gene family. In: Kumar A, ed. Eukaryotic gene expression. New York: Plenum Press, 1985:69–85.

6. McWilliams D, Boime I. Cytological localization of placental lactogen mRNA in syncytiotrophoblast layer of the human placenta. Endocrinology 1980; 107:761–5.

7. Cooke NE, Emery JG, Ray J, Urbanek M, Estes PA, Liebhaber SA. Placental expression of the human growth hormone-variant gene: a review. Troph Res 1991;5:61–74.

8. Clements J, Whitfeld P, Cooke NE, et al. Expression of the prolactin gene in human decidua-chorion. Endocrinology 1983;112:1133–4.

9. Soares MJ, Faria TN, Roby KF, Deb S. Pregnancy and the prolactin family of hormones: coordination of anterior pituitary, uterine, and placental expression. Endocr Rev 1991;12:402–23.

10. Ingraham HA, Chen R, Mangalam HJ, et al. A tissue-specific transcription factor containing a homeodomain specifies a pituitary phenotype. Cell 1988;55:519–29.

11. Bodner M, Castrillo J-L, Deerinck T, Ellisman M, Darin M. The pituitary-specific transcription factor GHF-1 is a homeobox-containing protein. Cell 1988;55:505–18.

12. Simmons DM, Voss JW, Ingraham HA, et al. Pituitary cell phenotypes involve cell-specific pit-1 mRNA translation and synergistic interactions with other classes of transcription factors. Genes Dev 1985;4:695–711.

13. Li S, Crenshaw EB, Rawson EJ, Simmons DM, Swanson LW, Rosenfeld MG. Dwarf locus mutants lacking three pituitary cell types result from mutations in the POU-domain gene pit-1. Nature 1990;347:528–33.

14. Pfäffle RW, DiMattia GE, Parks JS, et al. Mutation of the POU-specific domain of pit-1 and hypopituitarism without pituitary hypoplasia. Science 1992;257:1118–21.

15. Robins DM, Paek I, Seeburg PH, Axel R. Regulated expression of human growth hormone genes in mouse cells. Cell 1982;29:623–31.

16. Nickel BE, Kardami E, Cattini PA. Differential expression of human placental growth-hormone variant and chorionic somatomammotropin in culture. Biochem J 1990;267:653–8.

17. Lira SA, Crenshaw EB, Glass CK, Swanson LW, Rosenfeld MG. Identification of rat growth hormone sequences targeting pituitary expression in transgenic mice. Proc Natl Acad Sci USA 1988;85:4755–9.

18. Lira SA, Kalla KA, Glass CK, Drolet DW, Rosenfeld MG. Synergistic interactions between pit-1 and other elements are required for effective somatotroph rat growth hormone gene expression in transgenic mice. Mol Endocrinol 1993;7:694–701.

19. Hammer RE, Palmiter RD, Brinster RL. Partial correction of murine hereditary growth disorder by germline incorporation of a new gene. Nature 1984;311:65–7.
20. Dillon N, Grosveld F. Transcriptional regulation of multigene loci: multilevel control. Trends Genet 1993;9:134–7.
21. Morse RH. Transcribed chromatin. TIBS 1992;17:23–6.
22. Grosveld F, Blom van Assendelft G, Greaves DR, Kollias G. Position-independent, high-level expression of the human β-globin gene in transgenic mice. Cell 1987;51:975–85.
23. Stief A, Winter DM, Strätling WH, Sippel AE. A nuclear DNA attachment element mediates elevated and position-independent gene activity. Nature 1989;341:343–5.
24. Kellum R, Schedl P. A position-effect assay for boundaries of higher order chromosomal domains. Cell 1991;64:941–50.
25. Mirkovitch J, Mirault ME, Laemmli UK. Organization of the higher-order chromatin loop: specific DNA attachment sites on nuclear scaffold. Cell 1984;39:223–32.
26. Gross DS, Garrard WT. Nuclease hypersensitive sites in chromatin. Annu Rev Biochem 1988;57:159–97.
27. Forrester WC, Epner E, Driscoll MC, et al. A deletion of the human β-globin locus activation region causes a major alteration in chromatin structure and replication across the entire β-globin locus. Genes Dev 1990;4:1637–49.
28. Jones BK, Gerhard GS, Liebhaber SA, Cooke NE. High-level, pituitary specific expression of the human growth hormone gene in transgenic mice. 74th annu meet Endocr Soc, San Antonio, TX, June, 1992:446.
29. Jones BK, Monks B, Liebhaber SA, Cooke NE. Analysis of a locus control region for the human growth hormone gene cluster. 75th annu meet Endocr Soc, Las Vegas, NV, June, 1993:219.
30. MacLeod JN, Lee AK, Liebhaber SA, Cooke NE. Developmental control and alternative splicing of the placentally expressed transcripts from the human growth hormone gene cluster. J Biol Chem 1992;267:14219–26.
31. Braunstein GD, Rasor JL, Engvall E, Wade ME. Interrelationships of human chorionic gonadotropin, human placental lactogen, and pregnancy-specific β1-glycoprotein throughout normal human gestation. J Obstet Gynecol 1980;138:1205–13.
32. Frankenne F, Closset J, Gomez F, Scippo ML, Smal J, Hennen G. Expression of the growth hormone variant gene in human placenta. J Clin Endocrinol Metab 1987;64:635–7.
33. Seeburg PH. The human growth hormone gene family: nucleotide sequences show recent divergence and predict a new polypeptide hormone. DNA 1982;1:239–49.
34. Misra-Press A, Cooke NE, Liebhaber SA. The human chorionic somato-mammotropin-like (hCS-L) gene expresses a wide spectrum of alternatively spliced mRNAs. 75th annu meet Endocr Soc, Las Vegas, NV, June, 1993:1461.
35. Cooke NE, Ray J, Emery JG, Liebhaber SA. Two distinct species of human growth hormone-variant mRNA in the human placenta predict the expression of novel growth hormone proteins. J Biol Chem 1988;263:9001–6.

36. Estes PA, Cooke NE, Liebhaber SA. A difference in the splicing patterns of the closely related normal and variant human growth hormone gene transcripts is determined by a minimal sequence divergence between two potential splice-acceptor sites. J Biol Chem 1990;265:19863–70.

37. Estes PA, Cooke NE, Liebhaber SA. Native RNA secondary structure controls alternative splice-site selection and generates two growth hormone isoforms. J Biol Chem 1992;267:14902–8.

38. Liebhaber SA, Urbanek M, Ray J, Tuan RS, Cooke NE. Characterization and histological localization of human growth hormone-variant gene expression in the placenta. J Clin Invest 1989;83:1985–91.

39. Ray J, Jones BK, Liebhaber SA, Cooke NE. Glycosylated human growth hormone variant. Endocrinology 1989;125:566–8.

40. Lee AK, MacLeod JN, Ray J, Cooke NE, Liebhaber SA. The human growth hormone-variant gene encodes a novel membrane-associated protein product. Clin Res 1990:296a.

41. Ray J, Okamura H, Kelly PA, Cooke NE, Liebhaber SA. Human growth hormone-variant demonstrates a receptor binding profile distinct from that of the normal pituitary growth hormone. J Biol Chem 1990;265:7939–44.

42. Baumann G, Davila N, Shaw MA, Ray J, Liebhaber SA, Cooke NE. Binding of human growth hormone (GH)-variant (placental GH) to GH-binding proteins in human plasma. J Clin Endocrinol Metab 1991;73:1175–9.

43. MacLeod JN, Worsley I, Ray J, Friesen HG, Liebhaber SA, Cooke NE. Human growth hormone-variant is a biologically active somatogen and lactogen. Endocrinology 1991;128:1298–302.

44. Plaut K, Ikeda M, Vonderhaar BK. Role of growth hormone and insulin-like growth factor-I in mammary development. Endocrinology 1993;133:1843–8.

45. Feldman M, Ruan W, Cunningham BC, Wells JA, Kleinberg DL. Evidence that the growth hormone receptor mediates differentiation and development of the mammary gland. Endocrinology 1993;133:1602–8.

46. Jammes H, Gaye P, Belair L, Djiane J. Identification and characterization of growth hormone receptor mRNA in the mammary gland. Mol Cell Endocrinol 1991;75:27–35.

47. Postel-Vinay M-C, Belair L, Kayser C, Kelly PA, Djiane J. Identification of prolactin and growth hormone binding proteins in rabbit milk. Proc Natl Acad Sci USA 1991;88:6687–90.

48. Bchini O, Andres AC, Schubaur B, et al. Precocious mammary gland development and milk protein synthesis in transgenic mice ubiquitously expressing human growth hormone. Endocrinology 1991;128:539–46.

49. Daughaday WH, Trivedi B, Winn HN, Yan H. Hypersomatotropism in pregnant women, as measured by a human liver radioreceptor assay. J Clin Endocrinol Metab 1990;70:215–21.

50. Duquesnoy P, Sobrier ML, Anselem S, Goossens M. Defective membrane expression of human growth hormone (GH) receptor causes Laron-type GH insensitivity syndrome. Proc Natl Acad Sci USA 1991;88:10272–6.

51. Leung DW, Spencer CA, Cachianes G, et al. Growth hormone receptor and serum binding protein: purification, cloning and expression. Nature 1987; 330:537–43.

52. Urbanek M, MacLeod JN, Cooke NE, Liebhaber SA. Expression of a human growth hormone (hGH) receptor isoform is predicted by tissue-specific

alternative splicing of exon 3 of the hGH receptor gene transcript. Mol Endocrinol 1992;6:279–87.

53. Sobrier M-L, Duquesnoy P, Duriez B, Anselem S, Goossens M. Expression and binding properties of two isoforms of the human growth hormone receptor. FEBS Lett 1993;319:16–20.

54. Kliman HJ, Nestler JE, Sermasi E, Sanger JM, Strauss JF. Purification, characterization, and in vitro differentiation of cytotrophoblasts from human term placentae. Endocrinology 1986;118:1567–82.

55. Urbanek M, Russell JE, Cooke NE, Liebhaber SA. Functional characterization of the alternatively spliced, placental human growth hormone receptor. J Biol Chem 1993;268:19025–32.

12

Growth Hormone Economy in Menopausal Women: Effects of Age

MARK L. HARTMAN, JILL A. KANALEY, AND
ARTHUR WELTMAN

Growth hormone (GH) is secreted by the anterior pituitary gland in a pulsatile fashion under the regulation of two hypothalamic peptides: *GH releasing hormone* (GHRH) stimulates GH synthesis and secretion, while *somatostatin* (SRIH) inhibits GH release without affecting GH synthesis (1). *Insulin-like growth factor I* (IGF-I), synthesized under GH control in the liver and other tissues, mediates some of the metabolic effects of GH (2) and exerts a rapid negative feedback effect on GH release (3). GH secretion declines with aging in both men and women (4, 5). Since GH is an anabolic and lipolytic hormone, it has been hypothesized that decreasing GH secretion with aging may be responsible for some of the changes in body composition that occur with aging (6). In this brief review, the etiology and metabolic implications of the hyposomatotropism of aging are discussed; possible gender differences in these effects are highlighted. The changes in body composition that occur with aging and menopause are reviewed with an emphasis on the possible role for declining GH concentrations in mediating these changes.

Decline of GH Secretion with Aging

Figure 12.1 illustrates the decline in GH secretion that occurs in men and women progressively after age 20 (4). This decrease occurs predominantly before age 40–50 (4, 7) and affects both daytime and nocturnal GH secretion (8, 9). Although young eumenorrheic women secrete more GH than young men, no gender differences in GH release in older subjects have been documented (5). Serum concentrations of IGF-I decline with age in a pattern that closely corresponds to that seen with GH levels (8, 10). Provocative tests of GH secretion in older subjects have yielded

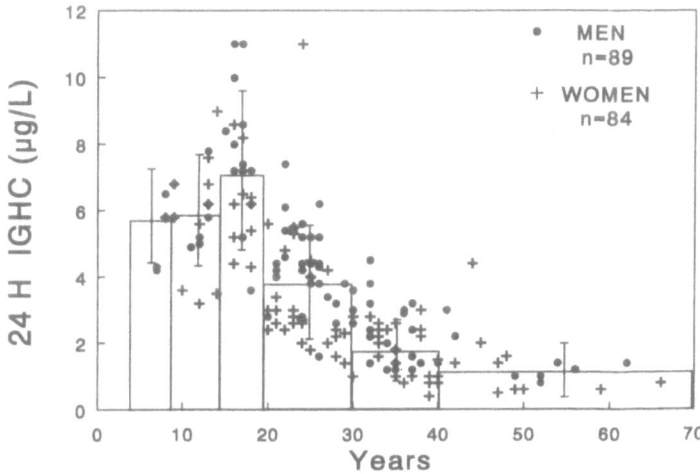

FIGURE 12.1. Relationship between 24-h IGHC and age of 89 male (solid circle) and 84 female (+) normal subjects. Reprinted with permission from Zadik, Chalew, McCarter, Meistas, and Kowarski (4), © The Endocrine Society, 1985.

conflicting results. Whereas some reports found no effect of age on the GH responses to hypoglycemia (11), arginine (12), or GHRH (13), others document diminished responses to hypoglycemia (14) and GHRH (15). Differences in the ages of the young control groups might account for some of these conflicting reports.

Etiology of Hyposomatotropism of Aging

It is not known whether the reduction in GH secretion with age results from a primary pituitary abnormality, or from decreased GHRH, or increased SRIH release. Several factors may contribute to the decline in GH secretion with age.

Pituitary Aging

In a pathologic study of 28 autopsied subjects between 16 and 90 years of age, both the number and size of immunoreactive somatotrophs decreased significantly with age; the greatest decrease occurred between young (ages 16–37) and middle-aged (ages 40–55) subjects, with no further significant decreases in the older (ages 60–90) group. This is consistent with the timing of the decrease in serum GH concentrations. The mechanisms responsible for this observation are not known, but could include diminished GHRH release by the hypothalamus (16).

Hypothalamic GHRH Synthesis

GHRH immunoreactivity and mRNA levels are reduced by approximately 50% in aged (24 months) compared to young (3 months) rats (17). Administration of a GHRH antagonist to healthy young men results in decreased pulsatile GH release; this results primarily from decreased GH pulse amplitudes, as is the case in aging (18). Since GH synthesis and secretion is dependent on GHRH, these observations suggest that declining levels of GH secretion with aging might result from a primary decrease in GHRH synthesis. This hypothesis is supported by the observation that administration of *subcutaneous* (sc) GHRH for 14 days to older men resulted in increased serum concentrations of GH and IGF-I (19).

Gonadal Steroids

The decline in serum GH concentrations with age in men and women correlates with changes in gonadal steroid levels. When the entire age range of men and women was compared, serum estradiol (but not testosterone) levels largely accounted for the differences in 24-h *integrated GH concentrations* (IGHC) (5). However, when men aged 21–71 years were studied, serum testosterone (not estradiol) was the best correlate of 24-h GH secretion (7). Oral estrogen replacement therapy increases 24-h spontaneous and GHRH-stimulated GH release in menopausal women, although serum IGF-I levels are decreased (20, 21). Animal studies suggest that estrogen likely stimulates GH release via effects on hypothalamic SRIH and GHRH secretion (22). Studies to date indicate that transdermal estrogen therapy may not increase serum GH concentrations or decrease IGF-I levels in menopausal women (21, 23). This suggests that stimulation of GH release by oral estrogens may be a consequence of lower serum IGF-I levels and reduced negative feedback by IGF-I on GH secretion (21).

The high concentrations of estrogen in portal blood following oral administration may inhibit hepatic IGF-I synthesis, as demonstrated in rats (24). Endogenous estrogen levels likely regulate GH secretion since serum GH concentrations in eumenorrheic women are maximal during the late follicular phase of the menstrual cycle when serum estradiol levels are highest (25). In summary, current evidence favors the hypothesis that the decline in estradiol concentrations after menopause removes a potent stimulus to GH secretion. The role of gonadal steroids in regulating GH secretion is reviewed in greater detail in other contributions to this symposium.

Sleep

Significant changes in the amount and pattern of sleep occur with aging. There is a reduction in slow wave sleep, increased nighttime wakefulness,

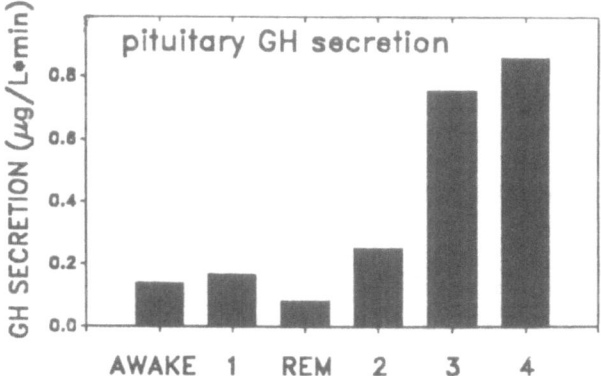

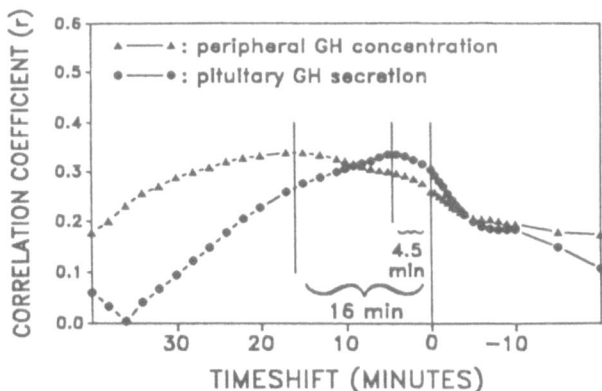

FIGURE 12.2. Relationship between GH secretion and sleep. In the top panel mean GH secretion rates calculated by deconvolution analysis are shown for each stage of sleep, as determined by concomitant electroencephalographic (EEG) monitoring and blood sampling at 30-sec intervals in young men. In the bottom panel the correlations between sleep stage and pituitary GH secretion (solid circle) or plasma GH concentrations (solid triangle) in relation to the time interval separating the 2 events are shown. A positive lag (left) indicates that hormone values occur after the EEG findings, and a negative lag (right) means that hormone values occur earlier than the EEG. The time lag corresponding to the maximal correlation coefficient is noted for each analysis. Reprinted with permission from Holl, Hartman, Veldhuis, Taylor, and Thorner (27), © The Endocrine Society, 1991.

and increased fragmentation of sleep by periods of wakefulness (26). In young subjects GH release is closely associated with slow wave sleep. By sampling blood at 30-sec intervals for measurement of GH during simultaneous electroencephalographic monitoring of sleeping subjects, we demonstrated that GH secretion rates were maximally correlated with slow wave sleep with a lag of 4.5 min (Fig. 12.2) (27).

It is not known whether slow wave sleep stimulates GH secretion or whether common neural inputs stimulate both slow wave sleep and GH secretion. If the former is true, then declining levels of slow wave sleep with aging may remove a potent stimulus to GH secretion. However, a recent study demonstrated that GHRH may have sleep-promoting effects. Administration of GHRH to young men at times of decreased sleep propensity (e.g., during the third period of *rapid-eye-movement* [REM] sleep) resulted in an 8-fold decrease in the duration of wakefulness and a 9-fold increase in slow wave sleep during the second hour after administration of GHRH (28). These data support the hypothesis that decreased GH secretion and fragmentation of sleep in older adults may be related to decreased GHRH synthesis in the central nervous system.

Adiposity and Physical Fitness

Adiposity

Both increasing age and adiposity are associated with decreased GH secretion. These effects may be difficult to separate since relative adiposity increases with age. In obese subjects the GH responses to hypoglycemia and GHRH are clearly blunted and are partially reversible with 72 h of fasting or with weight loss (29–31). Decreased GH concentrations observed in obese men have been attributed to decreased GH secretion rates and accelerated metabolic clearance of GH as determined by deconvolution analysis (32). Hyperinsulinemia associated with excess adiposity may reduce GH secretion by direct effects on the pituitary (33) or by enhancing negative feedback by IGF-I (3).

In a study of 21 healthy men aged 21–71, both age and *body mass index* (BMI) were found to be significant negative correlates of GH secretion and half-times of GH disappearance. Increasing age was associated with decreased GH secretory pulse frequency without significant effects on pulse amplitude. In contrast, increasing BMI was associated with decreased GH secretory pulse amplitude without significant effects on pulse frequency (7). This suggests that age and adiposity may influence pulsatile GH secretion by distinct mechanisms. However, since the BMI is not an accurate measure of adiposity, future studies of these relationships should employ more accurate methods for estimation of body fat, as will be detailed below.

Furthermore, since low-amplitude GH pulses (below 0.2 μg/L) were likely not detected in the *immunoradiometric assay* (IRMA) used in this study, it is possible that GH pulse frequency and amplitudes may have been underestimated and overestimated, respectively. These conclusions will need to be confirmed using new, more sensitive GH assays that are able to measure GH in all samples. For example, we have modified a new

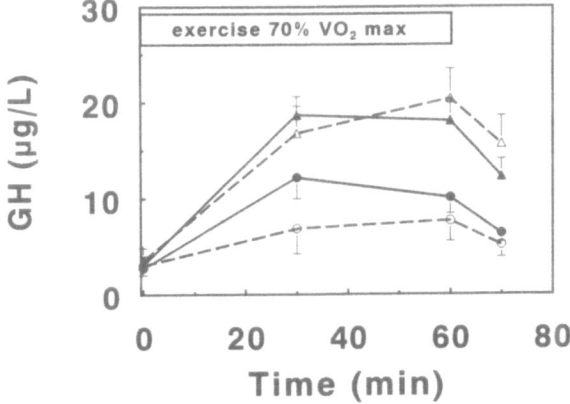

FIGURE 12.3. Plasma GH responses to 60 min of treadmill exercise in young (ages 20–32) and older (ages 60–70) men. Subjects exercised at a work rate that elicited 70% of the subjects' maximal oxygen consumption (VO$_2$ max [mL/kg^{-1}/ min^{-1}]). Young trained men (VO$_2$ max = 64.9 ± 5.3, n = 11) (solid line and triangle) and young sedentary men (VO$_2$ max = 46.4 ± 4.0, n = 13) (dotted line and open triangle) had similar GH responses that were greater than those observed in both groups of older subjects. Trained older men (VO$_2$ max = 50.0 ± 4.8, n = 11) (solid line and circle) had greater GH responses than sedentary older men (VO$_2$ max = 26.8 ± 2.4, n = 10) (dotted line and open circle). Adapted from data presented in Hagberg, Seals, and Yerg (36).

GH chemiluminescence immunoassay to improve its sensitivity to 0.002 µg/L, which enabled serum GH concentrations to be measurable even after suppression with an oral glucose load (34).

Physical Fitness

In our culture physical fitness levels tend to decline with aging. Acute exercise is a known stimulus of GH secretion, although significant inter-subject variability exists (35). The GH response to both aerobic and resistance exercise is reduced in older men (36, 37). As illustrated in Figure 12.3, the GH response to aerobic exercise is greater in aerobically fit compared to sedentary older subjects (36). We are unaware of studies relating 24-h GH release to measures of fitness in older men and women. It is not known whether chronic exercise training will result in increased 24-h GH release in older subjects.

Gender Differences

Since differences in body composition exist between men and women, we hypothesized that the strength of relationships between serum GH levels and age, body composition, and fitness would differ between premeno-pausal women and young men. These relationships were studied in 32 eumenorrheic women and 12 men aged 20–40 years (38). In men significant relationships existed between 24-h serum IGHC and age ($r = -0.79$, $P = 0.002$), percent body fat as calculated by hydrostatic weighing ($r = -0.75$, $P = 0.005$), and aerobic fitness as measured by peak oxygen consumption ($V_{O_2 peak}$: $r = 0.58$, $P = 0.05$), but not BMI ($r = -0.53$, $P = 0.08$). In women a significant relationship existed between 24-h IGHC and age ($r = -0.35$, $P = 0.05$), but not BMI ($r = -0.19$, $P = 0.29$); relationships between 24-h IGHC and percent body fat ($r = -0.29$, $P = 0.11$) and $V_{O_2 peak}$ ($r = 0.31$, $P = 0.08$) approached significance. Standardized regression coefficients revealed that for each standard deviation change in age, BMI, percent body fat, or $V_{O_2 peak}$, the associated change in 24-h IGHC was 1.9–2.6 times greater in men than women.

We conclude that age, percent body fat, and fitness are related to 24-h GH release in young adults and that these relationships are considerably stronger in men than women. These data demonstrate the importance of studying both men and women and highlight the need to use more precise measures of body composition than BMI. Relationships between accurate measures of body composition and 24-h GH secretion and clearance in postmenopausal women have not been reported to our knowledge. We are currently studying these relationships in older men and women.

Effects of Fasting

The relationships between GH secretion and adiposity or physical fitness may be obscured by the suppressive effects of nutrition. This is particularly true if subjects studied are within a relatively narrow range of percent body fat or $V_{O_2 peak}$. In one study we observed no significant correlation between BMI and 24-h GH production rates in 9 young, normal-weight men in the fed state, but after 2 days of fasting, a highly significant inverse correlation was observed ($r = -0.90$, $P < 0.001$) (39). In a second study nocturnal GH secretion (2200–0800 h) was correlated with percent body fat ($r = -0.75$, $P = 0.01$), but not with $V_{O_2 peak}$ ($r = 0.24$, $P = 0.51$) in the fed state. However, after 2 days of fasting, nocturnal GH secretion was significantly correlated with both percent body fat ($r = -0.69$, $P = 0.03$) and $V_{O_2 peak}$ ($r = 0.65$, $P = 0.04$) (40). Thus, fasting may be a useful physiologic stimulus to study these relationships in older men and women.

Changes in Body Composition with Aging and Menopause

Significant changes in body composition occur with aging. On average between the ages of 30 and 75, muscle mass decreases by 20%–50%, bone mass decreases by 20%, and the proportion of body fat increases by

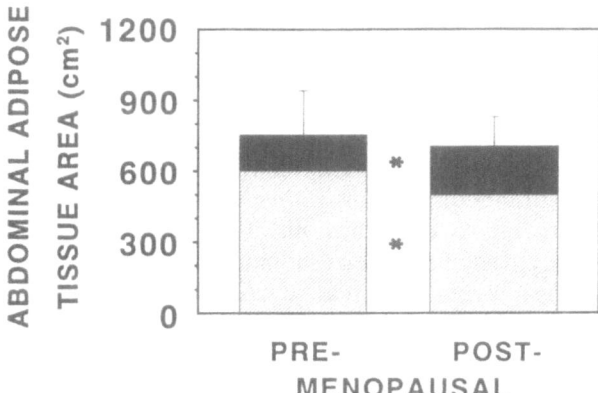

FIGURE 12.4. Abdominal adipose tissue areas (subcutaneous [hatched], intra-abdominal visceral [solid]), as determined by CT in 40 premenopausal (34.7 ± 11.1 years) and 17 postmenopausal (60.0 ± 7.8 years) obese women matched for BMIs (39.0 ± 6.3 and 37.4 ± 5.3 kg/m², respectively). (*$P < 0.01$.) Adapted from data presented in Zamboni, Armellini, Milani, et al. (46).

100% (6). In addition, a significant redistribution of fat from peripheral sc to intraabdominal adipose tissue depots occurs, as assessed by *computed tomography* (CT) (41). Total body water tends to decrease, although the aqueous fraction of the fat-free mass significantly increases (42). The changes in body water and bone introduce significant errors into the estimation of body fat by simple 2-compartment models, such as those employed with hydrostatic weighing. Thus, accurate evaluation of body composition of older subjects requires the use of more complex multi-compartment models (42, 43).

Although body fat distribution in women differs from that in men, recent studies have demonstrated that postmenopausal women also have increased intraabdominal visceral and decreased sc abdominal fat compared to premenopausal women matched for BMI (Fig. 12.4) (46). Regionally specific changes in adipocyte metabolism occur after menopause. In premenopausal women femoral adipocytes are larger, have lower rates of lipolysis, and greater lipoprotein lipase activity than abdominal adipocytes. After menopause the size of femoral adipocytes decreases, lipoprotein lipase activity in femoral adipocytes decreases, and the lipolytic activity of abdominal adipocytes decreases (47). These changes in adipocyte metabolic activity after menopause may favor the redistribution of fat from femoral-gluteal sites to abdominal depots, resulting in an increased abdominal obesity as occurs in men. A few recent studies suggest that estrogen replacement therapy may prevent age-related increases in body fat (48) and, specifically, accumulation of abdominal fat (49). Whether this reflects a direct effect of estrogen on adipocyte metabolism or an indirect effect via other hormonal changes, such as enhanced GH secretion, is unknown.

The changes in body composition with aging have several health implications. Decreased muscle strength and bone density increase the risk of fractures, particularly at sites where trabecular bone predominates, such as the spine and the proximal femur (6). Recent evidence suggests that increased intraabdominal fat, rather than abdominal or chest sc fat, is linked to the metabolic complications of upper-body or android obesity, including insulin resistance, hyperinsulinemia, glucose intolerance, type II diabetes mellitus, hyperlipidemia, decreased *high-density lipoprotein* (HDL) cholesterol concentrations, and hypertension (44, 45). Visceral fat depots are more sensitive to lipolytic stimuli and less sensitive to the inhibitory effects of insulin on lipolysis. This may result in elevated portal vein concentrations of *free fatty acids* (FFAs) that are known to inhibit hepatic extraction of insulin and stimulate hepatic gluconeogenesis and synthesis of *very low density lipoproteins* (VLDLs). Elevated systemic concentrations of FFA and insulin may contribute to insulin resistance, glucose intolerance, and hypertension (44, 45). Reduced HDL cholesterol in subjects with abdominal obesity may result from an increased exchange of cholesterol for *triglycerides* (TGs) with VLDL particles, as well as from enhanced activity of hepatic TG lipase (44). These hypothetical mechanisms are illustrated in Figure 12.5.

Effects of GH on Body Composition

The observation that GH treatment of GH-deficient children to promote linear growth also increased *lean body mass* (LBM) and decreased fat mass (50) raised the hypothesis that declining GH secretion with aging might have a role in the altered body composition of older persons (6). Recent clinical studies have demonstrated that treatment of GH-deficient adult men and women with *recombinant human GH* (rhGH) for 4–6 months decreased fat mass and increased LBM (51–53). In addition, increased thigh muscle volume and decreased adipose tissue volumes in the legs, trunk, abdomen, arms, and head/neck areas have been demonstrated by CT (51, 53, 54). As shown in Figure 12.6, the greatest decrease in adipose tissue volume occurred in the intraabdominal visceral depot (53). This is particularly noteworthy since increased visceral fat may be one mechanism by which cardiovascular mortality is increased in GH-deficient adults (55).

Preliminary observations suggest that GH may increase expression of hepatic receptors for *low-density lipoproteins* (LDLs) (56), possibly accounting for the observation that plasma cholesterol levels were lowered by rhGH treatment of GH-deficient subjects (52). The changes in skeletal muscle volume by rhGH treatment may also have functional importance.

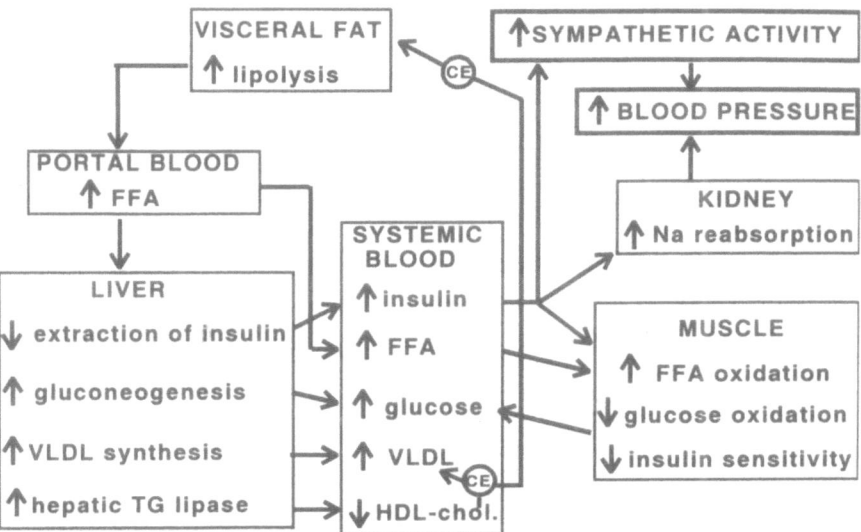

FIGURE 12.5. Hypothetical mechanisms for the adverse metabolic consequences associated with increased intraabdominal visceral fat. (FFA = free fatty acids; VLDL = very low density lipoproteins; TG = triglyceride; HDL = high-density lipoproteins; CE = cholesterol esters; Na = sodium.) For reviews, see references 44 and 45.

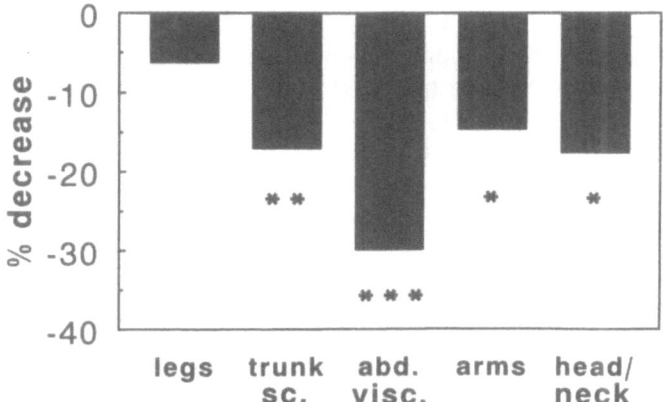

FIGURE 12.6. Changes in adipose tissue areas estimated by CT of different adipose tissue depots in 10 patients with adult-onset GH deficiency treated for 26 weeks with rhGH. The amount of adipose tissue in legs, sc trunk depot, intraabdominal visceral depot, arms, and head/neck areas decreased by 6.3%, 17.1%, 29.9%, 14.7%, and 17.7%, respectively. (*$P < 0.05$; **$P < 0.01$; ***$P < 0.001$.) Reprinted with permission from Bengtsson, Edén, Lönn, et al. (53), © The Endocrine Society, 1993.

Increased strength of the hip flexors and limb girdle muscles was documented in one study; other muscle groups were unaffected (54). A small
increase in exercise capacity was also demonstrated in two studies (51,
57). Although effects on muscle strength and exercise capacity were
small, it is important to note that these subjects were not trained. It is
possible that greater effects may be observed with a combination of rhGH
treatment and physical training.

Few data are available on the effects of rhGH treatment in healthy
older subjects. Short-term (7 days) rhGH treatment of men and women
over age 60 resulted in decreased urinary nitrogen excretion, decreased
plasma cholesterol levels, increased serum concentrations of parathyroid
hormone and calcitriol, and increases in markers of bone turnover (serum
osteocalcin and urinary hydroxyproline). Serum insulin concentrations
were also increased, although glucose intolerance did not occur except at
the highest dose of rhGH administered (58).

The only published study of long-term rhGH treatment in older persons
is that of Rudman et al. (59), who treated 21 men over age 60, selected
for low IGF-I levels, with 0.03-mg/kg rhGH ($n = 12$) or placebo ($n = 9$) 3
times per week for 6 months. This dose of rhGH raised serum IGF-I
concentrations into the range observed in healthy men 20–40 years old
and resulted in an 8.8% increase in LBM that was statistically significant
compared to the placebo-treated group. A 14.4% decrease in adipose
tissue mass and a 1.6% increase in lumbar vertebral bone density also
occurred in the rhGH-treated group; these changes were significant by
within-group comparisons, but not when compared to the placebo-treated
group. Nevertheless, taken together with the studies performed with GH-
deficient subjects, these findings support the hypothesis that the decline
of GH secretion with aging may contribute to the changes in body composition observed with aging.

It is important to note that no measurements of body fat distribution,
plasma lipids, muscle strength, or exercise capacity were performed in the
study by Rudman et al. (59). Thus, the significance of these findings for
the cardiovascular and musculoskeletal health of older men remains
uncertain. There are no published studies of the effect of rhGH treatment
in postmenopausal women. Since gender differences exist in body composition and the relationship between body composition, fitness, and GH
secretion, it is essential that future studies of rhGH treatment of older
subjects include women. At the University of Virginia, we are conducting
a placebo-controlled, randomized trial of rhGH treatment for 1 year in
men and women over age 60. Subjects are also being randomized to
aerobic exercise training, strength training, or no training to test the
hypothesis that physical training will enhance the effects of rhGH on
muscle, bone, and adipose tissue. Detailed assessments of body composition, body fat distribution, bone density, carbohydrate and lipid metabolism, metabolic rate, aerobic fitness, muscle strength, and balance are

being performed. The results of this study and similar studies being conducted at other centers may shed further light on the role of declining GH secretion in the aging process.

Reversing the Decline of GH Secretion with Aging

If ongoing trials of rhGH treatment of older men and women demonstrate beneficial health effects, then future treatment strategies may include pharmacologic or physiologic interventions to enhance endogenous GH secretion. Pharmacologic interventions could include administration of rhGH, GHRH (or analogs), nonpeptide GH secretagogues, and gonadal steroid replacement. These options have been reviewed in several other chapters in this volume.

Physiologic interventions, such as exercise training or dietary modifications, may also be of value in enhancing GH secretion in older persons. We reported recently that 1 year of moderately intense aerobic exercise training (above the lactate threshold) increased 24-h IGHC nearly 2-fold in eumenorrheic women between 18 and 40 years of age. This increase in GH release was accounted for by a nearly 2-fold increase in GH pulse amplitude and area with no change in GH pulse frequency, as assessed by the cluster algorithm. GH release was not enhanced in women who trained at a lower exercise intensity (at the lactate threshold) (60). Representative 24-h profiles of serum GH concentrations from subjects in the 3 groups in this study are shown in Figure 12.7. We are unaware of any studies investigating the effects of exercise training on 24-h GH release in older men or women.

The effects of chronic dietary modifications on GH secretion in older men and women are unknown. We studied the effects of short-term fasting on GH secretion in 4 postmenopausal women and 2 men (ages 55–81; BMI: 22–24 kg/m^2) (61). Blood was obtained every 5 min for 24 h on a control (fed) day and on day 2 of a fast. GH concentrations were analyzed with a multiple-parameter deconvolution method to simultaneously resolve endogenous GH secretory and clearance rates (62). Twenty-four-hour GH production rates increased more than 4-fold with the 2-day fast (38 ± 25 vs. 166 ± 42 µg/L of distribution volume; $P = 0.003$). This increase was accounted for by more than 2-fold increases in the number of detectable GH secretory bursts and mass of GH secreted per burst. The endogenous half-life of GH disappearance was not significantly altered by the 2-day fast.

In a similar study of 9 young men (ages 24–28; BMI: 21–25 kg/m^2), 24-h GH production rates increased by a similar order of magnitude with 2 days of fasting (78 ± 12 vs. 371 ± 57 µg/L of distribution volume; $P = 0.0001$), although the absolute levels of GH secretion were approximately 2-fold higher in these young men in both the fed and fasted states

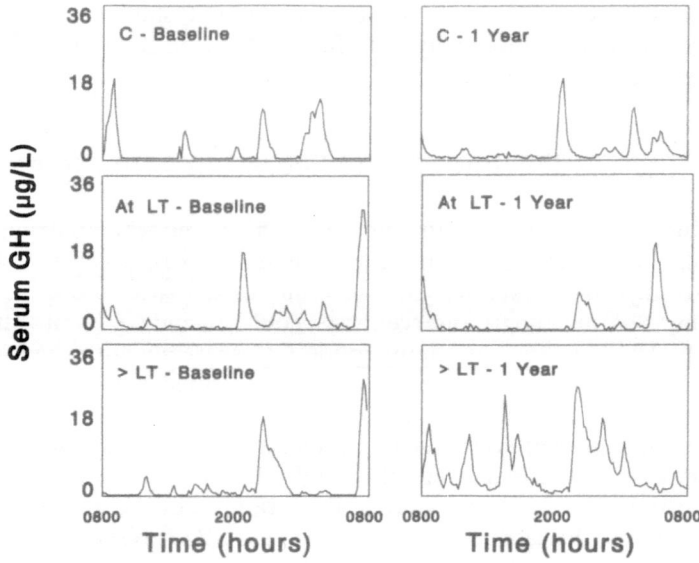

FIGURE 12.7. The 24-h serum GH concentrations for 3 eumenorrheic women before and 1 year after either no exercise training (C = control) (top panels) or aerobic exercise training at 1 of 2 intensity levels: at the lactate threshold (LT) (middle panels) or above the lactate threshold (>LT) (bottom panels). Note that 1 year of training at the higher intensity was associated with increased pulsatile GH release. Reprinted with permission from Weltman, Weltman, Schurrer, Evans, Veldhuis, and Rogol (60).

compared to the older subjects (39, 61). These data demonstrate that diminished pulsatile GH secretion associated with aging is partially reversible with a simple physiologic intervention, such as fasting. This suggests that modification of lifestyle factors, such as diet and/or exercise, may augment endogenous GH secretion in older subjects.

Conclusion

The decline of GH secretion with aging is likely related to several factors, including possible aging effects on the pituitary and hypothalamus, decreased circulating concentrations of gonadal steroids, increased adiposity, decreased physical fitness, and fragmented sleep. Lower serum GH concentrations may be related to some of the changes in body composition and metabolism that occur with aging. Although recent studies lend support to this hypothesis, it is not known whether rhGH treatment will improve the functional capacity, risk of cardiovascular disease, or risk of fractures of older persons. Ongoing clinical trials of rhGH treatment of older men and women may clarify whether such treatment is likely to have beneficial health effects. It is possible that the combination of rhGH treatment and exercise training may result in greater effects. Since fasting increases GH secretion in older subjects, lifestyle modifications such as changes in diet or exercise may increase serum GH concentrations. However, because fasting-enhanced GH secretion in older subjects remains less than half of that seen in young men, it seems likely that a primary age-related decline in hypothalamic GHRH synthesis and release occurs. Thus, restoration of serum GH concentrations in older subjects to levels observed in young subjects may require pharmacologic interventions.

References

1. Hartman ML, Veldhuis JD, Thorner MO. Normal control of growth hormone secretion. Horm Res 1993;40:37–47.
2. Daughaday WH, Rotwein P. Insulin-like growth factors I and II: peptide, messenger ribonucleic acid and gene structures, serum, and tissue concentrations. Endocr Rev 1989;10:68–91.
3. Hartman ML, Clayton PE, Johnson ML, et al. A low-dose euglycemic infusion of recombinant human insulin-like growth factor I rapidly suppresses fasting-enhanced pulsatile growth hormone secretion in humans. J Clin Invest 1993;91:2453–62.
4. Zadik Z, Chalew SA, McCarter RJ, Meistas M, Kowarski AA. The influence of age on 24-hour integrated concentration of growth hormone in normal individuals. J Clin Endocrinol Metab 1985;60:513–6.
5. Ho KY, Evans WS, Blizzard RM, et al. Effects of sex and age on the 24-hour profile of growth hormone in man: importance of endogenous estradiol concentrations. J Clin Endocrinol Metab 1987;64:51–8.
6. Rudman D. Growth hormone, body composition and aging. J Am Geriatr Soc 1985;33:800–7.
7. Iranmanesh A, Lizarralde G, Veldhuis JD. Age and relative adiposity are specific negative determinants of the frequency and amplitude of growth hormone (GH) secretory bursts and the half-life of endogenous GH in healthy men. J Clin Endocrinol Metab 1991;73:1081–8.
8. Rudman D, Kutner MH, Rogers CM, Lubin MF, Fleming GA, Bain RP. Impaired growth hormone secretion in the adult population. J Clin Invest 1981;67:1361–9.
9. Vermeulen A. Nyctohemeral growth hormone profiles in young and middle aged men: correlations with somatomedin-C levels. J Clin Endocrinol Metab 1987;64:884–8.
10. Clemmons DR, Van Wyk JJ. Factors controlling blood concentration of somatomedin-C. J Clin Endocrinol Metab 1984;13:113–43.
11. Kalk WJ, Vinik AI, Pimstone BL, Jackson WPU. Growth hormone response to insulin hypoglycemia in the elderly. J Gerontol 1973;28:431–3.
12. Dudl RJ, Ensinck JW, Palmer HE, Williams RH. Effect of age on growth hormone secretion in man. J Clin Endocrinol Metab 1973;37:11–6.

13. Pavlov EP, Harman SM, Merriam GR, Gelato MC, Blackman MR. Responses of growth hormone (GH) and somatomedin-C to GH releasing hormone in healthy aging. J Clin Endocrinol Metab 1986;62:595–600.
14. Muggeo M, Fedele D, Tiengo A, Molinari M, Crepaldi G. Serum growth hormone and cortisol responses to insulin stimulation in healthy aging. J Gerontol 1975;30:546–51.
15. Shibasaki T, Shizume K, Nakahara M, et al. Plasma growth hormone response to growth hormone-releasing factor in acromegalic patients. J Clin Endocrinol Metab 1984;58:212–4.
16. Sun YK, Xi YP, Fenoglio CM, et al. The effect of age on the number of pituitary cells immunoreactive to growth hormone and prolactin. Hum Pathol 1984;15:169–80.
17. De Gennaro Colonna V, Zoli M, Cocchi D, et al. Reduced growth hormone releasing factor (GHRF)-like immunoreactivity and GHRF gene expression in the hypothalamus of aged rats. Peptides 1989;10:705–8.
18. Jaffe CA, Friberg RD, Barkan AL. Suppression of growth hormone (GH) secretion by a selective GH-releasing hormone (GHRH) antagonist: direct evidence for involvement of endogenous GHRH in the generation of GH pulses. J Clin Invest 1993;92:695–701.
19. Corpas E, Harman SM, Pineyro MA, Roberson R, Blackman MR. Growth hormone (GH)-releasing hormone-(1-29) twice daily reverses the decreased GH and insulin-like growth factor-I levels in old men. J Clin Endocrinol Metab 1992;75:530–5.
20. Dawson-Hughes B, Stern D, Goldman J, Reichlin S. Regulation of growth hormone and somatomedin-C secretion in postmenopausal women: effect of physiological estrogen replacement. J Clin Endocrinol Metab 1986;63:424–32.
21. Weissberger AJ, Ho KKY, Lazarus L. Contrasting effects of oral and trans-dermal routes of estrogen replacement therapy on 24-hour growth hormone (GH) secretion, insulin-like growth factor I, and GH-binding protein in postmenopausal women. J Clin Endocrinol Metab 1991;72:374–81.
22. Wehrenberg WB, Giustina A. Basic counterpoint: mechanisms and pathways of gonadal steroid modulation of growth hormone secretion. Endocr Rev 1992;13:299–308.
23. Bellantoni MF, Harman SM, Cho DE, Blackman MR. Effects of progestin-opposed transdermal estrogen administration on growth hormone and insulin-like growth factor-I in postmenopausal women of different ages. J Clin Endocrinol Metab 1991;72:172–8.
24. Murphy LJ, Friesen HG. Differential effects of estrogen and growth hormone on uterine and hepatic insulin-like growth factor I gene expression in the ovariectomized hypophysectomized rat. Endocrinology 1988;122:325–32.
25. Faria ACS, Bekenstein LW, Booth RA, et al. Pulsatile growth hormone release in normal women during the menstrual cycle. Clin Endocrinol (Oxf) 1992;36:591–6.
26. Prinz PN, Vitiello MV, Raskind MA, Thorpy MJ. Geriatrics: sleep disorders and aging. N Engl J Med 1990;323:520–6.
27. Holl RW, Hartman ML, Veldhuis JD, Taylor WM, Thorner MO. Thirty-second sampling of plasma growth hormone in man: correlation with sleep stages. J Clin Endocrinol Metab 1991;72:854–61.

28. Kerkhofs M, Van Cauter E, Van Onderbergen A, Caufriez A, Thorner MO, Copinschi G. Sleep promoting effects of growth hormone-releasing hormone in normal men. Am J Physiol 1993;264:E594–8.

29. Ball MF, el-Khodary AZ, Canary JJ. Growth hormone response in the thinned obese. J Clin Endocrinol Metab 1972;34:498–511.

30. Williams T, Berelowitz M, Joffe SN, et al. Impaired growth hormone responses to growth hormone-releasing factor in obesity: a pituitary defect reversed with weight reduction. N Engl J Med 1984;311:1403–7.

31. Kelijman M, Frohman LA. Enhanced growth hormone responsiveness (GH) to GH-releasing hormone after dietary manipulation in obese and nonobese subjects. J Clin Endocrinol Metab 1988;66:489–94.

32. Veldhuis JD, Iranmanesh A, Ho KKY, Waters MJ, Johnson ML, Lizarralde G. Dual defects in pulsatile growth hormone secretion and clearance subserve the hyposomatotropism of obesity in man. J Clin Endocrinol Metab 1991; 72:51–9.

33. Yamashita S, Melmed S. Effects of insulin on rat anterior pituitary cells: inhibition of growth hormone secretion and mRNA levels. Diabetes 1986; 35:440–7.

34. Chapman IM, Hartman ML, Straume M, Johnson ML, Veldhuis JD, Thorner MO. Enhanced sensitivity GH chemiluminescence assay reveals lower post-glucose nadir GH concentrations in men than women. J Clin Endocrinol Metab 1994;78:1312–9.

35. Raynaud J, Capderou A, Martineaud JP, Bordachar J, Durand J. Intersubject variability in growth hormone time course during different types of work. J Appl Physiol 1983;55:1682–7.

36. Hagberg JM, Seals DR, Yerg JE, et al. Metabolic responses to exercise in young and older athletes and sedentary men. J Appl Physiol 1988;65:900–8.

37. Pyka G, Wiswell RA, Marcus R. Age-dependent effect of resistance exercise on growth hormone secretion in people. J Clin Endocrinol Metab 1992;75: 404–7.

38. Weltman A, Weltman JY, Hartman ML, et al. Relationship between age, percentage body fat, fitness and 24 hour growth hormone release in healthy young adults: effects of gender. J Clin Endocrinol Metab 1994;78:543–8.

39. Hartman ML, Veldhuis JD, Johnson ML, et al. Augmented growth hormone (GH) secretory burst frequency and amplitude mediate enhanced GH secretion during a two day fast in normal man. J Clin Endocrinol Metab 1992; 74:757–65.

40. Hartman ML, Haskvitz EM, Weltman A, Thorner MO. Percent body fat and VO_2 max are determinants of fasting enhanced growth hormone secretion. Med Sci Sports Exerc 1992;24:S53.

41. Borkan GA, Hults DE, Gerzof SG, Robbins AH, Silbert CK. Age changes in body composition revealed by computed tomography. J Gerontol 1983;38: 673–7.

42. Hewitt MJ, Going SB, Williams DP, Lohman TG. Hydration of the fat-free mass in children and adults: implications for body composition assessment. Am J Physiol 1993;265:E88–95.

43. Baumgartner RN, Heymsfield SB, Lichtman S, Wang J, Pierson RN. Body composition in elderly people: effect of criterion estimates on predictive equations. Am J Clin Nutr 1991;53:1345–53.

44. Després JP, Moorjani S, Lupien PJ, Tremblay A, Nadeau A, Bouchard C. Regional distribution of body fat, plasma lipoproteins, and cardiovascular disease. Arteriosclerosis 1990;10:497–511.
45. Björntorp P. Metabolic implications of body fat distribution. Diabetes Care 1991;14:1132–43.
46. Zamboni M, Armellini F, Milani MP, et al. Body fat distribution in pre- and post-menopausal women: metabolic and anthropometric variables and their inter-relationships. Int J Obes 1992;16:495–504.
47. Rebuffé-Scrive M, Eldh J, Hafstrom LO, Björntorp P. Metabolism of mammary, abdominal, and femoral adipocytes in women before and after menopause. Metabolism 1986;35:792–7.
48. Hassager C, Christiansen C. Estrogen/gestagen therapy changes soft tissue body composition in postmenopausal women. Metabolism 1989;38: 662–5.
49. Haarbo J, Marslew U, Gotfredsen, Christiansen C. Postmenopausal hormone replacement therapy prevents central distribution of body fat after menopause. Metabolism 1991;40:1323–6.
50. Frasier SD. Human pituitary growth hormone (hGH) therapy in growth hormone deficiency. Endocr Rev 1983;4:155–70.
51. Jørgensen JOL, Pedersen SA, Thuesen L, et al. Beneficial effects of growth hormone treatment in GH-deficient adults. Lancet 1989;1:1221–5.
52. Saloman F, Cuneo RC, Hesp R, Sönksen PH. The effects of treatment with recombinant growth hormone on body composition and metabolism in adults with growth hormone deficiency. N Engl J Med 1989;321:1797–803.
53. Bengtsson BA, Edén S, Lönn L, et al. Treatment of adults with growth hormone (GH) deficiency with recombinant human GH. J Clin Endocrinol Metab 1993;76:309–17.
54. Cuneo RC, Saloman F, Wiles CM, Hesp R, Sönksen PH. Growth hormone treatment in growth hormone-deficient adults, I. Effects on muscle mass and strength. J Appl Physiol 1991;70:688–94.
55. Rosén T, Bengtsson BA. Premature cardiovascular mortality in hypopituitarism—a study of 333 consecutive patients. Lancet 1990;336:285–8.
56. Rudling M, Norstedt G, Olivercona H, Reihner E, Gustafsson JA, Angelin B. Importance of growth hormone for the induction of hepatic low density lipoprotein receptors. Proc Natl Acad Sci USA 1992;89:6983–7.
57. Cuneo RC, Saloman F, Wiles CM, Hesp R, Sönksen PH. Growth hormone treatment in growth hormone-deficient adults, II. Effects on exercise performance. J Appl Physiol 1991;70:695–700.
58. Marcus R, Butterfield G, Holloway L, et al. Effects of short term administration of recombinant human growth hormone to elderly people. J Clin Endocrinol Metab 1992;70:519–27.
59. Rudman D, Feller AG, Nagraj HS, et al. Effects of human growth hormone in men over 60 years old. N Engl J Med 1990;323:1–6.
60. Weltman A, Weltman JY, Schurrer R, Evans WS, Veldhuis JD, Rogol AD. Endurance training amplifies the pulsatile release of growth hormone: effects of training intensity. J Appl Physiol 1992;72:2188–96.
61. Hartman ML, Pezzoli SS, Thorner MO. Diminished pulsatile growth secretion associated with aging is reversed by fasting. Clin Res 1991;39(2):165A.

62. Veldhuis JD, Carlson ML, Johnson ML. The pituitary gland secretes in bursts: appraising the nature of the glandular secretory impulses by simultaneous multiple-parameter deconvolution of plasma hormone concentration. Proc Natl Acad Sci USA 1987;84:7686–90.

13

Growth Hormone Economy in Hypogonadism: Effects of Sex Steroids

Ken K.Y. Ho, Andrew J. Weissberger, and John J. Kelly

The adolescent growth spurt is accompanied by a progressive rise in serum concentrations of gonadal steroids that occurs contemporaneously with an increase in spontaneous *growth hormone* (GH) release (1, 2). This observation suggests that gonadal steroids play an important role in determining GH secretion and in regulating body growth and composition. Evidence for androgens as a positive regulator of GH secretion is provided by the finding that testosterone is positively correlated with plasma IGF-I and 24-h mean GH concentration in males during puberty (3). There is evidence that estrogen may also be an important regulator of GH secretion in women, although this has been less well studied. Significant correlations have been reported between serum estradiol levels and GH secretion in pubertal girls (4). GH secretion varies during the menstrual cycle, with mean concentrations highest during the late follicular phase when estrogen concentration is highest (5). Mean 24-h GH and IGF-I levels are lower in postmenopausal women than in premenopausal women, suggesting that reduced activity of the somatotrophic axis in the menopause may be secondary to estrogen deficiency (6). The collective observations suggest that gonadal steroid deficiency may be associated with a reduction in GH secretion, which can be restored by sex steroid replacement.

In males GH secretion is reduced in hypogonadism and stimulated by androgen treatment, which also increases IGF-I levels (7, 8). Recent data suggest that the stimulatory effect of testosterone may in part be dependent on aromatization to estrogen (9). The consequences of hypogonadism and of estrogen replacement on GH economy in women are less clear. Girls with Turner syndrome appear to have reduced spontaneous GH release (10). While estrogen treatment augments spontaneous

GH secretion (11), the effects on IGF-I appear to be inconsistent, with reports of an increase (12) or no change (13) in IGF-I levels. Older studies utilizing higher doses of estrogens reported suppression of IGF-I activity in menopausal (14) and acromegalic women (15). Thus, the available data on estrogen effects in hypogonadal women appear confusing. In the last few years, we have undertaken studies to clarify the role of estrogen in the regulation of GH secretion and action using the postmenopausal state as a model of hypogonadism.

Estrogen Effect in the Menopause

We compared GH secretion obtained from 20-min blood measurements and serum IGF-I levels in weight-matched premenopausal and post-menopausal women and investigated in a prospective study design the effects of unopposed estrogens using a dosage previously reported to increase serum IGF-I in Turner patients (16). We also investigated possible effects on *GH binding protein* (GHBP) that had recently been identified in human serum, shown to bind substantial amounts of GH, and considered likely to modulate GH action (17, 18). We measured GHBP levels using a chromatography-based activity assay (17). In studying the effects of estrogen replacement, we considered that the route of administration may be important since the liver is the major source of circulating IGF-I (19), and orally administered estrogens exert nonphysiological effects on hepatic protein synthesis that are avoided when the nonoral route is used (20, 21). Accordingly, we undertook two regimens for estrogen replacement: ethinyl estradiol administered orally and 17β-estradiol administered transdermally (Estraderm TTS 100, Ciba Geigy).

We confirmed previous observations that mean 24-h GH concentrations and IGF-I levels were significantly lower in postmenopausal compared to younger cycling women. Both replacement regimens resulted in significant and comparable reductions in circulating levels of LH and FSH. Administration of oral ethinyl estradiol resulted in a 3-fold increase in mean 24-h GH concentrations, with the rise being characterized by an increase in mean pulse amplitude, but not frequency. In contrast, transdermal estrogen administration did not result in a significant change in mean 24-h GH concentrations.

Figure 13.1 shows representative 24-h GH profiles from a postmenopausal subject studied before treatment and during oral and transdermal estrogen therapy. Oral administration of ethinyl estradiol resulted in a uniform and significant reduction in mean IGF-I levels. In contrast, IGF-I levels showed a mild but significant increase when estrogen was administered via the transdermal route (Fig. 13.2). GHBP levels were not significantly different between pre- and postmenopausal women, but also revealed a route-dependent change with estrogen therapy. There was a

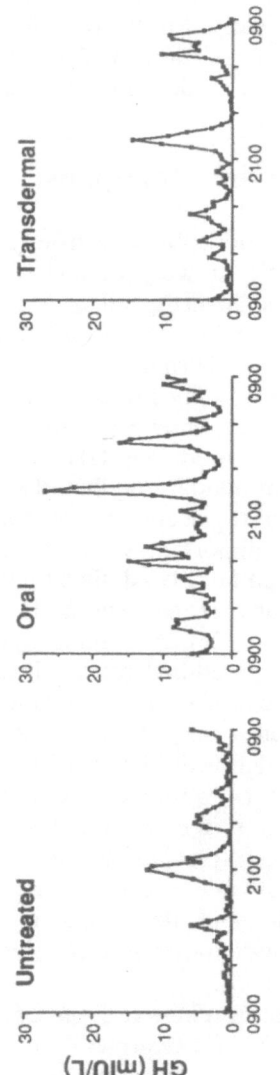

FIGURE 13.1. A 24-h GH profile from a menopausal subject studied before (left panel) and during estrogen replacement with oral ethinyl estradiol (middle panel) and transdermal 17β-estradiol (right panel). Reprinted with permission from Weissberger, Ho, and Lazarus (16), © The Endocrine Society, 1991.

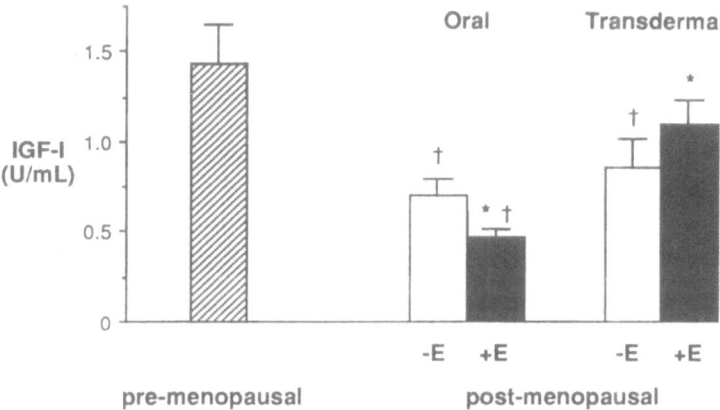

FIGURE 13.2. Mean (±SE) serum IGF-I concentrations in a group of premeno-pausal women ($n = 7$) and 2 groups of postmenopausal women ($n = 7$) before (−E) and after (+E) 12 weeks of oral ethinyl estradiol or transdermal 17β-estradiol treatment. (*$P < 0.05$ vs. pretreatment [−E]; $^†P < 0.05$ vs. premeno-pausal.) Modified with permission from Weissberger, Ho, and Lazarus (16), © The Endocrine Society, 1991.

significant increase in binding activity following the oral route, but no change following the transdermal route. We have more recently shown that the increase in binding activity arises from an increase in binding capacity and no change in affinity (22).

Our results show that estrogen treatment in the menopause has a significant effect on the GH-IGF-I axis, but demonstrate for the first time that the effects are dependent on the route of administration. Our finding of IGF-I suppression after ethinyl estradiol administration is supported by the results of two other studies using a similar dosage (23, 24). Since estrogens have been shown to inhibit hepatic IGF-I mRNA generation in the rat (25), we have postulated that the IGF-I suppressive effect of oral ethinyl estradiol is a consequence of a first-pass hepatic effect. This proposal is in keeping with the well-recognized observation that orally administered estrogens alter the synthesis of proteins of hepatic origin, such as angiotensinogen, clotting factors, and lipoproteins, effects that are avoided when a parenteral route is used. Because the fall in circulating IGF-I occurred in the setting of enhanced GH secretion, we further propose that the increase in GH concentration occurred as a result of reduced negative-feedback inhibition of IGF-I on GH secretion. A first-pass mechanism is also likely to explain the differing effects of oral and transdermal estrogens on serum GHBP since the liver is rich in *GH receptors* (GH-Rs) from which GHBP is thought to be derived.

We observed a small but significant increase in IGF-I levels following transdermal estrogen administration that was not accompanied by a change in GH levels. The mechanisms subserving these changes are unclear. In a study of transdermal estrogen administration to postmenopausal women, Bellantoni et al. (26) did not find a significant change in IGF-I levels, but observed a reduction in the peak GH response to GHRH. The latter observations suggest that the bioactivity of IGF-I may have been increased by transcutaneous estrogen administration. The reason for the discrepancy in immunoreactive IGF-I results between the two studies is not clear, but may have arisen from the different types of progestins used (norethisterone in our study vs. medroxyprogesterone acetate) or the timing of IGF-I measurements in relation to progesterone treatment.

Different Oral Estrogen Formulations

The estrogen types used in our study were not identical. Consequently, the data do not totally exclude the possibility that the contrasting effects of ethinyl estradiol and 17β-estradiol reflect intrinsic chemical differences, rather than the dissimilar routes of administration. It has been reported that induction of hepatic protein synthesis by ethinyl estradiol appears to be greater than its ability to suppress gonadotropin secretion when compared to other estrogen types and that its hepatic effects are not entirely eliminated when administered parenterally via the vaginal route (27, 28).

In order to address whether reduction of IGF-I and elevation of GH and GHBP are specific properties of ethinyl estradiol or intrinsic to the oral route of administration, we compared the effects of oral administration of ethinyl estradiol (20 μg), conjugated equine estrogen (premarin: 1.25 mg), and estradiol ester (estradiol valerate: 2 mg) (29). We used standard doses employed for hormone replacement therapy, and the dosage for estrogen type was determined from published data showing that they induced comparable systemic effects as indicated by gonadotropin suppression (27, 30). The 3 estrogen formulations were administered to 6 menopausal women in a randomized crossover design for a period of 4 weeks each, and 24-h studies were undertaken before and at the end of each treatment cycle during the estrogen-only phase, immediately before coadministration of medroxyprogesterone acetate.

The administration of all 3 estrogen formulations resulted in a significant reduction in IGF-I levels compared to baseline and corresponding elevations of mean 24-h GH and GHBP concentrations (Fig. 13.3). The percent increase in mean 24-h GH during treatment was significantly and inversely related to the percent decrease in IGF-I levels. All 3 estrogen formulations resulted in significant suppression of LH and FSH and in elevation of the hepatic proteins SHBG and angiotensinogen. GHBP increased in parallel with these hepatic proteins.

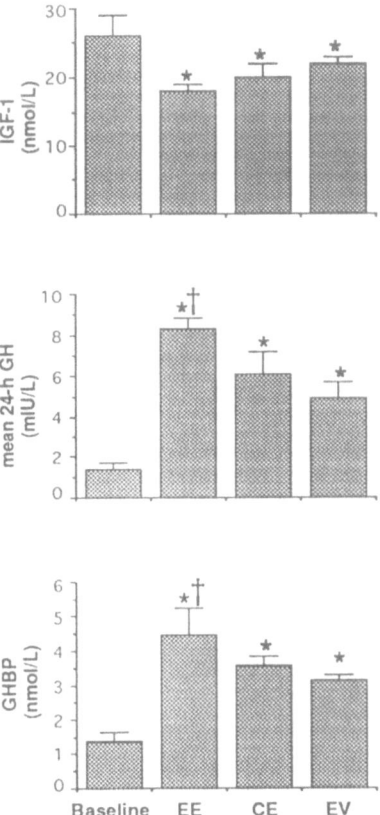

FIGURE 13.3. Mean (±SE) IGF-I (top), mean 24-h GH (middle), and GHBP (bottom) concentrations in 6 postmenopausal women before and during treatment with ethinyl estradiol (EE), conjugated equine estrogen (CE), and estradiol valerate (EV). (*$P < 0.05$ vs. baseline; †$P < 0.05$ vs. EV.) Reprinted with permission from Kelly, Rajkovic, O'Sullivan, Sernia, and Ho (29), © Blackwell Scientific Publications, 1993.

The uniform responses displayed by all 3 estrogen formulations administered by the oral route stand in contrast to those observed following transdermal delivery and strongly suggest that the reduction in IGF-I levels is an intrinsic effect of oral estrogens. The increase in GHBP concentration reflects another level of action of estrogen on the GH-IGF-I axis. GHBP shows immunologic identity to the extracellular domain of the GH-R, and in humans, it is believed to be derived from proteolytic cleavage of this receptor (31). All 3 estrogen formulations increased GHBP in parallel with elevations in SHBG and angiotensinogen, both recognized as estrogen-sensitive hepatic proteins. Together with our observations that transdermal estrogen delivery had no effect on GHBP, the data suggest that GHBP is an estrogen-sensitive hepatic protein and that the liver is a major source of GHBP in humans.

The inverse order of effect between treatment groups in IGF-I suppression and GH elevation provides further support for the notion that the increases in GH concentration may have arisen from reduced feedback inhibition by IGF-I on pituitary GH release. The mechanism of reduced IGF-I synthesis is unclear, but may involve attenuation of GH action by GHBP that was increased by estrogen treatment or a direct effect of estrogen on hepatic IGF-I synthesis. Irrespective of the mechanisms involved, the perturbation of the GH-IGF-I axis suggests that this sex steroid is likely to have significant biological and route-dependent effects on GH- and IGF-I-responsive tissues.

Biological Effects

To address whether estrogen-induced changes in GH and IGF-I levels might produce secondary biological effects, we studied changes in circulatory markers of connective and bone tissue metabolism during oral and transdermal estrogen treatment. Circulating concentrations of propeptides for collagen rise during GH treatment, reflecting stimulation of collagen synthesis (32, 33). Increases in the propeptide for *type I collagen*, the predominant protein matrix of bone, and *type III collagen*, the major structural protein in soft connective tissue, occur during GH treatment, indicating stimulation of bone and nonbone collagen synthesis. It is likely that these anabolic effects of GH are mediated by IGF-I since specific receptors for IGF-I are present in fibroblasts (34) and osteoblasts (35). Thus, increases in circulatory markers of connective tissue and bone tissue metabolism following GH treatment may reflect the biological effects of IGF-I. Because the route of administration determined whether IGF-I levels increased or decreased, we therefore measured propeptide levels of type I and type III collagen from our study comparison of oral and transdermal estrogen (36).

Transdermal estrogen therapy resulted in a significant increase in propeptide levels of type III and type I collagen. These changes stand in contrast to the effects observed following oral ethinyl estradiol treatment, which induced a fall in levels of type III and I procollagen (Fig. 13.4). Analysis of covariance revealed that the two modes of treatment had highly significantly different effects on IGF-I, procollagen III, and procollagen 1 levels, but not on LH and FSH levels, the latter indicating comparable systemic estrogenic effect.

These data show that estrogens have distinct effects on connective and skeletal tissue metabolism that are dependent on the route of administra-

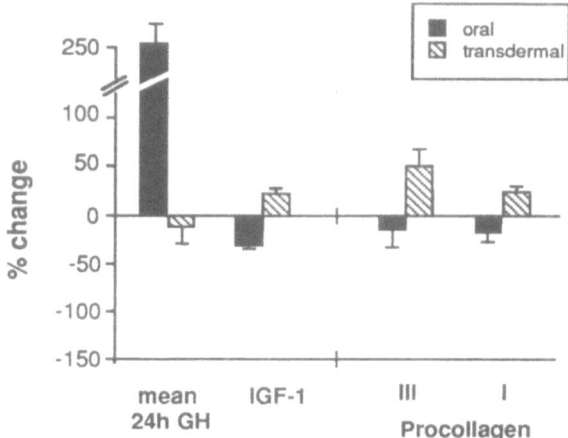

FIGURE 13.4. Changes in mean 24-h GH, IGF-I, procollagen III, and procollagen I levels following oral and transdermal estrogen treatment expressed as percentage change from pretreatment values. Reprinted with permission from Ho and Weissberger (36).

tion, with the changes in biochemical markers of fibroblast and osteoblast function being related to changes in circulating IGF-I. The dissociation of GH and IGF-I induced by oral estrogen and the resultant biological effects that we observed present a useful model to assess the relative contribution of the autocrine-paracrine action of IGF-I compared to the endocrine mode of IGF-I action. The importance of an endocrine role of IGF-I has been challenged by accumulating evidence that GH acts directly on a variety of tissues previously thought to be stimulated in vivo by IGF-I produced from the liver (37, 38). The observations that many extrahepatic tissues, such as fibroblasts, chondrocytes, and osteoblasts (39–41), are able to synthesize IGF-I under GH stimulation in culture suggest that the impact of GH on tissue growth may be as dependent on local IGF-I generation as on the liver-derived endocrine source. In our study, oral estrogen administration led to a 3-fold increase in GH concentrations that was not accompanied by a concomitant stimulation of connective or bone tissue metabolism. This observation strongly suggests that the dominant IGF-I effect on the cells is endocrine rather than paracrine-autocrine.

The finding of peripheral GH resistance in the oral estrogen study is interesting and intriguing. Since GHBP levels were also increased by oral estrogen treatment, one possible explanation is that GHBP may have attenuated GH action, as has been shown in in vitro studies (42). The physiological significance of GHBP—and in particular its impact on modulation of GH action in vivo—is unclear. It has recently been reported that GHBP enhances GH action when coadministered with GH to hypo-physectomized rats (43). It has also been proposed by some investigators that GHBP in blood may be a marker of tissue GH-R status, and if this is true, then tissue responsiveness to GH might be greater when GHBP levels are higher (44–46). The stimulation of GH secretion by oral estrogen therapy may expose the body to some of the direct (non-IGF-I-mediated) actions of GH, such as insulin resistance (47, 48) and sodium retention (49). The impact of concomitant elevation in GHBP is unknown, but likely to affect the actions of GH on body metabolism and composition.

Conclusion

Hypogonadism in the menopause is characterized by reduced GH and IGF-I levels. The effects of estrogen on GH economy in the menopause are complex and dependent on the route of delivery. Oral administration reduces circulating IGF-I and increases GH and GHBP levels, while transdermal administration causes no change in GH or GHBP, but a small rise in IGF-I levels. These diverse effects on each of the components of the somatotrophic axis suggest that estrogens are likely to have quite different metabolic effects that will be determined by the route by which they are administered. There is increasing evidence that circulating IGF-I plays an important endocrine role in the regulation of anabolic processes. The chronic suppression of circulating IGF-I that occurs with the oral route may have long-term negative effects on IGF-I-responsive tissues. The biological sequelae of elevated GH and GHBP levels are not known and are currently under investigation. Nevertheless, the findings of distinct route-dependent effects of estrogen on the somatotrophic axis may have important implications for postmenopausal health, particularly in relation to body metabolism and composition.

Acknowledgments. We gratefully acknowledge and thank Kay Cooper for excellent secretarial assistance. We thank Ciba Geigy Australia for provision of Estraderm. This work was supported in part by the National Health and Medical Research Council of Australia.

References

1. Rose SR, Municchi G, Barnes KM, et al. Spontaneous growth hormone secretion increases during puberty in normal girls and boys. J Clin Endocrinol Metab 1991;73:428–35.
2. Martha PM Jr, Gorman KM, Blizzard RM, Rogol AD, Veldhuis JD. Endogenous growth hormone secretion and clearance rates in normal boys, as determined by deconvolution analysis: relation to age, pubertal status and body mass. J Clin Endocrinol Metab 1992;74:336–44.

3. Martha PM Jr, Rogol AD, Veldhuis JD, Kerrigan JR, Goodman DW, Blizzard RM. Alterations in the pulsatile properties of circulating growth hormone concentration during puberty in boys. J Clin Endocrinol Metab 1989;69:563–70.
4. Wennick JMB, Waal HAD, Schoemaker R, Blaau G, van den Braken C, Schoemaker J. Growth hormone secretion patterns in relation to LH and estradiol secretion throughout normal female puberty. Acta Endocrinol (Copenh) 1990;124:129–35.
5. Faria ACS, Beckenstein LW, Booth RA, et al. Pulsatile growth hormone release in normal women during the menstrual cycle. Clin Endocrinol (Oxf) 1992;36:591–6.
6. Ho KY, Evans WS, Blizzard RM, et al. Effects of sex and age on the 24 hour secretory profile of GH secretion in man: importance of endogenous estradiol concentrations. J Clin Endocrinol Metab 1987;64:51–8.
7. Link K, Blizzard R, Evans W, Kaiser D, Parker M, Rogol A. The effect of androgens on the pulsatile release and the 24-hour mean concentration of growth hormone in peripubertal males. J Clin Endocrinol Metab 1986;62: 159–64.
8. Liu L, Merriam GR, Sherins RJ. Chronic sex steroid exposure increases mean plasma growth hormone concentration and pulse amplitude in men with isolated hypgonadotropic hypogonadism. J Clin Endocrinol Metab 1987;64: 58.
9. Weissberger AJ, Ho KKY. Activation of the somatotropic axis by testosterone in adult males: evidence for the role of aromatization. J Clin Endocrinol Metab 1993;76:1407–12.
10. Ross JL, Meyerson Long L, Loriaux DL, Cutler GB. Growth hormone secretory dynamics in Turner syndrome. J Pediatr 1985;106:202–6.
11. Mauras N, Rogol AD, Veldhuis JD. Specific, time-dependent actions of low-dose ethinyl estradiol administration on the episodic release of growth hormone, follicular-stimulating hormone and luteinizing hormone in pre-pubertal girls with Turner's syndrome. J Clin Endocrinol Metab 1989; 69:1053–8.
12. Ross JL, Cassorla FG, Skerda MC, Valk IM, Loriaux DL, Cutler GB. A preliminary study of the effect of estrogen dose on growth in Turner's syndrome. N Engl J Med 1983;309:1104–6.
13. Copeland KC. Effects of acute high dose and chronic low dose estrogen on plasma somatomedin C and growth in patients with Turner's syndrome. J Clin Endocrinol Metab 1988;66:1278–82.
14. Wiedemann E, Schwartz E, Frantz A. Acute and chronic estrogen effects upon serum somatomedin activity, growth hormone, and prolactin in man. J Clin Endocrinol Metab 1976;42:942–52.
15. Clemmons DR, Underwood LE, Ridgway EG, Kliman B, Kjellberg RN, Van Wyk JJ. Estradiol treatment of acromegaly: reduction of immunoreactive somatomedin C and improvement of metabolic status. Am J Med 1980;69: 571–5.
16. Weissberger AJ, Ho KY, Lazarus L. Contrasting effects of oral and trans-dermal routes of estrogen replacement therapy on 24-hour growth hormone (GH) secretion, insulin-like growth factor 1 and GH binding protein in postmenopausal women. J Clin Endocrinol Metab 1991;72:374–81.

17. Herington AC, Ymer S, Stevenson J. Identification and characterization of specific binding proteins for growth hormone in normal human sera. J Clin Invest 1986;77:1817–23.
18. Baumann G, Stolar MW, Amburn K, Barsano CP, DeVries BC. A specific growth hormone-binding protein in human plasma: initial characterization. J Clin Endocrinol Metab 1986;62:134–41.
19. D'Ercole AJ, Stiles AD, Underwood LE. Tissue concentrations of soma-tomedin C: further evidence for multiple sites of synthesis and paracrine or autocrine mechanisms of action. Proc Natl Acad Sci USA 1984;81:935–9.
20. de Lignieres B, Basdevant A, Thomas G. Biological effects of 17β estradiol in postmenopausal women: oral versus percutaneous adminstration. J Clin Endocrinol Metab 1986;62:536–41.
21. Chetkowski R, Meldrum D, Steingold K. Biological effects of transdermal estradiol. N Engl J Med 1986;314:1615–20.
22. Ho KY, Valiontis E, Waters MJ, Rajkovic IA. Regulation of growth hormone binding protein in man: comparison of gel chromatography and immuno-precipitation methods. J Clin Endocrinol Metab 1993;76:302–8.
23. Dawson-Hughes B, Stern D, Goldman J, Reichlin S. Regulation of growth hormone and somatomedin-C secretion in postmenopausal women: effect of physiological estrogen replacement. J Clin Endocrinol Metab 1986;63:424–32.
24. Duursma S, Bijlsma J, Van Paassen H, van Buul-Offers S, Skottner-Lundin A. Changes in serum somatomedin and growth hormone concentrations after 3 weeks oestrogen substitution in post-menopausal women; a pilot study. Acta Endocrinol (Copenh) 1984;106:527–31.
25. Murphy LJ, Freisen HG. Differential effects of estrogen and growth hormone on uterine and hepatic insulin-like growth factor-1 expression in the ovariec-tomized hypophysectomized rat. Endocrinology 1988;122:325–32.
26. Bellantoni MF, Harman M, Cho DE, Blackman MR. Effects of progestin-opposed transdermal estrogen administration on growth hormone and insulin-like growth factor-1 in postmenopausal women. J Clin Endocrinol Metab 1991;72:172–8.
27. Maschak CA, Lobo R, Dozono TR, et al. Comparison of pharmacokinetic properties of various estrogen formulations. Am J Obstet Gynecol 1982;144:511–8.
28. Goebelsmann U, Maschak CA, Mishell DR. Comparison of hepatic impact of oral and vaginal administration of ethinyl estradiol. Am J Obstet Gynecol 1985;151:868–77.
29. Kelly JJ, Rajkovic IA, O'Sullivan AJ, Sernia C, Ho KKY. Effects of different oestrogen formulations on insulin-like growth factor-1, growth hormone and growth hormone binding protein in post-menopausal women. Clin Endocrinol (Oxf) 1993.
30. Mandel FP, Geola FL, Lu J, et al. Biological effects of various doses of ethinyl estradiol. Am J Obstet Gynecol 1982;59:673–9.
31. Leung DW, Spencer SA, Cachaines G, et al. Growth hormone receptor and serum binding protein: purification, cloning and expression. Nature 1987;330:537–43.
32. Carey D, Goldberg B, Ratzan S, Rubin K, Rowe D. Radioimmunoassay for type 1 procollagen in growth hormone-deficient children before and during treatment with growth hormone. Pediatr Res 1985;19:8–11.

33. Tapanainen P, Ristelli L, Knip M, Kaar M, Ristelli J. Serum aminoterminal propeptide of type III procollagen: a potential predictor of the response to growth hormone therapy. J Clin Endocrinol Metab 1988;67:1244–9.
34. Rechler MM, Nissley SP, Podskalny JM, Moses AC, Fryklund L. Identification of a receptor for somatomedin-like polypeptides in human fibroblasts. J Clin Endocrinol Metab 1977;44:820–7.
35. Bennett A, Chen T, Feldman D, Hintz R, Rosenfeld R. Characterization of insulin-like growth factor 1 receptors on cultured rat bone cells: regulation of receptor concentration by glucocorticoids. Endocrinology 1984;115:1577–83.
36. Ho KKY, Weissberger AJ. Impact of short-term estrogen administration on growth hormone secretion and action: distinct route-dependent effects on connective and bone tissue metabolism. J Bone Miner Res 1992;7:821–7.
37. Daughaday W, Rotwein P. Insulin-like growth factors I and II: peptide, messenger ribonucleic acid and gene structures, serum, and tissue concentrations. Endocr Rev 1989;10:68–91.
38. Isaksson O, Eden S, Jansson J-O. Mode of action of pituitary growth hormone on target cells. Annu Rev Physiol 1985;47:483–99.
39. Cook JJ, Haynes KM, Werther GA. Mitogenic effects of growth hormone in cultured human fibroblasts. J Clin Invest 1988;81:206–12.
40. Barnard R, Ng K, Martin T, Waters M. Growth hormone (GH) receptors in clonal osteoblast-like cells mediate a mitogenic response. Endocrinology 1991;128:1459–64.
41. Lindahl A, Isgaard J, Nilsson A, Isaksson O. Growth hormone potentiates colony formation of epiphyseal chondrocytes in suspension culture. Endocrinology 1986;5:1843–8.
42. Lim L, Spencer SA, McKay P, Waters MJ. Regulation of growth hormone (GH) bioactivity by a recombinant human GH-binding protein. Endocrinology 1990;127:1287–91.
43. Clark RG, Cunningham B, Moore JA, et al. Growth hormone binding protein enhances growth promoting activity of GH in the rat [Abstract]. Prog 73rd meet Endocr Soc, Washington, DC, 1991:433.
44. Baumann G, Shaw MA, Merimee TJ. Low levels of high-affinity growth hormone binding protein in African pigmies. N Engl J Med 1989;320:1705–9.
45. Baruch Y, Amit TPH, Enat R, Youdim MBH, Hochberg Z. Decreased serum growth hormone-binding protein in patients with liver cirrhosis. J Clin Endocrinol Metab 1991;73:777–80.
46. Daughaday WH, Trivedi B, Andrews BA. The ontogeny of serum GH binding protein in man: a possible indicator of GH receptor development. J Clin Endocrinol Metab 1987;65:1072–4.
47. Bratusch-Marrain P, Smith D, DeFronzo R. The effect of growth hormone on glucose metabolism and insulin secretion in man. J Clin Endocrinol Metab 1982;55:973–82.
48. Davidson M. Effect of growth hormone on carbohydrate and lipid metabolism. Endocr Rev 1987;8:115–31.
49. Ho KY, Weissberger AJ. The antinatriuretic action of biosynthetic human growth hormone in man involves activation of the renin-angiotensin system. Metabolism 1990;39:133–7.

Part III

Growth Hormone:
The Ovarian Connection

14

Ovary as a Site of Growth Hormone Reception and Action

Eli Y. Adashi

Although specific [^{125}I] human *growth hormone* (GH) binding to the 15,000 × g fraction of homogenized luteinized rat ovaries has been described (1), it is recognized that the *human GH* (hGH) ligand is capable of binding to both lactogenic and somatogenic murine receptor sites (2). Thus, although these findings are compatible with the presence of somatogenic receptor sites in the luteinized rat ovary, no firm conclusions can be drawn in this regard. The above notwithstanding, the rat ovary has been shown to be a site of GH action as assessed by its response to GH receptor-selective ligands of ovine origin. These findings suggest, by inference, the existence of specific ovarian *GH receptor* (GH-R) sites.

These conclusions have recently been strengthened immunohisto-chemically by Lobie et al. (3, 4). Specifically, use was made of immuno-histochemistry to localize GH-R/*binding protein* (BP) in the adult female reproductive tract. Ovaries from neonatal animals were also examined to determine if GH-R/BP expression is developmentally regulated. Localization of the GH-R/BP was noted in both the nucleus and cytoplasm of positively staining cells, thereby reaffirming previous reports of a nuclear GH-R. Intense GH-R/BP immunoreactivity was noted in the germinal epithelium. Strong immunoreactivity was also exhibited by scattered oocytes and lutein cells of the corpus luteum. Moderate immunoreactivity was evident in scattered oocytes, granulosa cells, theca interna, and theca externa. Ovarian granulosa cells from 10-day postnatal (unlike adult) rats displayed strong immunoreactivity.

Taken together, these findings suggest widespread distribution of the GH-R/BP in the rat ovary, suggesting functions above and beyond those previously described. Further confirmation of these conclusions was recently provided by Tiong and Herington, whose observations revealed transcripts corresponding to both the GH-R (4.5 kb) and the *GH binding protein* (GHBP) (1.2 kb) in the adult rat ovary (5).

First to demonstrate the apparent dependence of rat ovarian function on GH were Advis et al., whose observations revealed that the in vitro ovarian *progesterone* (P) response to both hCG and hFSH was distinctly increased by prior in vivo GH treatment (6). This effect of GH was not reproduced either by the in vivo administration of LH (at a dose calculated by *radioimmunoassay* [RIA] to be contaminating the GH preparation) or by FSH (at a dose that induced a marked increase in aromatase activity in the same ovaries). GH treatment of hypophysectomized rats failed to affect either aromatase activity or hCG-induced *estradiol* (E_2) release, indicating that GH does not directly facilitate the production of estradiol by the ovary.

To investigate possible direct effects of GH on the differentiation of ovarian cells, granulosa cells from immature, hypophysectomized, estrogen-primed rats were cultured for 72 h in the presence of FSH, with or without *ovine GH* (oGH) (7). Interestingly, concomitant treatment with oGH increased the FSH-inducible granulosa cell LH/hCG receptor binding capacity (but not affinity) by enhancing the action of low doses of FSH (FSH sensitivity) without substantial increases in the maximal response (FSH efficacy). Importantly, this synergistic interaction between FSH and GH could not be accounted for by an increase in the granulosa cell number in that the plated viable cell mass remained unaltered by any and all hormonal treatment groups. Expectedly, the LH/hCG receptors so induced were functional in nature, as indicated by their ability to mediate LH/hCG-stimulated cAMP generation. In addition to its effects on the LH/hCG receptor content, GH also augmented the FSH-stimulated accumulation of P and 20α-dihydroprogesterone in a dose-dependent manner. In contrast, GH treatment did not significantly affect FSH-stimulated estrogen production. The reason(s) underlying this dichotomy remains uncertain, and subsequent studies (8) have failed to document such selectivity of action.

Given the possibility that GH action may be associated with perturbation of intracellular cAMP economy, consideration was given to the possibility that GH-enhanced FSH hormonal action may involve an increase in stimulatable adenylate cyclase activity (7). Indeed, the ability of GH to augment FSH hormonal action was associated with an enhancement of FSH-stimulated cAMP generation. To examine the possible effect of GH on stimulatable adenylate cyclase activity independent of *FSH receptor* (FSH-R) involvement, use was also made of forskolin, a potent stimulant of the catalytic subunit of adenylate cyclase. Expectedly, exposure to forskolin produced significant increments in granulosa cell LH/hCG receptor binding capacity, an effect further augmented by the addition of GH. In addition, the concurrent application of GH substantially augmented cAMP analog (8-bromo-cAMP)-promoted LH/hCG receptor formation and P-production. Taken together, these observations suggest that the ability of GH to amplify FSH hormonal action may involve a site(s) of action both proximal and distal to cAMP generation.

In related studies, Hutchinson et al. reported on the ability of oGH to promote *pregnant mare's serum gonadotropin* (PMSG)-supported estrogen and P-biosynthesis by cultured rat granulosa cells (8). Importantly, oGH proved capable of enhancing estrogen and P-biosynthesis to a point beyond that associated with maximally stimulating doses of PMSG (an FSH-like principle). Interestingly, use was made of GH concentrations as high as 200 ng/mL, a concentration documented in the serum of early prepubertal female rats. These findings suggest that the in vitro findings reported may be of physiological relevance under corresponding in vivo circumstances. These studies contrast with those of Jia et al. (7), who failed to observe an effect of GH on granulosa cell aromatase activity. Although the reasons underlying the apparent discrepancies between the studies in question remain uncertain, they are likely to be due to subtle (albeit uncertain) differences in the experimental conditions involved.

Given ovaries from immature hypophysectomized rats, Usuki et al. (9) and Usuki and Shioda (10) were able to show that the in vitro provision of oGH results in a significant increase in *deoxyribonucleic acid* (DNA) polymerase α, but not β, activity. It has therefore been suggested that oGH is capable of inducing DNA synthesis in immature rat ovaries. Comparable results were observed under in vivo circumstances, wherein immature hypophysectomized rats were subjected to systemic treatment with oGH.

Studied at the level of the immature porcine granulosa cell, oGH proved to be a relatively ineffective stimulator of P-secretion (11). Indeed, treatment with oGH by itself increased the accumulation of P only 2.6-fold relative to controls as compared with a 7.4-fold increase observed following treatment with E_2 plus FSH. However, combined treatment with all 3 agonists produced a synergistic interaction yielding P-values 33 times those observed in the untreated state. These findings suggest that GH has a direct stimulatory action at the level of the porcine granulosa cell. However, compared to E_2 and FSH, established stimulators of porcine granulosa cells, GH appears less effective as a promoter of steroidogenesis. Nevertheless, GH dramatically enhances the effects of E_2 and FSH as assessed at the level of P-secretion.

Studied at the level of the human ovary, hGH (100–500 ng/mL) was without effect on P-accumulation by cultured highly luteinized (*in vitro fertilization* [IVF] cycle-derived) human granulosa cells (11). Likewise, concentrations of hGH of up to 300 ng/mL proved incapable of altering human granulosa cell IGF-I, IGF-II, or cholesterol side-chain cleavage gene expression (12). In contrast, IGF-II release was potently stimulated in a dose-dependent fashion (13). More recent observations by Mason et al. indicate that hGH may well exert a direct stimulatory effect on FSH-supported E_2 biosynthesis by the human ovary (14). Moreover, hGH also appeared to have a direct gonadotropic action on its own, even in the absence of added FSH (14).

Yet more recent observations by Lanzone et al. revealed that 1 mg/mL of hGH stimulated basal P-biosynthesis by cultured highly luteinized (IVF cycle-derived) human granulosa cells (15). More importantly, concomitant treatment with generally noneffective doses of hCG (6 and 12 ng/mL) and hGH (250 and 500 ng/mL) enhanced P-production to levels comparable to those seen with the highest doses of hGH ($\geq$1 mg/mL) or hCG (25–50 ng/mL) alone. Although the reasons underlying the apparent differences between the above-mentioned studies remain unknown, it is likely that the nature of the detection techniques and differences in the respective culture systems are at play. Consideration must also be given to the heterogenous nature of the granulosa cells employed, the origins of which can be traced to different follicles of divergent sizes.

Preliminary studies from this laboratory documented the highly luteinized human granulosa cell as a site of GH reception. Moreover, treatment of cultured human granulosa cells with forskolin, an established activator of adenylate cyclase, resulted in substantial up-regulation of granulosa cell GH-Rs, as assessed by the steady state levels of the corresponding transcript. Given that FSH hormonal action is mediated via cAMP, it is tempting to speculate that the ability of forskolin to up-regulate granulosa cell GH-Rs may have a bearing on the ability of FSH to effect a similar action. Accordingly, it is tempting to speculate that rising FSH levels during the proliferative phase may, in fact, result in progressive increments in granulosa cell GH-R levels. In contrast, no GH-Rs were noted at the level of the theca-interstitial cell when assessed under comparable in vitro circumstances.

More recently, additional efforts have been reported relevant to the localization of GH-Rs in human granulosa cells. Specifically, use was made of granulosa cells isolated either from natural cycles or from gonadotropin-stimulated cycles. Total RNA was hybridized with a [32P]-labeled rat GH-R cRNA probe. Importantly, note was made of a single major transcript with an estimated size of 4.5 kb, along with a minor transcript estimated at 1.3 kb. Conventional radio ligand receptor assays proved confirmatory. It is also of note that treatment with GH augmented basal as well as FSH-supported steroidogenesis in granulosa cells obtained from patients with natural cycle. However, the response to GH stimulation displayed considerable variation. All told, these data were interpreted to suggest the existence of functional GH-Rs in human granulosa cells (16).

Complementing the preceding in vitro observations is a limited body of information relevant to the gonadotropic property of GH in the human under in vivo circumstances. In this connection, Jorgensen et al. (17) examined the impact of systemic treatment with hGH on the reproduction of female rats. Specifically, rats were treated with hGH 0.3, 1.0, and

3.3 IU/kg daily for 2 weeks before and throughout mating (17). Treatment with 1.0 and 3.3 IU/kg of hGH produced a significant prolongation of the estrous cycle. As a consequence, the number of mating days was more than doubled as compared to placebo-treated controls. The number of implantation sites and corpora lutea was significantly higher in hGH-treated rats relevant to placebo-treated controls. The prolongation of the estrous cycle length was to an average of 10.1 days, and the plasma P-levels tended to be higher in those rats in which cyclic patterns had been most deranged. Although of interest, the present observations may reflect a lactogenic (rather than somatogenic) effect of hGH at the level of the rodent ovary (2).

Note must also be made of the inability of Kulin and associates to demonstrate a gonadotropic effect of hGH, as assessed by its in vivo impact on the steroidogenic response of the prepubertal human ovary to the acute administration of hCG (18). Specifically, short-term hCG stimulation tests were performed before, during, and after 1 year of GH therapy in 7 prepubertal GH-deficient girls. These findings confirm the functional capability of the prepubertal ovary when exposed acutely to LH/hCG, but argue against a role for GH in the stimulation of gonadal steroid production in the prepubertal female.

The expression of hGH in female transgenic mice is accompanied by sterility, whereas females expressing the *bovine GH* (bGH) gene are fertile. To obtain further insight into these experimental phenomena, detailed morphologic evaluation of the corresponding ovaries was undertaken (19). These observations revealed that the ovary, although not enlarged in either case, was affected by chronic exposure to the heterologous GH. Bovine GH, which in the mouse exhibits isolated somatotrophic activity, reduced the morphological signs of atresia. In contrast, hGH, which in the mouse has additional lactotrophic activity, produced complex alterations, including the acceleration of follicular development, increased atresia, and massive degeneration of interstitial cells. These findings were taken to mean that the expression of the hGH transgene leads to accelerated aging of the mouse ovary and that this effect is likely to be due to the combination of somatotrophic and lactotrophic activity of hGH in this species.

More recently, Gong and associates have undertaken to investigate the possible effect of *recombinant bGH* (rbGH) on ovarian folliculogenesis and ovulation rate (20). Heifers received daily 25-mg injections of either rbGH or vehicle for a period of 2 estrous cycles until slaughter. Importantly, only single ovulation was documented in all heifers. Treated animals displayed significantly more antral follicles as compared with controls. No effect of treatment was noted on FSH or LH binding to granulosa and theca cells or on LH binding to corpora lutea. As expected, treated heifers displayed significantly higher circulating levels of GH and IGF-I throughout the treatment period. However, no significant differences could be detected in the circulating concentrations of E_2 or P. Likewise, no significant differences could be detected regarding gonadotropin economy. All told, these observations indicated that treatment with GH may increase the population of antral follicles 2–5 mm in diameter in mature heifers. This effect did not appear to be mediated through changes in circulating gonadotropin concentrations or gonadotropin receptor levels. Although it is possible that GH may act via increased peripheral IGF-I concentrations, a direct effect of GH at the ovarian level could not be excluded.

Following with the preceding observations, Tapanainen and associates have examined the effects of treatment with GH in combination with an ultrashort-term GnRH agonist/hMG/hCG stimulation regimen (21). The prospective, randomized, placebo-controlled study involved 54 normally cycling women enrolled in an IVF program. Recombinant hGH (24 IU) or placebo were given intramuscularly on alternate days starting on cycle day 4 until the day of the last hMG injection. Although no dramatic differences were noted in the circulating levels of key reproductive hormones, note was made that the levels of 3β-hydroxysteroid dehydrogenase and aromatase mRNA were significantly higher in granulosa cells isolated from patients who received GH treatments than in patients receiving placebo. These observations could be taken to suggest that GH administration modifies ovarian steroidogenic response to gonadotropins in IVF patients, suggesting a role for GH in the regulation of human ovarian function.

Taken together, the accumulating information strongly suggests that the ovary is in fact a site of GH reception and action. However, additional formal demonstrations of ovarian GH-R gene expression and cellular localization would appear desirable. Although the exact role of GH in ovarian physiology remains to be determined, it may well act as a cogonadotropin—that is, as an amplifier of gonadotropin hormonal action.

References

1. Davies TF, Katikineni M, Chan V, Harwood JP, Dufau ML, Catt KJ. Lactogenic receptor regulation in hormone-stimulated steroidogenic cells. Nature 1980;283:863–5.
2. Ranke MB, Stanley CA, Tenore A, Rodbard D, Bongiovanni AM, Parks JS. Characterization of somatogenic and lactogenic binding sites in isolated rat hepatocytes. Endocrinology 1976;99:1033–45.
3. Usuki S, Shioda M. Growth hormone elevates DNA polymerase-α activity related to DNA synthesis in ovaries of hypophysectomized immature rat. Horm Metab Res 1989;21:455–6.
4. Lobie PE, Garcia-Aragon J, Bosco SW, Baumbach WR, Waters MJ. Cellular localization of the growth hormone binding protein in the rat. Endocrinology 1992;130(5):3057–65.

5. Tiong TS, Herington AC. Tissue distribution, characterization, and regulation of messenger ribonucleic acid for growth hormone receptor and serum binding protein in the rat. Endocrinology 1991;129:1628–34.
6. Advis JP, White SS, Ojeda SR. Activation of growth hormone short loop negative feedback delays puberty in the female rat. Endocrinology 1981;108:1343–52.
7. Jia XC, Kalmijn J, Hsueh AJW. Growth hormone enhances follice-stimulating hormone-induced differentiation of cultured rat granulosa cells. Endocrinology 1986;118:1401–9.
8. Hutchinson LA, Findlay JK, Herington AC. Growth hormone and insulin-like growth factor-I accelerate PMSG-induced differentiation of granulosa cells. Mol Cell Endocrinol 1988;55:61–9.
9. Usuki S, Kubota S, Miyakawa S. Growth hormone elevates deoxyribonucleic acid polymerase α activity in conjunction with deoxyribonucleic acid synthesis in immature rat ovaries. Biomed Res 1989;10:267–73.
10. Usuki S, Shioda M. Growth hormone elevates DNA polymerase-α activity related to DNA synthesis in ovaries of hypophysectomized immature rat. Horm Metab Res 1989;21:455–6.
11. Hsu CJ, Hammond JM. Concomitant effects of growth hormone on secretion of insulin-like growth factor I and progesterone by cultured porcine granulosa cells. Endocrinology 1987;121:1343–8.
12. Voutilainen R, Miller WL. Coordinate tropic hormone regulation of mRNAs for insulin-like growth factor II and the cholesterol side-chain-cleavage enzyme, P450scc, in human steroidogenic tissues. Proc Natl Acad Sci USA 1987;84:1590–4.
13. Ramasharma K, Li CH. Human pituitary and placental hormones control human insulin-like growth factor II secretion in human granulosa cells. Proc Natl Acad Sci USA 1987;84:2643–7.
14. Mason HD, Martikainen H, Beard RW, Anyaoku V, Franks S. Direct gonadotrophic effect of growth hormone on oestradiol production by human granulosa cells. J Endocrinol 1990;126:R1–4.
15. Lanzone A, Di Simone N, Castellani R, Fulghesu AM, Caruso A, Mancuso S. Human growth hormone enhances progesterone production by human luteal cells in vitro: evidence of a synergistic effect with human chorionic gonadotropin. Fertil Steril 1990;57:92–6.
16. Carlsson B, Bergh C, Bentham J, et al. Expression of functional growth hormone receptors in human granulosa cells. Hum Reprod 1991;7:1205–9.
17. Jorgensen KD, Svendsen O, Agergaard N, Skydsgaard K. Effect of human growth hormone on the reproduction of female rats. Pharmacol Toxicol 1991;68:14–20.
18. Kulin HE, Samojlik E, Demers LM, Santner SJ. Response of the prepubertal ovary to acute chorionic gonadotropin administration: absence of modulation by growth hormone. J Clin Endocrinol Metab 1985;60:208–11.
19. Mayerhoffer A, Weis J, Bartke A, Yun JS, Wagner TE. Effects of transgenes for human and bovine growth hormones on age-related changes in ovarian morphology in mice. Anat Rec 1990;227:175–86.
20. Gong JG, Bramley T, Webb R. The effect of recombinant bovine somato-tropin on ovarian function in heifers: follicular populations and peripheral hormones. Biol Reprod 1991;45:941–9.

21. Tapanainen J, Martikainen H, Voutilainen R, Orava M, Ruokonen A, Ronnberg L. Effect of growth hormone administration on human ovarian function and steroidogenic gene expression in granulosa-luteal cells. Fertil Steril 1991;58:726–32.

15

Ovarian IGF System: Interface with the Gonadotropic and Somatotrophic Axes

James M. Hammond, Susan Samaras, Randall Grimes, Daniel Hagen, and David Guthrie

This brief chapter outlines the several components of the ovarian *insulin-like growth factor* (IGF) system and their physiological regulation. Collectively, these data support a critical role for this system in the development and selection of ovarian follicles. Although the role of this system in luteal function is less secure, current evidence also supports a potential role in that ovarian tissue. These data involve descriptive studies in vivo or in vitro that fall short of compelling proof of an obligatory role for locally secreted IGFs in ovarian function in vivo. However, IGF-I gene disruption experiments have now been conducted in transgenic mice (1); further analysis of effects on ovarian function should provide critical missing information in this regard.

Although less direct, the experiments reviewed here have revealed much about the operation of the ovarian IGF system and its interface with the hormonal signals of the reproductive cycle (transmitted by variations in gonadotropin secretion), as well as the impact of *growth hormone* (GH) on this system. Our findings suggest that the impact of gonadotropins and GH on this system differ in important ways. We hypothesize that these differences account for some of the distinct differences in the effects of gonadotropins and GH on other, more classical indices of ovarian function in large animals; for example, steroidogenesis and ovulation rate. This chapter emphasizes the data collected in our own system (the porcine ovary) and our own studies in particular. Other authors in this volume (see Findlay et al. below) take a broader species perspective in dealing with the action of GH on the ovary. In addition, these issues have been reviewed in more comprehensive fashion by other recent authors (2). Finally, in each of the subsections below, we have summarized areas in which our data may differ significantly from those in other species.

Ovarian Secretion of the IGFs

The ability of porcine ovarian cells to secrete IGFs has been shown by studies measuring such factors in ovarian *follicular fluid* (FF), ovarian cell-conditioned media, and by analysis of the expression of mRNA for such peptides. IGFs were easily measured in porcine FF (3). IGF-I and -II were both present (3, 4); the concentration of IGF-I in FF increased with follicle development (3), but that of IGF-II did not (4).

Histochemically, there is heavy expression of mRNA for IGF-I in medium-sized follicles, as well as some expression in corpora lutea (5); in these follicles the mRNA was heavily enriched in granulosa versus theca (Fig. 15.1). It remains unclear whether the modest activity encountered in porcine theca derived from the degree of granulosa contamination of these preparations or from expression in theca. In dissected ovarian follicles mRNA for IGF-I was preferentially expressed in growing follicles and increased during the follicular phase until the LH surge, with a strong correlation with intrafollicular *estradiol* (E_2) (6) (Fig. 15.2). In contrast, mRNA levels for IGF-II did not vary strikingly or correlate with these physiological parameters (6). A role for FSH in these changes was shown by increases in ovarian IGF-I levels in *pregnant mare serum gonadotropin*

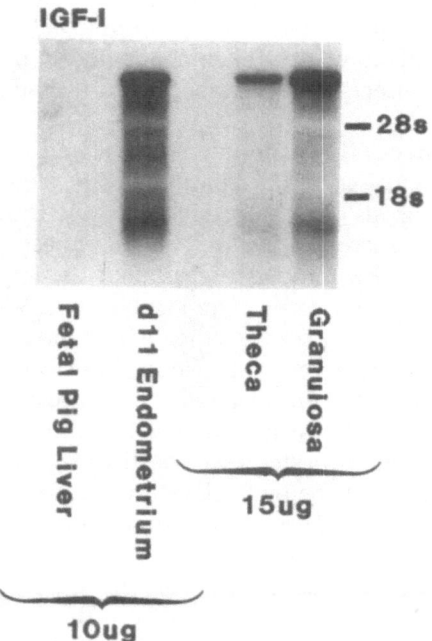

FIGURE 15.1. Northern blot of granulosa and theca expression of IGF-I mRNA.

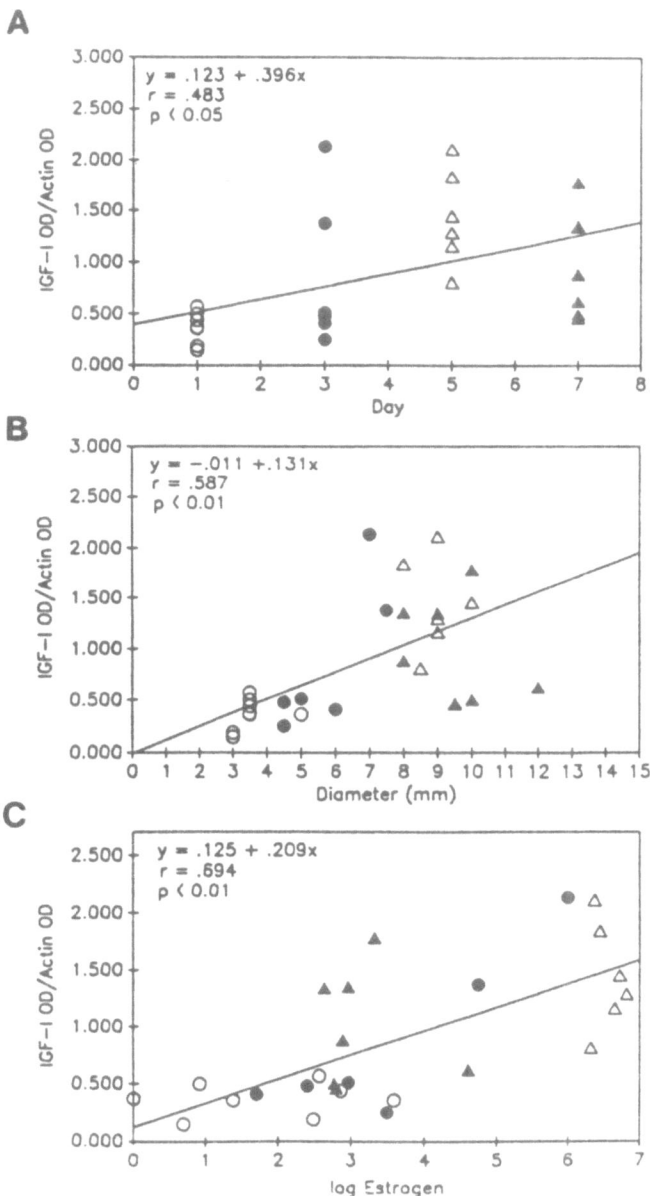

FIGURE 15.2. Expression of IGF-I mRNA in individual growing porcine follicles; correlation with day of follicular phase, size, and E_2 levels. Reprinted with permission from Samaras, Guthrie, Barber, and Hammond (6), © The Endocrine Society, 1993.

(PMSG)-treated animals (4, 7). With cultured porcine granulosa cells, IGF secretion could be maintained for 8–10 days in culture and was responsive to pituitary gonadotropins, cAMP, and E_2 (8), as well as *epidermal growth factor* (EGF) (9).

These data point to meaningful physiological control of the expression of IGF-I in the porcine ovary and are consistent with the notion that this expression is involved in the process of follicular growth, differentiation, and selection. Further, these findings emphasize IGF-I as the most important intrafollicular ligand, as it appears to be in the rat, whereas data in the human suggest that IGF-II is the granulosa cell-secreted IGF and the most responsive to hormonal and physiological stimuli (10, 11). Data derived from in situ hybridization in the rat (12) also suggest that IGF-I is the dominant peptide and that it is physiologically regulated during follicular development in vivo. However, experiments to demonstrate that IGF-I expression is gonadotropin dependent in rodent have been unsuccessful to date (13).

Expression of IGF Binding Proteins

Our early studies of IGF levels in FF and conditioned media (3, 14) indicated high-molecular weight components that bound IGF tracer in addition to the IGFs themselves. However, neither we nor other workers in the field anticipated the explosion of information that has occurred regarding the nature and complexity of this family of abundant ovarian proteins. Shown in Figure 15.3 is a schematic representation of a ligand blot in which these proteins have been resolved by electrophoresis and recognized by their ability to bind IGF-II tracer.

Based on these studies, as well as immunoprecipitation, immunoblotting, and deglycosylation, the principal *IGF binding proteins* (IGFBPs) in porcine FF and granulosa-conditioned media have been identified as IGFBP-2 through -5. In human FF the overall binding protein profile is similar (2). However, IGFBP-1, not detectable in the pig, is a minor but significantly regulated component (7). To our knowledge, levels in FF have not been systematically studied in rodent species.

In Figure 15.4 major changes in FF IGFBP profiles during follicular enlargement and development are shown in the densitometric tracing of ligand blots for large and small follicles. These studies and saturation analyses (15) indicated that small follicles have about twice the IGFBP activity of large follicles and that the difference is accounted for principally by differences in the low-molecular weight binding proteins IGFBP-2 and -4, which predominate in immature follicles. IGFBP-6 is also present in porcine FF (8) and expressed in the porcine ovary (Hammond et al., unpublished data), but the levels encountered have been too low to assess physiological regulation.

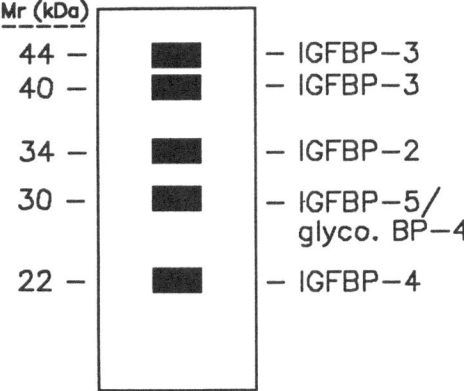

FIGURE 15.3. Schematic diagram of IGFBP activity as visualized by the ligand blot technique in porcine FF or granulosa cell-conditioned media (see reference 15).

We have also examined the IGFBP mRNA expression in individual developing porcine follicles in comparison to the expression of IGFs and steroid contents in the same follicles (16, 19). IGFBP-2, -4, and -5 were highly expressed by these follicles. In contrast, IGFBP-3 mRNA was considerably lower, with dominant expression in the corpus luteum. The expression of IGFBP-2 mRNA dropped quite substantially, with follicular growth and differentiation correlating inversely with follicle size, day of the follicular phase, and intrafollicular steroid levels (Fig. 15.5).

The mRNAs for IGFBP-4 and -5 were also abundant in these follicles (19). In the initial portion of the follicular phase, there was a substantial decline in the expression of these mRNAs; however, in the preovulatory follicle after the LH surge, there was increased expression in many follicles (19). These data are interesting because they differ from the levels of the IGFBP-3, -4, and -5 peptides as measured in our studies (15) and those of Howard et al. (20). These studies indicate high levels of IGFBP-3 and a substantial decline in all low-molecular weight IGFBPs during the preovulatory period. These data, along with additional information (discussed below), indicate the potential for posttranscriptional regulation of the levels of these proteins.

Cultured porcine granulosa cells can also secrete each of the IGFBPs found in porcine FF, although the profile of the IGFBPs found in such samples differs depending on the ovarian origin of the granulosa cell, the duration and conditions of culture, and treatment with hormones and growth factors. A schematic representation of data from multiple studies is illustrated in Figure 15.6, in which the effects of hormones and growth

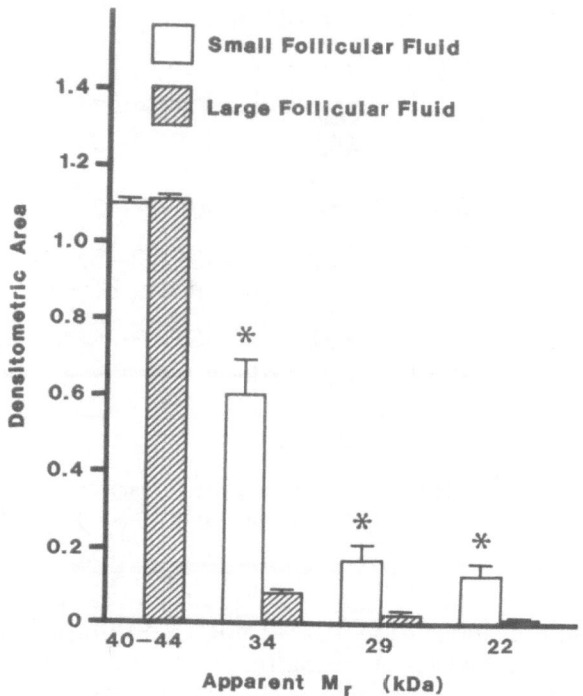

FIGURE 15.4. Densitometric tracing of ligand blot reflecting IGF binding activity of porcine FF; 4 samples pooled from large (preovulatory) and small (immature) follicles were examined. The bands indicated in order of decreasing molecular size are IGFBP-3, -2, -5, and -4 (see Fig. 15.2). Reprinted with permission from Mondschein, Etherton, and Hammond (15).

factors on levels of the several IGFBPs in culture medium are illustrated. Major stimulatory effects have been encountered with IGF-I, and inhibitory effects have been encountered with gonadotropins and other cAMP-dependent agonists.

Our data regarding the regulation of IGFBP levels in vitro are quite similar to those encountered in comparably detailed studies with cultured rat granulosa cells (24, 25). However, the IGFBP profile is apparently different. Porcine granulosa cells secrete abundant IGFBP-3, which we ascribe to a process of partial luteinization, as well as IGFBP-2 (16). In contrast, cultured rat granulosa cells secrete predominantly the low-molecular weight binding proteins IGFBP-4 and -5 (26). In porcine granulosa cells, changes in IGFBP levels in culture correlate, for the most part, with changes in steady state levels of mRNA (16, 21–23). However, posttranslational processing of the IGFBPs can be readily demonstrated

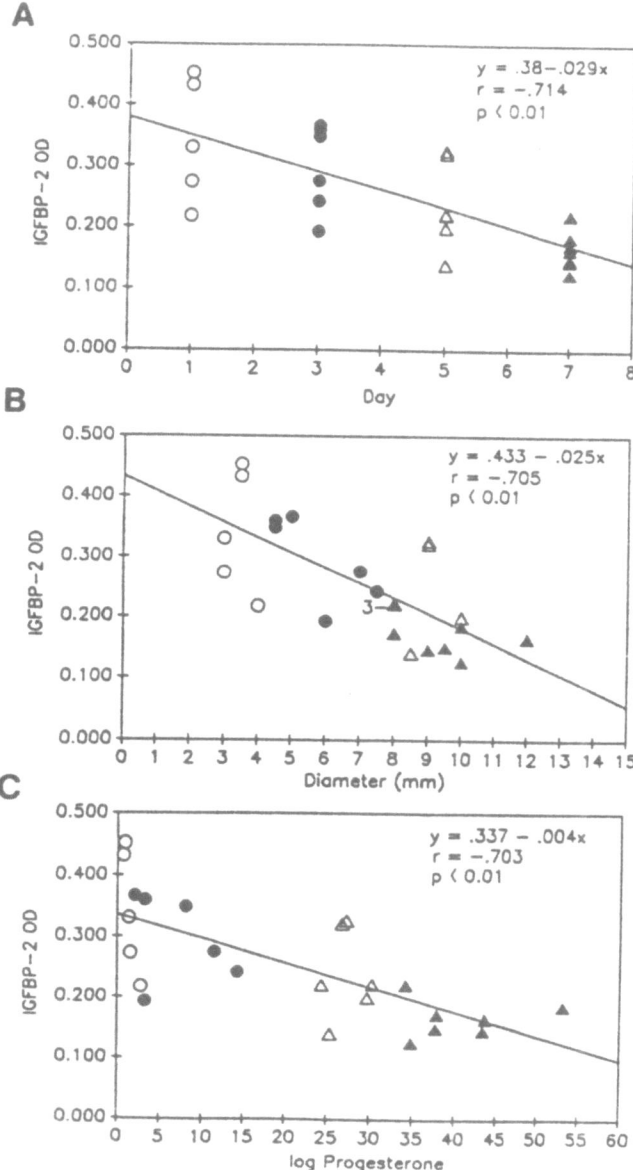

FIGURE 15.5. Expression of IGFBP-2 mRNA in individual follicles from follicular phase swine; correlation with day of the follicular phase, follicle diameter, and progesterone levels in FF. Reprinted with permission from Samaras, Guthrie, Barber, and Hammond (6), © The Endocrine Society, 1993.

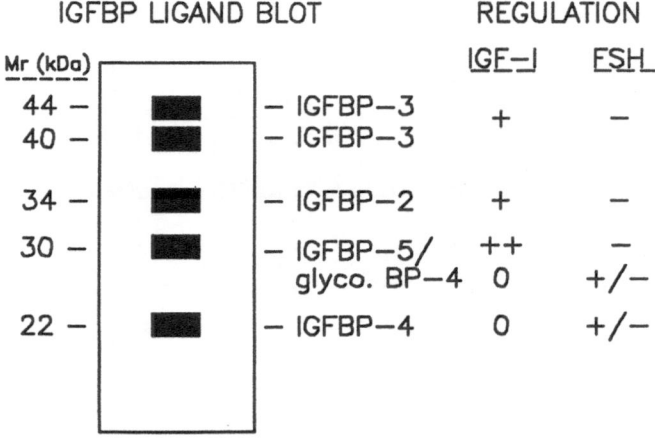

FIGURE 15.6. Schematic summary of regulation of IGFBP secretion in cultured porcine granulosa cells (see references 16 and 21–23). Reprinted with permission from Mondschein, Smith, and Hammond (16), © The Endocrine Society, 1990.

and impacts on these levels as well. Our own studies have emphasized IGFBP-3 (27), which is stabilized in culture by IGF treatment through inhibition of a serine protease that normally governs the half-life of this protein. Comparable data for IGFBP-4 and -5 have been presented in cultured rat granulosa cells, where proteolytic degradation appears to be a major control locus (26, 28). In these cultures both IGF-I (inhibitory) and FSH (stimulatory) seem important for proteolytic processing.

Actions of IGFs and IGFBPs

IGF Action

The action of the IGFs on ovarian cells has been investigated for more than 10 years, and the data generated can only be summarized here. A more comprehensive review of these issues has recently been published (2). The action of IGFs in the ovary, as in other tissues, depends on binding to cell-surface receptors and activation of a subsequent phosphorylation cascade that leads to either cell replication or differentiation. Current opinion favors the mediation of IGF effects through the type I receptor, but type II receptors and insulin receptors are also present on ovarian cells, and it remains possible that these binding sites may be involved in some actions of the several insulin-like peptides.

In cultured porcine ovarian cells, which actively replicate in vitro, important mitogenic actions of the IGFs can be readily demonstrated.

However, these effects are only pronounced in the presence of other growth factors that also support replication; for example, EGF and/or *transforming growth factor* α (TGFα) (29, 30). In the presence of gonadotropins, differential effects of the IGFs are more pronounced than their growth-stimulating activities. These actions have been investigated in great detail and involve the induction of gonadotropin receptors, amplification of gonadotropin-dependent cAMP generation, and cAMP action resulting in induction of steroidogenic enzymes with an increase in steady state mRNA for such enzymes.

Critical issues that are currently being defined include the interaction of the tyrosine kinase-dependent IGF transduction system and gonadotropin-stimulated protein kinase A system on critical intracellular phosphorylation substrates (31) and/or the interaction of the two pathways at the genomic level. While several of these mechanic details remain to be worked out, a large number of data uniformly support a critical stimulatory role for IGFs in ovarian cellular function in culture.

Actions of IGFBPs in the Ovary

As discussed earlier, data on the expression of the IGFBPs at the mRNA and/or peptide level suggest that these proteins are regulated independently of the IGFs; during spontaneous and hormone-induced ovarian cellular differentiation, they appear to be regulated in a fashion roughly opposite to the IGFs themselves. These data are consistent with IGFBPs as counterbalanced antagonistic principles to the IGFs. This possibility has been supported by a number of studies exploring the action of IGFBPs on cultured ovarian cells.

Our own studies to date (32) have emphasized the role of IGFBP-3, which is the most abundant IGFBP in the porcine ovary. As previously demonstrated in the rat ovary (34, 35), this protein was a potent inhibitor of IGF- and gonadotropin-dependent differentiation in porcine granulosa cells (32). Attempts to manipulate culture conditions by experimental paradigms involving preincubation and "washouts" to emphasize cell-associated IGFBP-3 have resulted in the enhancement of IGF-I effects in other cell culture systems (36). However, our own experiments have shown only inhibitory actions of the IGFBP-3 under these circumstances (Samaras and Hammond, unpublished data). Additional studies in cultured rat or human granulosa cells have also shown inhibitory effects of IGFBP-1, -2, -4, and -5 (26, 34, 37). Thus, the actions of exogenous IGFBPs on cultured ovarian cells reported to date have been consistently inhibitory on ovarian cellular function in vitro.

We (21) and others (38) have also examined the action of derivatized forms of IGF-I that are designed to reduce the affinity of these ligands for the IGFBPs. In our hands, the potency of such agonists is an order of magnitude greater than that of native IGF-I (21). These results suggest that the effects of the endogenous IGFBPs secreted by ovarian target cells are inhibitory of these IGF-I actions. In summary, data to date indicate that the IGFBPs are expressed and act in a manner that is consistent with a function as inhibitory agents modulating, in a negative fashion, ovarian cellular function in vivo and in vitro.

While these actions of the IGFBPs seem likely to be important and may prove crucial to understanding ovarian follicular development and selection, these data fail to provide an adequate explanation for the multiplicity of the IGFBPs, their geographical localization, and differential physiological control. These data, coupled with a variety of mechanistic studies in other systems, have led us and others to project a much more dynamic model of the role of IGFBPs in which these proteins may serve as major determinants of the bioavailability and targeting of the IGFs. In addition, recent data have suggested that these proteins may have signaling capabilities independent of the IGFs, generally serving to inhibit cell growth (39, 40).

A working model of the possible role of IGFBPs in targeting the actions of IGFs is shown in Figure 15.7. Central to this model is the

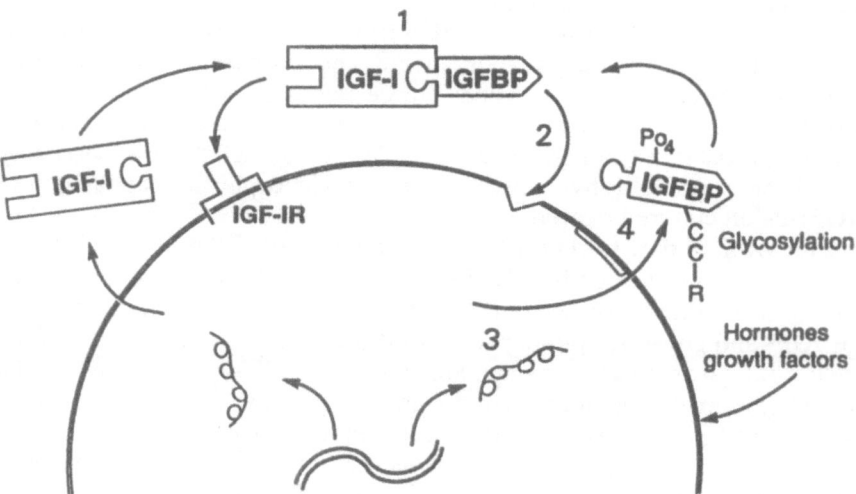

FIGURE 15.7. Model of IGF-IGFBP interaction: the IGF-I autocrine system in porcine granulosa cells. Several aspects of this system are illustrated, including IGF-I biosynthesis and action through the IGF-I receptor (IGF-IR). Points of emphasis relate to extracellular processing and targeting of IGF-I through the IGFBPs. Shown are the IGF-IGFBP complex in the extracellular space (1); the binding site for IGFBP on cell surface or extracellular matrix (2); posttranslational modification of IGFBPs (3); and putative cellular processing site for IGFBPs, e.g., the proteolytic enzyme phosphatase (4). Reprinted with permission from Hammond, Samaras, Grimes, et al. (5).

existence of an extracellular pool of IGFs complexed to the IGFBPs that serves as a reservoir of potentially active IGFs. With processing of this complex, the IGFs can be presented to receptors on target cells. This model has not been fully substantiated in any system, but data for the existence of activity at each of the control loci depicted have been developed in one or more systems (41, 42). These include binding of the IGFBPs to attachment sites on the cell glycocalyx (integrins or related attachment sites) or to mucopolysaccharides in the extracellular matrix and posttranslational modification of the IGFBPs by phosphorylation, glycosylation, or proteolytic processing that may modify the affinity of these proteins for the IGFs and, hence, delivery of the IGFs to the receptors.

Our own studies suggest that IGFBP-3 is proteolytically processed by porcine granulosa cells, probably by a serine protease, and that this action is inhibited by IGF-I (27). Similar effects can also be shown for IGFBP-5 (5), for which a striking induction of IGFBP-5 by IGF-I is associated with a diminution in its low-molecular weight processing variant. Further studies of this sort, as well as of the critical steps in the interaction of IGF-I-IGFBP complexes at the ovarian cell surface, need to be performed before the full importance of these mechanisms can be understood. These effects may be important in controlling the overall set point of the ovarian IGF system.

Contrasting Effects of Somatotropin and Gonadotropins on the Porcine Ovary

A central hypothesis advanced in preceding sections has been the notion of the IGF system as a local amplification mechanism for gonadotropin action. This central effect is achieved through a number of interlocking mechanisms, including gonadotropin induction of local IGF secretion, facilitory interactions between IGFs and gonadotropins on steroidogenesis, and gonadotropin inhibition of the secretion of inhibitory IGFBPs.

Using parallel investigative strategies, we have examined the effects of GH on ovarian cellular function in vitro and in vivo. Our initial studies in cultured porcine granulosa cells (43) indicated that GH effects paralleled and enhanced those of FSH on IGF-I and progesterone secretion. Studies with a monoclonal antibody to IGF-I suggested that the actions of GH, like those of FSH, were mediated at least in part by locally secreted IGF-I (44). A notable exception to the gonadotropin-somatotropin synergism was manifested on IGFBP secretion (Fig. 15.8). Instead of mimicking the inhibitory effects of FSH on IGFBP secretion, GH exhibited a modest stimulatory action on IGFBP levels that antagonized the effects of gonadotropins. The agonist actions of GH on these proteins were less consistent than those of IGF-I, but they were fairly dramatic in individual

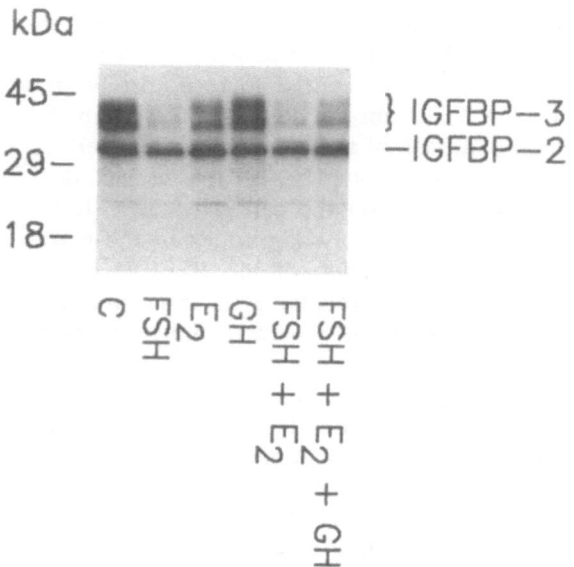

FIGURE 15.8. Ligand blot of secreted IGFBPs from porcine granulosa cells treated with pituitary and ovarian hormones.

experiments (45). These data may be important in understanding some of the paradoxical inhibitory actions of GH on the porcine ovary in vivo.

The actions of GH on ovarian function in swine have been inferred from several investigative strategies. Animals with reduced GH and IGF-I levels achieved by active immunization against *GH releasing hormone* (GHRH) generally have a reduced ovulation rate early in reproductive life (46), suggesting a supportive or permissive role for GH on ovarian function. On the other hand, animals treated with pharmacological doses of GH (4, 47) or rendered transgenic for the human GH gene (48) manifest a variety of negative effects on ovarian function. These include reduced ovarian androgen and estrogen levels (4, 47, 48) and delayed or impaired estrus (4).

As illustrated in Figures 15.9 and 15.10, exogenous GH can result in diminished androgen and estrogen levels in FF despite a substantial increase in intrafollicular IGF-I levels and preserved (or enhanced) progesterone levels (4, 48). Levels of IGFBPs in the FF of these animals are shown in Figures 15.11 and 15.12. In animals treated with GH alone, there was a significant increase in IGFBP-3; in animals cotreated with PMSG, which lowers IGFBP-2 levels, the gonadotropin effects on this protein were antagonized by GH (4). These studies leave a number of mechanistic questions unanswered, but they, as well as studies from other laboratories (48–50), clearly illustrate the potential for inhibitory actions

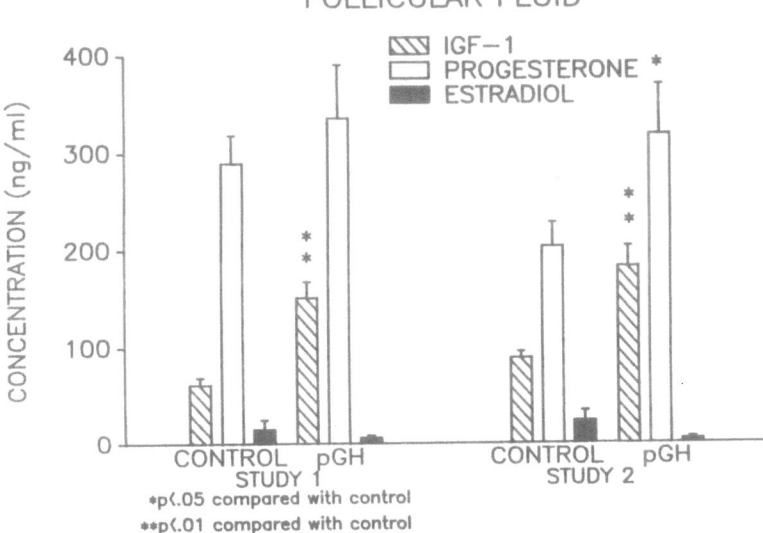

FIGURE 15.9. Concentrations of IGF-I, progesterone, and E_2 in aspirated FF from swine treated for 30 days with pGH. Reprinted with permission from Bryan, Hammond, Canning, et al. (47).

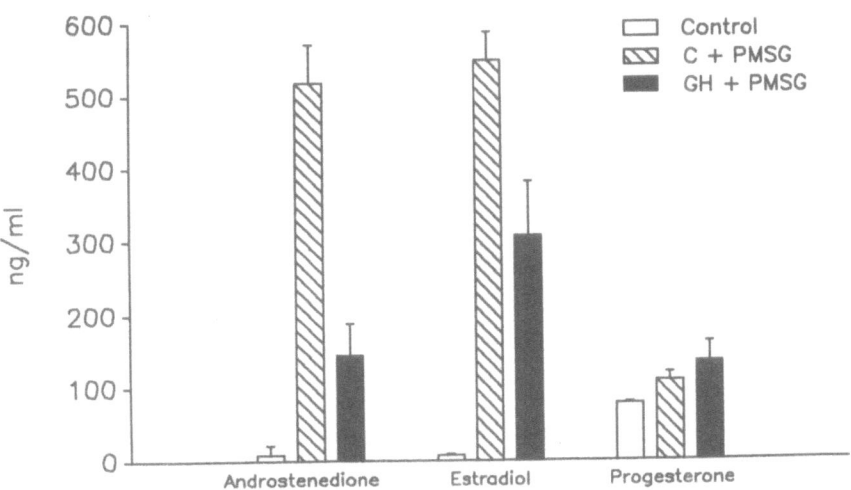

FIGURE 15.10. Effects of pGH pretreatment on steroid response to PMSG. Reprinted with permission from Samaras, Hagen, Bryan, Mondschein, Canning, and Hammond (4).

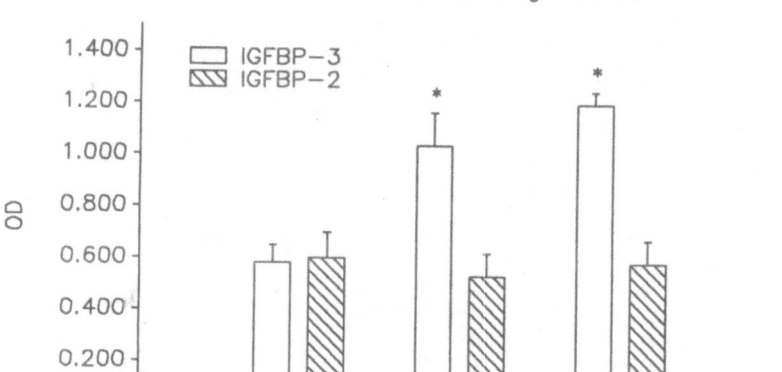

FIGURE 15.11. Effect of GH treatment on levels of IGFBP-2 and -3 in porcine FF. Reprinted with permission from Samaras, Hagen, Bryan, Mondschein, Canning, and Hammond (4).

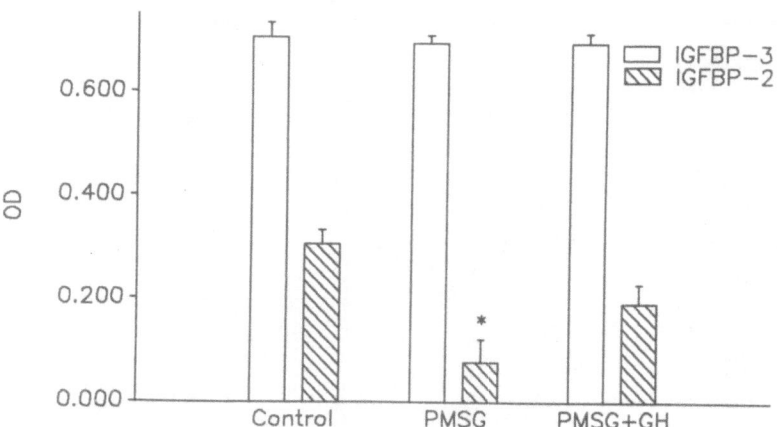

FIGURE 15.12. Effect of GH and PMSG on levels of IGFBP-2 and -3 in porcine FF. Reprinted with permission from Samaras, Hagen, Bryan, Mondschein, Canning, and Hammond (4).

of GH in the ovary of this species. The potential inhibitory roles for ovarian IGFBPs reviewed previously, along with the observed increase in these proteins, may explain the bipotential effects of GH on the ovary of some species.

Summary and Conclusions

This chapter has focused on outlining the current data on the several components of the ovarian IGF system—the IGFs themselves, the IGFBPs, and their cellular transduction systems—emphasizing the abundant information in the porcine model. Based on these data, we can piece together a construct of a highly integrated, mutually facilitory mechanism that could support selective growth and development of critical ovarian components.

The synergistic interaction of the IGFs and gonadotropins is well understood and persuasive, although some details remain to be established. In general, the IGFBPs appear to inhibit these processes; these actions probably derive from limiting the bioavailability of ovarian IGFs. However, the possibility of these proteins serving additional shuttle or targeting functions seems likely based on data in nonovarian systems. In the course of these investigations, a substantial number of observations have accumulated concerning the actions of GH on the porcine ovary. These data indicate the possibility of inhibitory ovarian actions of GH—possibly mediated by IGFBPs. These latter data also need to be incorporated into the ovarian IGF concept and probably into plans for the use of GH as a growth-promoting or fertility-inducing agent in farm animals, humans, and other species.

Acknowledgments. Research on the IGF system in the authors' laboratory is supported by Grant HD-24565 (J.M.H.). Critical technical support from J.A. Barber and S.F. Canning and the secretarial skills of L. Doster are gratefully acknowledged.

References

1. Liu J-P, Baker J, Perkins AS, Robertson EJ, Efstratiadis A. Mice carrying null mutations of the genes encoding insulin-like growth factor I (IGF-I) and type I IGF receptor (IGFIr). Cell 1993;75:59–72.
2. Giudice LC. Insulin-like growth factors and ovarian follicular development. Endocr Rev 1992;13:641–69.
3. Hammond JM, Baranao JLS, Skaleris D, Knight AB, Romanus JA, Rechler MM. Production of insulin-like growth factors by ovarian granulosa cells. Endocrinology 1975;117:2553–5.

4. Samaras SE, Hagen DR, Bryan KA, Mondschein JS, Canning SF, Hammond JM. Effects of growth hormone and gonadotropin on the insulin-like growth factor system in the porcine ovary. Biol Reprod 1994;50:178–86.

5. Hammond JM, Samaras SE, Grimes R, et al. Intraovarian regulation in the porcine ovary: the role of insulin-like growth factors and epidermal growth factor-related peptides. J Reprod Fertil Suppl 1993;48:117–25.

6. Samaras S, Guthrie D, Barber J, Hammond JM. Changes in mRNAs for the insulin-like growth factors and their binding proteins in individual porcine ovarian follicles. Endocrinology 1993;133:2395–8.

7. Hammond JM, Hsu C-J, Klindt J, Tsang BK, Downey BR. Gonadotropins increase concentrations of immunoreactive insulin-like growth factor-I in porcine follicular fluid in vivo. Biol Reprod 1988;38:304–8.

8. Hsu C-J, Hammond JM. Gonadotropins and estradiol stimulate immunoreactive insulin-like growth factor-I production by porcine granulosa cells in vitro. Endocrinology 1987;120:198–207.

9. Mondschein JS, Hammond JM. Growth factors regulate immunoreactive insulin-like growth factor-I production by cultured porcine granulosa cells. Endocrinology 1988;123:463–8.

10. Ramasharma K, Li CH. Human pituitary and placental hormones control human insulin-like growth factor II secretion in human granulosa cells. Proc Natl Acad Sci USA 1987;94:2643–7.

11. Voutilainen R, Miller WL. Coordinate tropic hormone regulation of mRNAs for insulin-like growth factor II and the cholesterol side-chain-cleavage enzyme P450scc, in human steroidogenic tissues. Proc Natl Acad Sci USA 1987;84:1590–4.

12. Oliver JE, Aitman TJ, Powell JF, Wilson CA, Clayton RN. Insulin-like growth factor I gene expression in the rat ovary is confined to the granulosa cells of developing follicles. Endocrinology 1989;124:2671–9.

13. Botero LF, Roberts CT Jr, LeRoith D, Adashi EY, Hernandez ER. Insulin-like growth factor I gene expression by primary cultures of ovarian cells: insulin and dexamethasone dependence. Endocrinology 1993;132:2703–8.

14. Hammond JM, Veldhuis JD, Seal TW, Rechler MM. Intraovarian regulation of granulosa-cell replication. In: Channing CP, Segal SJ, eds. Intraovarian control mechanisms, New York: Plenum Press, 1982:341–56.

15. Mondschein JS, Etherton TD, Hammond JM. Characterization of insulin-like growth factor-binding proteins of porcine ovarian follicular fluid. Biol Reprod 1991;44:315–20.

16. Mondschein JS, Smith SA, Hammond JM. Production of insulin-like growth factor binding proteins (IGFBPs) by porcine granulosa cells; identification of IGFBP-2 and -3 and regulation by hormones and growth factors. Endocrinology 1990;127:2298–307.

17. Seppala M, Wahlstrom T, Koskimies AI, et al. Human preovulatory follicles and corpus luteum contain placental protein 12. J Clin Endocrinol Metab 1984;58:505–10.

18. Shimasaki S, Ling N. Identification and molecular characterization of insulin-like growth factor binding proteins (IGFBP-1, -2, -3, -4, -5, and -6). Prog Growth Fact Res 1991;3:243–66.

19. Hammond JM, Barber JA, Canning SF, Guthrie HD. Expression of insulin-like growth factor binding proteins-4 and -5 during the growth of porcine ovarian follicles [Abstract]. Biol Reprod 1992;46(suppl 1):85.
20. Howard HJ, Ford JJ. Relationships among concentrations of steroids, inhibin, insulin-like growth factor-I (IGF-I), and IGF-binding proteins during follicular development in weaned sows. Biol Reprod 1992;47:193–201.
21. Grimes RW, Hammond JM. Insulin and insulin-like growth factors (IGFs) stimulate production of IGF-binding proteins by ovarian granulosa cells. Endocrinology 1992;131:553–8.
22. Grimes RW, Samaras SE, Barber JA, Shimasaki S, Ling N, Hammond JM. Gonadotropin and cAMP modulation of IGF binding protein production in ovarian granulosa cells. Am J Physiol 1992;262:3497–503.
23. Grimes RW, Barber JA, Shimasaki S, Ling N, Hammond JM. Porcine ovarian granulosa cells secrete insulin-like growth factor (IGF)-binding proteins-4 and -5 and express their mRNAs: regulation by FSH and IGF-I. Biol Reprod 1994;50:695–701.
24. Adashi EY, Resnick CE, Hernandez ER, Hurwitz A, Rosenfeld RG. Follicle-stimulating hormone inhibits the constitutive release of insulin-like growth factor binding proteins by cultured rat ovarian granulosa cells. Endocrinology 1990;126:1305.
25. Adashi EY, Resnick CE, Tedeschi C, Rosenfeld RG. A kinase-mediated regulation of granulosa cell-derived insulin-like growth factor binding proteins (IGFBPs): disparate response sensitivities of distinct IGFBP species. Endocrinology 1993;132:1463–8.
26. Liu X-J, Malkowski M, Guo Y, Erickson GF, Shimasaki S, Ling N. Development of specific antibodies to rat insulin-like growth factor-binding proteins (IGFBP-2 to -6): analysis of IGFBP production by rat granulosa cells. Endocrinology 1993;132:1176–83.
27. Grimes RW, Hammond JM. Proteolytic degradation of insulin-like growth factor-binding protein-3 by porcine ovarian granulosa cells in culture: regulation by IGF-I. Endocrinology 1994;134:337–43.
28. Fielder PJ, Pham H, Adashi EY, Rosenfeld RG. Insulin-like growth factors (IGFs) block FSH-induced proteolysis of IGF-binding protein-5 (BP-5) in cultured rat granulosa cells. Endocrinology 1993;133:415–8.
29. Hammond JM, English HF. Regulation of DNA synthesis in cultured porcine granulosa cells by growth factors and hormones. Endocrinology 1987;120:1039–46.
30. May JV, Frost JP, Schomberg DW. Differential effects of epidermal growth factor, somatomedin-C/insulin-like growth factor I, and transforming growth factor-beta on porcine granulosa cell deoxyribonucleic acid synthesis and cell proliferation. Endocrinology 1988;123:168–79.
31. Chakravorty A, Joslyn M, Davis JS. Synergistic interactions among the adenylate cyclase (LH/cAMP) and tyrosine kinase (insulin/IGF-I) systems in bovine luteal cells [Abstract #165]. Biol Reprod 1992;46(suppl 1):92.
32. Samaras SE, Hammond JM. Inhibitory actions of recombinant IGFBP-3 on the function of cultured porcine granulosa cells [Abstract]. Prog IX Ovarian Workshop on Ovarian Cell Interactions: Genes to Physiology, Chapel Hill, NC, 1992.

33. Ui M, Shimonaka M, Shimasaki S, Ling N. An insulin-like growth factor binding protein in ovarian follicular fluid blocks follicle-stimulating hormone-stimulated steroid production by ovarian granulosa cells. Endocrinology 1989; 125:912–6.
34. Bicsak TA, Shimonaka M, Malkowski M, Ling N. Insulin-like growth factor-binding protein (IGFBP) inhibition of granulosa cell function: effect on cyclic adenosine 3',5'-monophosphate, deoxyribonucleic acid synthesis, and comparison with the effect of an IGF-I antibody. Endocrinology 1990;126:2184–9.
35. Bicsak TA, Ling N, DePaolo LV. Ovarian intrabursal administration of insulin-like growth factor-binding protein inhibits follicle rupture in gonadotropin-treated immature rats. Biol Reprod 1991;44:599–603.
36. Conover CA, Ronk M, Lombana F, Powell DR. Structural and biological characterization of bovine insulin-like growth factor binding protein-3. Endocrinology 1990;127:2795–803.
37. Mason HD, Willis D, Holly JM, Cwyfan-Hughes SC, Seppala M, Franks S. Inhibitory effects of insulin-like growth factor-binding proteins on steroidogenesis by human granulosa cells in culture. Mol Cell Endocrinol 1992;89: R1–4.
38. Adashi EY, Resnick CE, Ricciarelli E, et al. Granulosa cell-derived insulin-like growth factor (IGF) binding proteins are inhibitory to IGF-I hormonal action: evidence derived from the use of a truncated IGF-I analogue. J Clin Invest 1992;90:1593–9.
39. Oh Y, Muller HL, Lamson G, Rosenfeld RG. Insulin-like growth factor (IGF)-independent action of IGF-binding protein-3 in Hs578T human breast cancer cells: cell surface binding and growth inhibition. J Biol Chem 1993; 268:14964–71.
40. Cohen P, Lamson G, Okajima T, Rosenfeld RG. Transfection of the human insulin-like growth factor binding protein-3 gene into balb/c fibroblasts inhibits cellular growth. Mol Endocrinol 1993;7:380–6.
41. Clemmons DR. Insulin-like growth factor binding proteins: roles in regulating IGF physiology. J Dev Physiol 1991;15:105–10.
42. Rechler MM. Insulin-like growth factor binding proteins. Vitam Horm 1993; 47:1–114.
43. Hsu C-J, Hammond JM. Concomitant effects of growth hormone on secretion of insulin-like growth factor-I and progesterone by cultured porcine granulosa cells. Endocrinology 1987;121:1343–8.
44. Mondschein JS, Canning SF, Miller DO, Hammond JM. Insulin-like growth factors (IGFs) as autocrine/paracrine regulators of granulosa cell differentiation and growth: studies with a neutralizing monoclonal antibody to IGF-I. Biol Reprod 1989;40:79–85.
45. Hammond JM, Mondschein JS, Samaras SE, Smith SA, Hagen DR. The ovarian insulin-like growth factor system. J Reprod Fertil 1991;43(suppl): 199–208.
46. Armstrong JD, Britt JH, Flowers WL, Campbell RM, Heimer EP. Effect of immunization against growth hormone releasing factor (GRFi) on onset of puberty and ovulation rate in gilts [Abstract #271]. Biol Reprod 1993; 48(suppl 1):126.
47. Bryan KA, Hammond JM, Canning S, et al. Reproductive and growth responses of gilts to exogenous porcine pituitary growth hormone. J Anim Sci 1989;67:196–205.

48. Guthrie HD, Pursel VG, Bolt DJ, Cooper BS. Expression of a bovine growth hormone transgene inhibits pregnant mare's serum gonadotropin-induced follicle maturation in prepuberal gilts. J Anim Sci 1993;71:3409–13.
49. Spicer LH, Klindt J, Bunomo FC, Maurer R, Yen JT, Echternkamp SE. Effect of porcine somatotropin on number of granulosa cell luteinizing hormone/human chorionic gonadotropin receptors, oocyte viability, and concentrations of steroids and insulin-like growth factors I and II in follicular fluid of lean and obese gilts. J Anim Sci 1992;70:3149–57.
50. Kirkwood RN, Thacker PA, Korchinski RS, Laarveld B. The influence of exogenous pituitary growth hormone on the ovulatory responses to PMSG and hCG and the LH response to estradiol benzoate in prepubertal gilts. Domest Anim Endocrinol 1988;55:61–9.

16

Is the Intraovarian IGF System a Mediator of Growth Hormone Action?

J.K. FINDLAY

There is a considerable body of evidence, reviewed in this volume and elsewhere (1–3), that supports the ability of *growth hormone* (GH) to influence the function of the ovary. It has been proposed that GH is a cogonadotropin that synergizes with *follicle stimulating hormone* (FSH) and *luteinizing hormone* (LH) in the promotion of ovarian function (3). The actions of GH on the ovary could be manifest in two ways, not necessarily mutually exclusive. GH could act via its receptors, resulting in direct modulation of the actions of gonadotropins on ovarian somatic cells. This implies an interaction between the second-messenger systems within the target cell subserving each of the pituitary hormones. Alternatively, GH could act via its receptors to stimulate the production of *insulin-like growth factor I* (IGF-I) that in turn could have autocrine or paracrine actions on the ovarian somatic cells to modify the actions of FSH and LH. Implicit in this second possibility is the presumption that the ovarian expression of the IGF-I gene and the intraovarian actions of IGF-I are either partially or totally GH dependent. The evidence for these propositions is discussed below.

It should be kept in mind that there is no compelling evidence to date that the putative GH-IGF axis within the ovary is indispensable for ovarian function. The evidence to be reviewed suggests that GH is not obligatory for fertility, underlining the possibility that compensatory mechanisms may exist within the ovary to ensure timely production of fertilizable oocytes.

GH and IGF Gene Expression

There is evidence for regulation of IGF-I gene expression in ovarian cells by GH based on both in vivo and in vitro studies, but there are some inconsistencies and gaps in our knowledge.

In Vivo Studies

It was shown that treatment of hypophysectomized female rats with estrogen and ovine GH resulted in an increased ovarian content of immunoreactive IGF-I 8 h later that could not be accounted for by serum contamination (4). GH given in vivo has been shown to result in an increase in intrafollicular concentrations of IGF-I in gilts (5) and women (6), although the results of such studies are not always consistent. For example, Volpe et al. (7) found no difference in the *follicular fluid* (FF) concentrations of IGF-I from women given *human menopausal gonadotropin* (hMG) plus GH versus those given hMG alone, although there was a significant linear correlation between intrafollicular GH and IGF-I. The problem of interpreting this and similar studies when GH is given in vivo is delineating between the direct effects of GH on ovarian production of IGF-I and the effects of GH on hepatic production of IGF-I that is reflected in serum and FF. An animal model suitable for investigating this question is the sheep with the ovary transplanted to the neck, allowing direct and easy access to the arterial and venous supplies of the ovary. Downing et al. (8) have used this model recently to demonstrate a direct effect of GH on LH-induced secretion of estradiol. Unfortunately, IGF-I secretion rates were not measured in these experiments.

One study has examined the in vivo effects of GH on steady state IGF-I mRNA levels in the ovary (9). In immature hypophysectomized rats treated with GH, there was a marked *decrease* in ovarian IGF-I mRNA in contrast to the liver, where IGF-I mRNA levels increased. Treatment of the rats with estrogen alone increased ovarian IGF-I mRNA, but decreased hepatic IGF-I mRNA. This underlines the tissue-specific regulation of IGF-I gene expression and also suggests that in the ovary it is estrogen, rather than GH, dependent.

In Vitro Studies

GH has been shown to stimulate the production of immunoreactive IGF-I by porcine granulosa cells in vitro (10). Attempts to repeat this observation using granulosa cells from rats and women have not been successful (11–13) despite significant effects of GH on steroidogenesis (12–14). Furthermore, it has not been possible to demonstrate a significant in vitro effect of GH on steady state levels of IGF-I mRNA in rat granulosa cells (11). While equine chorionic gonadotropin—*pregnant mare serum*

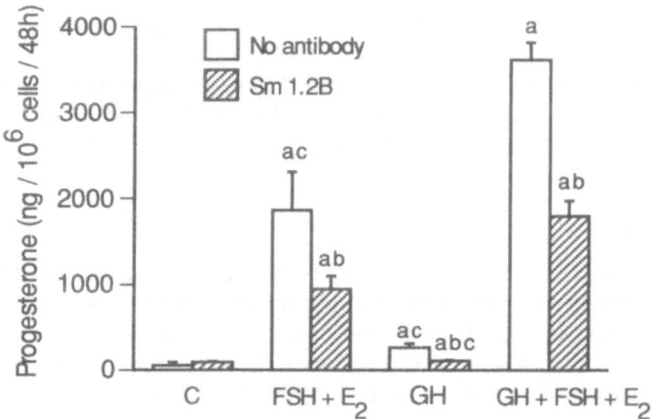

FIGURE 16.1. Effects of a MAb to IGF-I (Sm 1.2B) on hormone-stimulated progesterone production. Immature porcine granulosa cells were cultured in serum-free medium. Treatments with FSH (200 ng/mL), estradiol (E_2) (1 µg/mL), GH (100 ng/mL), and the MAb (1:500) were begun at the day 3 medium change and continued through day 7. Spent medium from days 5–7 was assayed for progesterone; data were normalized on the basis of cell counts determined on day 7. Data are the means ± SE of 6 cultures per treatment group. (The letter a denotes $P < 0.05$ vs. control; b: $P < 0.05$ vs. no MAb; c: $P < 0.05$ vs. FSH + E_2 + GH.) Reprinted with permission from Mondschein, Canning, Miller, and Hammond (15).

gonadotropin (PMSG)—treatment of rat granulosa cells in vitro resulted in an approximate 2-fold increase in IGF-I mRNA levels over basal levels, no further increase was observed with the addition of GH to PMSG. Voutilainen and Miller (13) were unable to demonstrate an effect of GH on IGF-I mRNA levels in human granulosa cells.

Perhaps the best, albeit indirect, evidence for an involvement of IGF-I in the actions of GH on granulosa cells comes from two studies that used a *monoclonal antibody* (MAb) to human IGF-I to block endogenous IGF-I produced in response to GH and monitored the effects on steroidogenesis. Mondschein et al. (15) used porcine granulosa cells cultured in serum-free medium and showed that the effects of FSH, GH, estradiol, or combinations thereof on progesterone production were inhibited by the addition of the IGF-I MAb (Fig. 16.1). It was shown that the inhibitory action of the MAb was unlikely to be due to nonspecific toxic effects because the MAb did not influence the ability of insulin to stimulate progesterone production under similar conditions. The authors concluded that the actions of GH on progesterone production were mediated at least in part by IGF.

We also used the IGF-I MAb in similar studies with rat granulosa cells (16, 17). As shown in Figure 16.2, the effect of GH on FSH (PMSG)-stimulated aromatase activity or progesterone production was completely

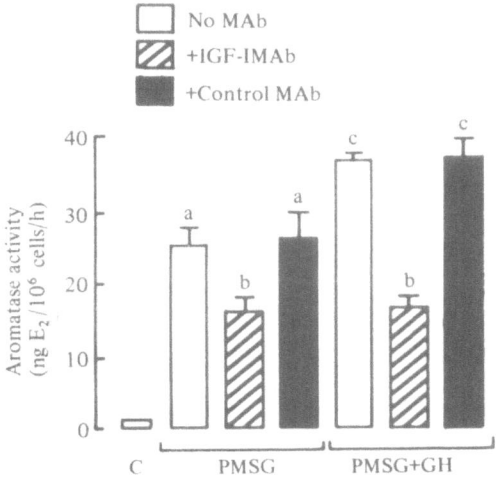

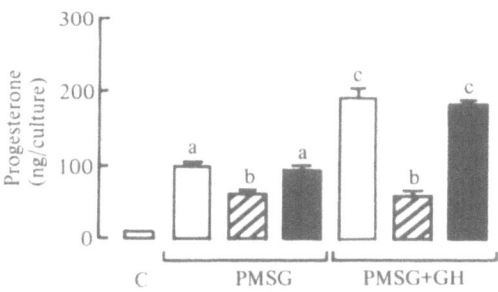

FIGURE 16.2. Aromatase activity and progesterone production by rat granulosa cells in a 3-day culture in the presence of PMSG (100 mU/mL) and/or GH (200 ng/mL) and the presence or absence of an anti-IGF-I MAb (1:1000 dilution). The control antibody used was an antihuman IgG MAb devoid of anti-IGF-I activity. (The letters a, b, and c denote significant differences; $P < 0.01$.) Reprinted with permission from Findlay, Zhiwen, Hutchinson, Carson, Burger, and Herington (17), © Cambridge University Press, 1990.

blocked by the addition of the IGF-I MAb. The same effect was observed when the MAb was added to cells treated with PMSG alone. There were no effects of a control MAb, nor were there any effects of the IGF-I MAb on stimulation of steroidogenesis by insulin (Fig. 16.3). Furthermore, the effects of the IGF-I MAb could not be accounted for by neutralization of IGF-I contamination of PMSG or GH. Therefore, in this study, the actions of GH could be accounted for by the endogenous production of IGF-I. Furthermore, the actions of FSH (PMSG) alone

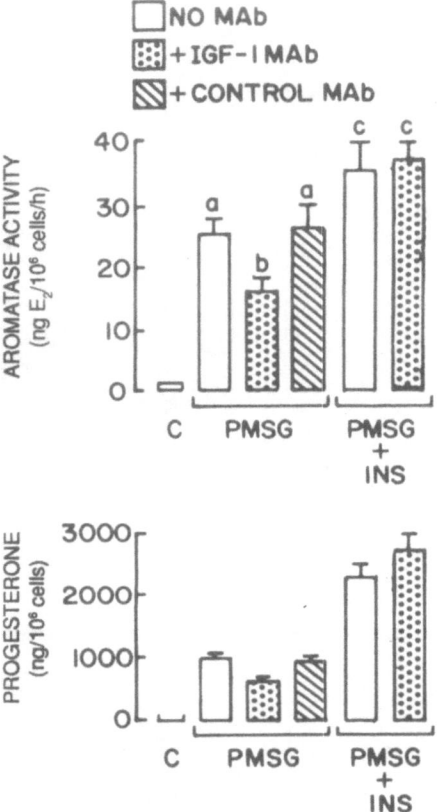

FIGURE 16.3. Aromatase activity and progesterone synthesis levels in rat granulosa cells after a 3-day culture with PMSG alone (100 mU/mL) or PMSG + insulin (8 μg/mL) in the absence (open bars) or presence (hatched and dotted bars) of 1:1000 final dilution of control or IGF-I MAb. Values are mean ± SD of triplicate cultures. (The letters a through c [a vs. b vs. c] denote significant differences; $P < 0.01$.) Drawn from data of Hutchinson, Herington, and Findlay (16).

were not completely blocked by the IGF-I MAb, supporting the conclusion that FSH alone has effects on granulosa cells that are both dependent and independent of IGF-I.

GH and IGF-I Actions

If IGF-I is responsible for mediating some or all of the actions of GH on ovarian somatic cells, then many of the effects of GH on granulosa cell differentiation should be mirrored by the addition of IGF-I. Examples of this can be found in a number of studies in which both GH and IGF-I amplify the effects of FSH action on granulosa cells (see 3). For example, in our studies using rat granulosa cells, both IGF-I and GH had similar time- and dose-dependent effects on PMSG induction of aromatase activity and progesterone production (14). Neither IGF-I nor GH had any detectable effects in the absence of PMSG.

However, there are instances where the in vitro actions of GH and IGF-I on granulosa cells are divergent. He and Herington (18) investigated the effects of human GH and IGF-I on PMSG-induced progesterone production by rat granulosa cells in vitro under different culture conditions. Specifically, they varied the content of transferrin, cortisol, and insulin in the serum-free medium. There was a marked difference between IGF-I and GH in their effects on PMSG-induced steroidogenesis in the presence of cortisol and transferrin. In the presence of cortisol, the effect of IGF-I was inhibited, while the stimulatory effect of GH was amplified. Furthermore, transferrin facilitated the action of GH and inhibited the effect of IGF-I. The authors concluded that at least some of the effects of GH on the differentiation of granulosa cells are direct and not mediated by IGF-I and that there exist different and independent pathways for the effects of IGF-I and GH on granulosa cells influenced by gonadotropins. These conclusions should be tempered by the fact that human GH was used, which has both somatotrophic and lactogenic activities, so some of the effects observed on granulosa cells could involve lactogenic sites.

A second example of the divergent effects of GH and IGF-I on granulosa cells comes from a recent study of the mitogenic actions of these hormones on bovine granulosa cells in vitro (19). Recombinant *bovine somatotropin* (BST) treatment inhibited the incorporation of [^{3}H]-thymidine into granulosa cells from large (>10 mm in diameter) follicles in a dose-dependent manner by up to 40%. In contrast, there were no significant effects of BST on thymidine incorporation into granulosa cells from small-sized (<5 mm) or medium-sized (5–10 mm) follicles. There was no interaction between BST and either FSH or LH on thymidine incorporation by granulosa cells from all 3 size classes. In contrast to BST, recombinant human IGF-I stimulated thymidine incorporation into granulosa cells from all 3 size classes in a dose-dependent manner up to the highest dose tested (300 ng/mL). Furthermore, there was a synergistic effect of IGF-I and FSH or LH on granulosa cells from small follicles, but not from medium or large follicles, where the effects were additive. In this study the medium consisted of Medium 199 with Earle's salts, glutamine, and antibiotics, but no insulin, transferrin, or cortisol, and the plates were coated with donor calf serum. The study supports the conclusion that GH and IGF-I can have direct, independent actions on granulosa cells in vitro.

The mechanisms by which GH and IGF-I exert these independent effects on granulosa cells are unclear. He et al. (20) have recent evidence that the effects of IGF-I on PMSG-induced granulosa cell differentiation may require, at least in part, an interaction with a G-protein-mediated signaling mechanism. The possibility remains that in vivo the actions of GH may involve products derived from theca cells. In this regard, *IGF binding protein 3* (IGFBP-3) is a product of the rat theca cell (21). Treatment of hypophysectomized rats with FSH or estrogen decreased the abundance of transcripts for ovarian IGFBP-3, whereas treatment with GH produced a >3-fold increase in steady state levels of IGFBP-3 mRNA levels. If these transcripts are also reflected in increased levels of IGFBP-3 protein, this suggests that GH could indirectly influence the actions of IGF-I within the ovarian follicle in a negative fashion by increasing the levels of IGFBP-3, which would neutralize the biological activity of IGF-I.

Taken together, the observations lead one to conclude that there remains a degree of uncertainty about the role of GH in the regulation of expression of the IGF-I gene in the ovary. That being the case, the relative importance in vivo of GH versus other hormones, particularly FSH, known to regulate IGF-I production by ovarian cells becomes an important question.

Importance of GH and IGF-I for Fertility

There is a considerable amount of data, particularly from women being treated with gonadotropin for infertility, that GH can augment the response and improve the success rate (2, 3). However, it appears that circulating IGF-I and/or GH are not an absolute requirement for ovarian function and fertility because Laron-type dwarfs, who have defective GH receptors and barely detectable serum and FF IGF-I, are fertile (see 2). It is possible that these subjects have sufficient intraovarian IGF-I production that is regulated independently of GH; for example, by FSH. It is worth noting at this point that there is no compelling evidence to show that IGFs are obligatory for normal ovarian function.

It has been shown recently that a strain of *spontaneous dwarf rats* (SDR) carries an autosomal recessive mutation in the GH gene such that the pituitary content of GH mRNA is only 3%–6% of controls (22). This truncated mRNA is not translated so that GH cannot be detected in the pituitary gland of SDR. While the homozygotes are deficient in GH, they do have low IGF-I levels in serum and have a reduced body weight compared with heterozygotes and wild-type controls. The SDR are fertile, although their litter size is reduced (5.2 ± 0.5) compared to controls (8.6 ± 1.1) ($P = 0.008$). These data also support the conclusion that GH is not obligatory for fertility if gonadotropin levels are adequate.

In a broader context, one of the striking things about the effects that many of the growth factors and cytokines have on ovarian cells is their similarity. In vitro studies using rat granulosa cells show that a number of different factors can increase steroidogenesis, cAMP production, and ovarian peptide secretion when added in the presence of gonadotropin. For example, we have shown that IGF-I, TGFβ, and activin all increase steroid and inhibin output by rat granulosa cells in the presence of FSH (PMSG) (14, 23, 24). This raises two questions. First, is the regulation of ovarian cells such that there are backup systems involving other chemokines that come into play if one of the systems becomes inadequate? Second, what is unique about the actions of these factors that might indicate an obligatory role for each of them during folliculogenesis?

To answer the first question will require considerably more data, particularly using selective ablation of the various factors at the ovarian level once this transgenic technology is developed. With respect to the second question, data already exist that differentiate between the actions of activin and TGFβ on the one hand and IGF-I on the other hand on receptors for FSH on rat granulosa cells. Both activin and TGFβ can increase the number of FSH receptors (25–27), whereas IGF-I cannot mimic these actions (28). Further work along these lines will be important in establishing an indispensable role for these putative local regulatory factors in ovarian folliculogenesis.

Acknowledgments. My thanks to Faye Coates and Sue Panckridge for help with preparation of the manuscript. I thank the following colleagues for providing me with data in press: Drs. He Hong, Peter Roupas, Adrian Herington, J.G. Gong, and Robert Webb. The financial support of the National Health and Medical Research Council of Australia is gratefully acknowledged.

References

1. Hammond JM, Mondschein JS, Samaras SE, Smith SA, Hagen DR. The ovarian insulin-like growth factor system. J Reprod Fertil 1991;43:199–208.
2. Giudice LC. Insulin-like growth factors and ovarian follicular development. Endocr Rev 1992;13:641–69.
3. Katz E, Ricciarelli E, Adashi EY. The potential relevance of growth hormone to female reproductive physiology and pathophysiology. Fertil Steril 1993; 59:8–34.
4. Davoren JB, Hsueh AJ. Growth hormone increases ovarian levels of immunoreactive somatomedin C/insulin-like growth factor I in vivo. Endocrinology 1986;118:888–90.
5. Spicer LJ, Klindt J, Maurer R, Buonomo FC, Echternkamp SE. Effect of porcine somatotropin (pST) on numbers of granulosa cell LH/hCG receptors, oocyte viability, and concentrations of progesterone (p) and insulin-like growth factor-I (IGF-I) in follicular fluid (FFL) of lean and obese gilts [Abstract 410]. J Anim Sci 1990:68.

6. Owen EJ, Torresani T, West C, Mason BA, Jacobs HS. Serum and follicular fluid insulin like growth factors I and II during growth hormone co-treatment for in-vitro fertilization and embryo transfer. Clin Endocrinol (Oxf) 1991; 35:327–34.

7. Volpe A, Coukos G, Barreca A, et al. Ovarian response to combined growth hormone-gonadotropin treatment in patients resistant to induction of super-ovulation. Gynecol Endocrinol 1989;3:125–33.

8. Downing JA, Scaramuzzi RJ, Joss J. Steroid secretion in ewes following an ovarian arterial infusion of sheep growth hormone [Abstract 43]. Proc 25th annu conf Australian Soc Reprod Biol, Dunedin, New Zealand, August 23–25, 1993.

9. Hernandez ER, Roberts CT Jr, LeRoith D, Adashi EY. Rat ovarian insulin-like growth factor I (IGF-I) gene expression is granulosa cell-selective: 5'-untranslated mRNA variant representation and hormonal regulation. Endocrinology 1989;125:572–4.

10. Hsu CJ, Hammond JM. Concomitant effects of growth hormone on secretion of insulin-like growth factor I and progesterone by cultured porcine granulosa cells. Endocrinology 1987;121:1343–8.

11. Herington AC, Zhang Z, Hutchinson L, Burger HG, Findlay JK. Growth factors and the differentiation of rat ovarian granulosa cells. In: Li W-X, Chen H-C, Hahn DW, McGuire JL, eds. The role of growth factors, onco-genes, and gonadal polypeptides. Proc Int Symposium on Frontiers in Repro-duction Research, Beijing, China, July 28–30, 1988. NIH:139–64.

12. Mason HD, Martikainen H, Beard RW, Anyaoku V, Franks S. Direct gonadotrophic effect of growth hormone on oestradiol production by human granulosa cells in vitro. J Endocrinol 1990;126:R1–4.

13. Voutilainen R, Miller WL. Coordinate tropic hormone regulation of mRNAs for insulin-like growth factor II and the cholesterol side-chain-cleavage enzyme P450 scc, in human steroidogenic tissues. Proc Natl Acad Sci USA 1987; 84:1590–4.

14. Hutchinson LA, Findlay JK, Herington AC. Growth hormone and insulin-like growth factor-I accelerate PMSG-induced differentiation of granulosa cells. Mol Cell Endocrinol 1988;55:61–9.

15. Mondschein JS, Canning SF, Miller DQ, Hammond JM. Insulin-like growth factors (IGFs) as autocrine/paracrine regulators of granulosa cell differentia-tion and growth: studies with a neutralizing monoclonal antibody to IGF-I. Biol Reprod 1989;41:79–85.

16. Hutchinson LA, Herington AC, Findlay JK. IGF-I mediates growth hormone- and FSH-stimulated steroidogenesis in cultured rat granulosa cells [Abstract No. 833]. Proc 69th annu meet Endocr Soc, Indianapolis, IN, June 10–12, 1987.

17. Findlay JK, Zhiwen Z, Hutchinson LA, Carson RS, Burger HG, Herington AC. Intrafollicular roles of inhibin and interferon-α related peptides. In: Haseltine F, Findlay JK, eds. Growth factors in fertility regulation. Cambridge University Press, 1990:195–208.

18. He H, Herington AC. Differentiation between the effects of IGF-1 and GH on PMSG-induced progesterone production by rat granulosa cells. Growth Reg 1991;1:65–71.

19. Gong JG, McBride D, Bramley TA, Webb R. Effects of recombinant bovine somatotrophin, insulin-like growth factor-I and insulin on the proliferation of bovine granulosa cells in vitro. J Endocrinol 1993;139:67–75.
20. He H, Herington AC, Roupas P. Involvement of G proteins in the effect of insulin-like growth factor-I on gonadotropin-induced rat granulosa cell differentiation. Growth Reg 1993.
21. Ricciarelli E, Hernandez ER, Tedeschi C, et al. Rat ovarian insulin-like growth factor binding protein-3: a growth hormone-dependent theca-interstitial cell-derived antigonadotropin. Endocrinology 1992;130:3092–4.
22. Gargosky SE, Nanto-Salonen K, Tapanainen P, Rosenfeld RG. Pregnancy in growth hormone-deficient rats: assessment of insulin-like growth factors (IGFs), IGF-binding proteins (IGFBPs) and IGFBP protease activity. J Endocrinol 1993;136:479–89.
23. Zhang ZW, Findlay JK, Carson RS, Herington AC, Burger HG. Transforming growth factor β enhances basal and FSH stimulated inhibin production by rat granulosa cells in vitro. Mol Cell Endocrinol 1988;58:161–6.
24. Xiao S, Findlay JK. Interactions between activin and FSH suppressing protein and their mechanisms of action on cultured rat granulosa cells. Mol Cell Endocrinol 1991;79:99–107.
25. Hasegewa Y, Miyamoto K, Abe Y, et al. Induction of follicle stimulating hormone receptor by erythroid differentiation factor on rat granulosa cell. Biochem Biophys Res Commun 1988;156:668–74.
26. Xiao S, Robertson DM, Findlay JK. Effects of activin and FSH-suppressing protein/follistatin on FSH receptors and differentiation of cultured rat granulosa cells. Endocrinology 1992;131:1009–16.
27. Gitay-Goren H, Kim I-C, Miggans ST, Schomberg DW. Transforming growth factor β modulates gonadotropin receptor expression in porcine and rat granulosa cells differently. Biol Reprod 1993;48:1284–9.
28. Adashi EY, Resnick CE, Hernandez ER, et al. Insulin-like growth factor-I as an amplifier of follicle-stimulated hormone action studies on mechanism(s) and site(s) of action in cultured rat granulosa cells. Endocrinology 1988;122:1583–91.

17

Evidence for a Role of Growth Hormone-Insulin-Like Growth Factor System in the Polycystic Ovary Syndrome (PCOS)

S.S.C. Yen, G.A. Laughlin, and A.J. Morales

The reproductive axis (GnRH-gonadotropin-ovarian system) is vulnerable to impacts of centrally initiated neuroendocrine aberrations, as well as peripherally instigated endocrine-metabolic derangements (1). Emerging evidence strongly supports the existence of a reciprocal regulatory system between insulin and the *growth hormone* (GH)-*insulin-like growth factor I* (IGF-I) axis with physiological implications of a cooperative control of metabolic fuels during feeding and fasting (2–4).

In *the polycystic ovary syndrome* (PCOS), the presence of *insulin resistance* (IR) and hyperinsulinemia has been viewed to be pivotal in the maintenance of hyperandrogenic chronic anovulation (5–7). That the GH-IGF-I system may be involved in this setting is indicated by the ability of insulin to inhibit *IGF binding protein 1* (IGFBP-1) and *sex hormone binding globulin* (SHBG) production that, respectively, modulate the bioavailability of IGF-I and sex steroids to target tissues (8, 9). While it is well recognized that these effects are amplified in the presence of obesity (5–7), the relative contribution of PCOS per se remains to be elucidated.

GH-IGF System

IGF-I and IGF-II are multifunctional polypeptides with mitogenic and insulin-like activities. They appear to have both endocrine (circulating compartment) and paracrine/autocrine (locally produced) modes of action. Circulating IGF-I is derived from the liver, and its synthesis and secretion are GH dependent (2, 3). IGFs are also synthesized in multiple extrahepatic tissues, including the ovary where they are under the control of FSH and play autocrine and paracrine roles in folliculogenesis (10, 11). Unlike insulin, circulating and tissue IGFs are associated with high-affinity IGFBPs that serve to modulate the bioavailability of IGFs at receptor sites.

Six IGFBPs that differ in molecular size, hormonal control, and functional significance have been characterized (12). Among them, IGFBP-1 and -3 are the best studied. IGFBP-3, a major binding moiety in serum, possesses the highest binding affinity for IGFs and is in a saturated state. Circulating concentrations of IGFBP-3 are regulated by GH and IGF-I levels (13). IGFBP-1 is a relatively low affinity binding protein and is unsaturated. Increments of circulating IGFs that occur are bound by IGFBP-1. Thus, IGFBP-1 serves as an acute modulator of bioactivity of IGFs, whereas IGFBP-3 functions as a reservoir for long-term monitoring of IGF levels (reviewed in 2, 3, 13).

Insulin is a major regulator of hepatic IGFBP-1 production, and it enhances the magnitude of IGFBP-1 translocation to extravascular space (2, 3, 13). In humans rapidly fluctuating serum insulin concentrations in response to meals result in inverse changes in serum IGFBP-1 levels, but not serum IGFBP-2 or -3 levels (14). This insulin-mediated diurnal rhythm for IGFBP-1 levels in serum and its extravascular translocation are consistent with the multicompartmental distribution and double experiential disappearance rate ($\alpha = 7.5\,min$, $\beta = 120\,min$) (2). These dynamics of serum IGFBP-1 and its relative unsaturated state account for most of the binding activity of newly delivered free IGF in circulation (2, 3). IGFBP-1 sequesters free IGF-I and inhibits its metabolic (insulin-like activity) and mitogenic actions. Thus, insulin regulation of hepatic IGFBP-1 production may coordinate insulin and IGF action with nutritional signals.

Insulin and the IGF System in PCOS

The presence of IR and secondary hyperinsulinemia in association with PCOS is now well recognized (15–17). The IR in PCOS appears to involve a novel mechanism of marked defects in glucose transport sensitivity without significant alterations in receptor dynamics (16, 17). The degree of IR varies markedly between lean and obese PCOS; using pharmacological challenges (i.e., hyperinsulinemic euglycemic clamp), IR can be revealed in both lean and obese PCOS, with the latter group having more IR and being associated with fasting hyperinsulinemia (15–17). It is inferred that a spectrum of IR and hyperinsulinemia exists in PCOS. Its metabolic impact during the intermittent feeding and physiological fasting of the daily life cycle may also vary according to the degree of compensatory hypersecretion of insulin. The relative involvement of the IGF-IGFBP-1 system should, therefore, be concordant with varying degrees of overexpression of insulin action on hepatic production of IGFBP-1 and SHBG.

IR in PCOS: Incipient Versus Overt

Investigations were conducted in 32 women of 4 distinct groups ($N = 8$ each, ages 18–36) characterized on the basis of the presence or absence of obesity for PCOS and their respective matched controls with regular menstrual cycles. Obese women had a *body mass index* (BMI) >30 kg/m^2, and lean women had a BMI <23 kg/m^2. Such distinct BMIs were chosen to highlight the impact of obesity and to discriminate the expression of IR in PCOS in response to metabolic demands during the daily feeding/fasting cycle.

Responses to Acute Glucose Challenge

A rapid *intravenous* (IV) *glucose tolerance test* (GTT) was performed on all subjects. Glucose (300 mg/kg) was administered IV within 2 min, and blood samples were obtained at 2, 4, 8, 19, 22, 30, 40, 50, 70, 90, and 180 min. At 20 min, regular insulin at 0.03–0.05 μ/kg (depends on BMI) was injected IV as a bolus. Serum insulin and plasma glucose levels were analyzed in all samples, and the insulin *sensitivity index* (S_i) was determined by using the MIN-/MOD computer program (©, R.N. Bergman) run on an IBM-compatible personal computer.

As shown in Figure 17.1, the S_i is significantly and progressively decreased in *lean PCOS* (LPCO), *obese control* (OC) and *obese PCOS* (OPCO) as compared to *lean control* (LC). This finding indicates that IR is present in PCOS as well as in obesity, and an additive effect is present in OPCO. This conclusion is highly consistent with hyperinsulinemic euglycemic clamp studies by Dunaif et al. (15) and insulin action studies of adipocytes in vitro by Ciaraldi et al. in PCOS with varying degrees of adiposity (17).

Responses to Physiologic Challenge

The *insulin* (I) and *glucose* (G) responses to the 3 standard meals during the feeding hours of the day (0800–2000 h) were measured and expressed as G : I ratio, which provides a peripheral index of IR. As shown in Figure 17.1, IR is no longer evident in LPCO, whereas OPCO exhibited marked IR with hyperinsulinemia during fasting and in response to meals. Again, OC independent of PCOS showed significant IR ($P < 0.001$) as compared to LC, and an additive effect of obesity and OPCO is again evident. Thus, hyperinsulinemia seen under physiologic conditions in OPCO, but

Acute glucose challenge   **Response to meals**

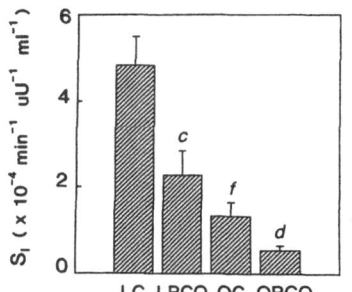
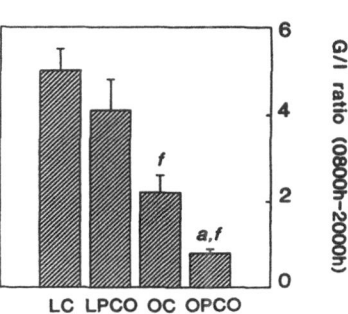

FIGURE 17.1. Insulin S_i as determined by rapid IV GTT and glucose:insulin ratio (G/I) during feeding phase of the day in lean control (LC), lean PCOS (LPCO), obese control (OC), and obese PCOS (OPCO). Significant levels between PCOS and control are indicated by the letters a ($P < 0.05$), b ($P < 0.01$), and c ($P < 0.001$); and between obese and lean groups, by d ($P < 0.05$), e ($P < 0.01$), and f ($P < 0.001$).

not LPCO, may inhibit hepatic production of IGFBP-1 (2, 3) and SHBG (8) by increasing the bioavailability of IGFs and sex steroids, both of which may exert impacts on the H-P-O axis.

24-Hour Metabolic Clock Studies

Assessments of the impact of obesity, hyperinsulinemia, and PCOS as independent variables on the IGF system during the 24-h metabolic clock revealed in LPCO, as in LC, normal excursions of glucose and insulin in response to meals, as well as a normal diurnal rhythm of IGFBP-1, with a decline during the feeding phase and a rise during nocturnal fasting. In contrast, OCs were hyperinsulinemic with normoglycemia, but attenuated IGFBP-1. In OPCO, the hyperinsulinemia occurs both during the day (feeding) and night (fasting) in association with hyperglycemia and a marked reduction of IGFBP-1 levels. In all groups 24-h integrated levels of IGF-I, IGF-II, and IGFBP-3 were unaltered (14). The integrated (24-h) data are presented in Figure 17.2. The reduction of IGFBP-1 levels is entirely insulin dependent (Fig. 17.3). As a consequence, the insulin-dependent increases in the IGF-I:IGFBP-1 ratio enhance the bioavailability of IGFs, and an endocrine mode of free IGFs with insulin-like action on target cells may prevail in OPCO (Fig. 17.4).

A direct endocrine action of circulating free IGFs and insulin on ovarian thecal-stromal cells and pituitary cells, where receptors for insulin and IGFs are present (18, 19), may function to augment LH action on androgen biosynthesis and secretion (20, 21) and on pituitary LH release (22,

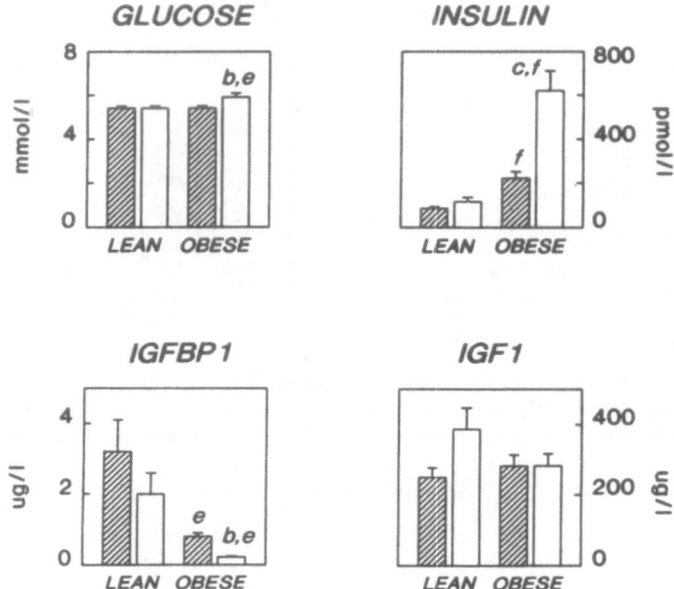

FIGURE 17.2. Integrated 24-h mean (± SE) concentrations obtained during feeding and fasting phases of the day for plasma glucose, serum insulin, IGFBP-1, and IGF-I in lean and obese PCOS (open bars) and their respective normal controls (hatched bars). Significant levels between PCOS and control are indicated by the letters a ($P < 0.05$), b ($P < 0.01$), and c ($P < 0.001$); and between obese and lean groups, by d ($P < 0.05$), e ($P < 0.01$), and f ($P < 0.001$).

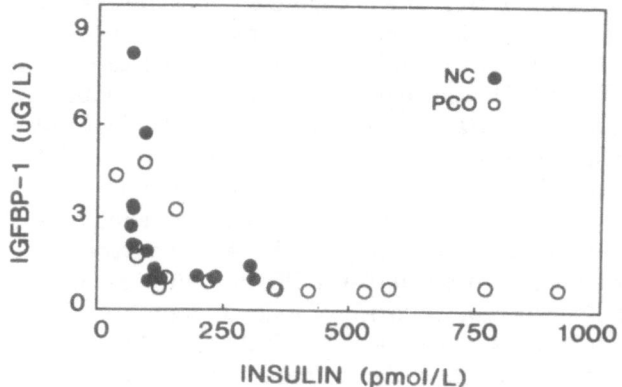

FIGURE 17.3. Apparent determination of IGFBP-1 levels by the relative insulin concentration in both normal controls (NC) and subjects with PCOS.

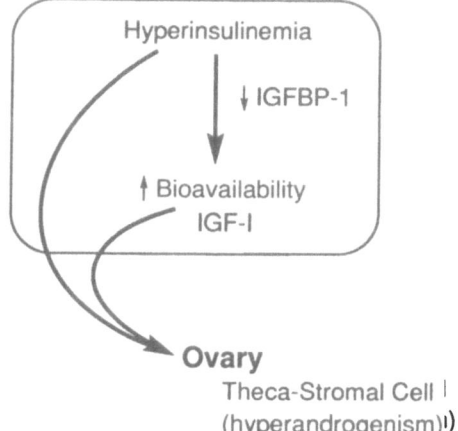

FIGURE 17.4. Diagrammatic illustration depicting the role of IR-hyperinsulinemia and its inhibitory effect on hepatic production of IGFBP-1. This event leads to an increased bioavailability of free IGF-I to the ovary, thereby enhancing LH-stimulated androgen biosynthesis in the thecal-stromal cell in PCOS.

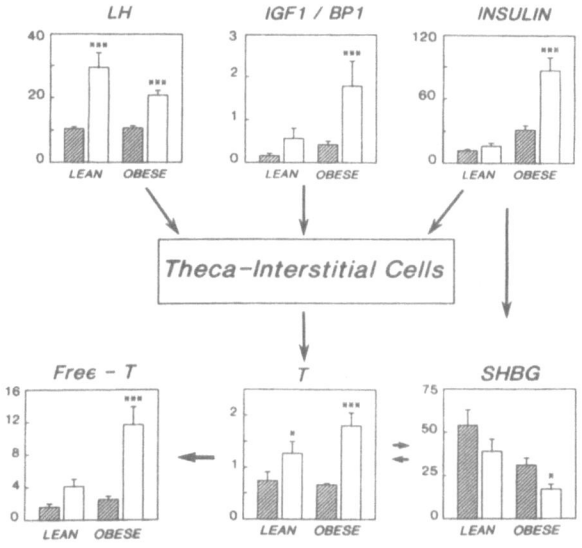

FIGURE 17.5. Integrated (24-h mean ± SE) endocrine input of LH (IU/L), IGF-I:IGFBP-1 (ratio × 100), and insulin (pmol/L) to thecal-stromal cells and responses of testosterone (T) (nmol/L), SHBG (nmol/L), and free T (T:SHBG ratio × 100) levels in LPCO and OPCO (open bars) and their respective normal controls (hatched bars). Significant levels between PCOS and control are indicated by the letters a ($P < 0.05$), b ($P < 0.01$), and c ($P < 0.001$); and between obese and lean groups, by d ($P < 0.05$), e ($P < 0.01$), and f ($P < 0.001$).

23). The resulting hyperandrogenemia, both within and outside the ovary, may inhibit follicular maturation and serve as a substrate for peripheral conversion to estrogen. It is suggested that the endocrine input to the ovarian thecal-stromal cells are remarkably different in PCOS with or without hyperinsulinemia (Fig. 17.5). Thus, in OPCO insulin and IGF-I, as well as LH, are involved, which in combination with the decreased SHBG level, results in a greater degree of hyperandrogenism in both free and bound forms. In LPCO, in contrast, the androgenesis by the thecal-stromal compartment can be accounted for solely by a greater elevation of LH than that in OPCO in the absence of alterations of circulating insulin and IGF-I levels. These differences in endocrine-metabolic inputs observed between LPCO and OPCO serve to fill an important gap for future elucidations of their pathophysiologic events.

Conclusion

In PCOS the IGF system is involved with increased bioavailable IGF-I to target cells. The occurrence of this change is highly dependent on the presence of obesity and is entirely insulin dependent. As a consequence, hepatic production of IGFBP-1 and SHBG is reduced. These changes, coupled with the synergistic action of IGF-I and insulin on LH-induced androgenesis in thecal-stromal cells, account for a more severe hyperandrogenic state in OPCO than in LPCO.

References

1. Yen SSC. Reproductive strategy in women: neuroendocrine basis of endogenous contraception. In: Roland R, ed. Neuroendocrinology of reproduction. Amsterdam: Excerpta Medica, 1988:231–9.
2. Holly JMP. The physiological role of IGFBP-1. Acta Endocrinol (Copenh) 1991;124:55–62.
3. Lee PDK, Conover CA, Powell DR. Regulation and function of insulin-like growth factor-binding protein-1. Proc Soc Exp Biol Med 1993;204:4–29.
4. Conover CA, Lee PDK, Kanaley JA, et al. Insulin regulation of insulin-like growth factor binding protein-1 in obese and nonobese humans. J Clin Endocrinol Metab 1992;74:1355–60.
5. Nestler JE, Clore JN, Blackard WG. The central role of obesity (hyperinsulinemia) in the pathogenesis of the polycystic ovary syndrome. Am J Obstet Gynecol 1989;161:1095.
6. Pasquali R, Casimirri F. The impact of obesity on hyperandrogenism and polycystic ovary syndrome in premenopausal women. Clin Endocrinol (Oxf) 1993;39:1–16.
7. Conway GS, Jacobs HS. Clinical implications of hyperinsulinemia in women. Clin Endocrinol (Oxf) 1993;39:623–32.

8. Nestler JE, Powers LP, Matt DW, et al. A direct effect of hyperinsulinemia on serum sex hormone-binding globulin in obese women with the polycystic ovary syndrome. J Clin Endocrinol Metab 1991;72:83–9.

9. Suikkari A-M, Koivisto VA, Rutanen E-M, et al. Insulin regulates the serum levels of low molecular weight insulin-like growth factor-binding protein. J Clin Endocrinol Metab 1988;66:266–72.

10. Adashi EY, Rohan M. Intraovarian regulation peptidergic signaling systems. TEM 1992;3:243–8.

11. Giudice LC. Insulin-like growth factors and ovarian follicular development. Endocr Rev 1992;13:641–69.

12. Shimasaki S, Ling N. Identification and molecular characterization of insulin-like growth factor binding proteins (IGFBP-1,-2,-3,-4,-5,-6). Prog Growth Factor Res 1991:243–66.

13. Clemmons DR. Insulin-like growth factor-binding proteins. TEM 1990;Nov/Dec:412–7.

14. Yen SSC, Laughlin GA, Morales AJ. Interface between extra- and intra-ovarian factors in polycystic ovarian syndrome. Ann NY Acad Sci 1993;687:98.

15. Dunaif A, Segal KR, Futterweit W, Dobrjansky A. Profound peripheral insulin resistance, independent of obesity, in polycystic ovary syndrome. Diabetes 1989;38:1165–74.

16. Dunaif A, Segal KR, Shelly DR, et al. Evidence for distinctive and intrinsic defects in insulin action in polycystic ovary syndrome. Diabetes 1992;41:1257–66.

17. Ciaraldi TP, El-Roeiy A, Madar Z, et al. Cellular mechanisms of insulin resistance in polycystic ovarian syndrome. J Clin Endocrinol Metab 1992;75:577–83.

18. El-Roeiy A, Chen X, Roberts VJ, Yen SSC. Localization of mRNAs encoding IGF-I, IGF-II, and receptors of IGF-I (IGF-Ir), IGF-II (IGF-IIr) and insulin (Ir) in normal human ovaries. J Clin Endocrinol Metab 1993;77:1411–8.

19. Bach MA, Bondy CA. The insulin-like growth factor system in the human pituitary. J Endocrinol 1993;1:187–91.

20. Barbieri RL, Makris A, Randall RW, et al. Insulin stimulates androgen accumulation in incubations of ovarian stroma obtained from women with hyperandrogenism. J Clin Endocrinol Metab 1986;62:904–10.

21. Poretsky L, Kalin MF. The gonadotropic function of insulin. Endocr Rev 1987;8:132–41.

22. Kanematsu T, Irahara M, Miyake T, et al. Effect of insulin-like growth factor-I on gonadotropin release from the hypothalamus-pituitary axis in vitro. Acta Endocrinol (Copenh) 1991;125:227–33.

23. Soldani R, Murphy A, Cagnacci A, Yen SSC. Insulin-like growth factors (IGFs) augment GnRH-mediated LH secretion from rat anterior pituitary cells in vitro. Acta Endocrinologica (in press).

18

Evidence Against a Role for the Growth Hormone-Insulin-Like Growth Factor System in the Polycystic Ovary Syndrome

STEPHEN FRANKS, DEBBIE WILLIS, DIANA HAMILTON-FAIRLEY, DAVINIA M. WHITE, AND HELEN D. MASON

Evidence that has accumulated from both clinical and in vitro studies suggests that each of the elements of the *growth hormone* (GH)-*insulin-like growth factor I* (IGF-I) system can affect the function of the human ovary (1). When administered in vivo, GH exerts a gonadotropic action on ovarian folliculogenesis; this effect is accompanied, although not necessarily mediated, by changes in circulating IGF-I (1–3). Studies in vitro have shown that GH has a direct, stimulatory effect on granulosa cell steroidogenesis (1, 4, 5), supporting the concept that it has the potential to act as a *cogonadotropin* under physiological conditions (1).

IGF-I is a potent stimulant of steroid production by human follicular cells and, under certain conditions, interacts synergistically with gonadotropins in this regard (6, 7). Recently, it has been shown that, as in the rat follicle, the *IGF binding proteins* (IGFBPs) IGFBP-1 and IGFBP-3 neutralize IGF-activated steroidogenesis by human granulosa cells (8). Intriguingly, both IGFBPs were able to block, partially, the effect of FSH alone (Fig. 18.1). The mechanism of this phenomenon remains unclear, but one possibility is that sequestration of endogenous IGFs is involved.

Significance of the GH-IGF System to the Human Ovary

There seems little doubt, therefore, that the GH-IGF-I system could play a part in human ovarian function, probably by modulation of the action of gonadotropins. It is an attractive proposition to consider that this may

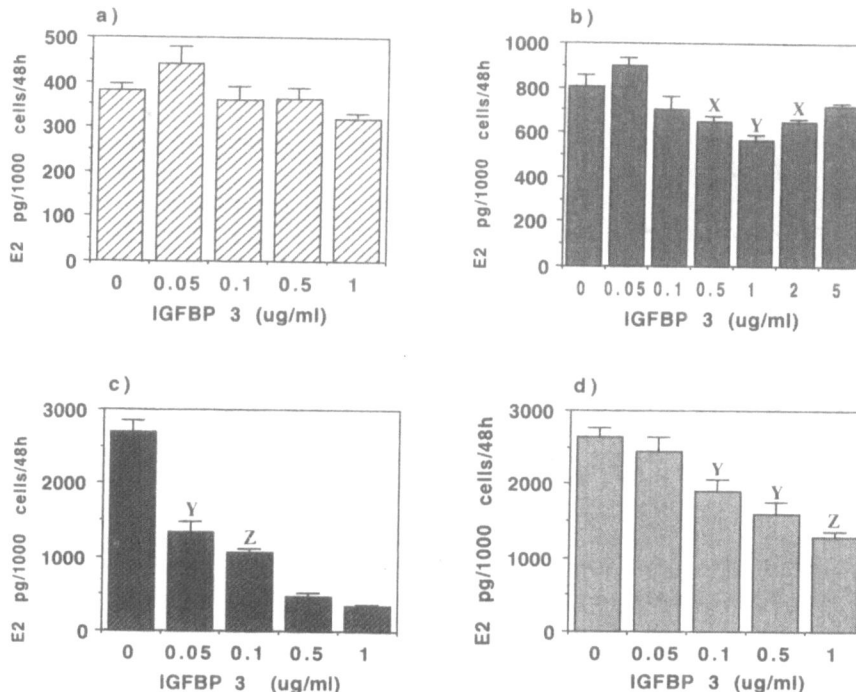

FIGURE 18.1. Estradiol (E_2) production in granulosa cells cultured in the presence or absence of a range of doses of IGFBP-3 in medium containing testosterone alone (*a*), 5 ng/mL FSH (*b*), 10 ng/mL IGF-I (*c*), or FSH + IGF-I (*d*). All cells were from one 18-mm follicle in a woman with regular menses. Results are mean (SE) of triplicate wells. (X = $P < 0.05$; Y = $P < 0.01$; Z = $P < 0.001$.) Reprinted with permission from Mason, Willis, Holly, Cwyfan-Hughes, Seppala, and Franks (8), © Elsevier Science Publishers, 1992.

be of physiological significance, especially during puberty when these effects could provide the means for the necessarily close coordination of somatic growth and reproductive development. There may also be a role for this system in "fine-tuning" gonadotropin action during the menstrual cycle. Excitingly, recent data suggest that the intraovarian IGF-IGFBP system may subserve the fundamental functions of folliculogenesis and atresia that are not necessarily dependent on cyclical gonadotropin activity (1, 9).

It is tempting, therefore, to suggest that abnormalities of the GH-IGF-I axis may play a part in the mechanism of disordered ovarian function, particularly in *polycystic ovary syndrome* (PCOS), which in the absence of a clear etiological mechanism, is a prime target for such speculation. The questions that need to be answered are (i) What is the evidence for abnormalities of the GH-IGF system in PCOS? and, more importantly (ii) are such disorders of primary pathogenetic importance, or do they merely reflect a metabolic and/or reproductive disturbance, the principal cause of which lies elsewhere? It will be argued that although derangements of the GH-IGF system have been identified in PCOS, these phenomena are secondary to abnormalities of glucose/insulin homeostasis and that hyperinsulinemia, per se, is a primary component of the mechanism of anovulation in hyperandrogenemic women with PCOS.

Abnormalities of the GH-IGF System in PCOS

Serum GH levels in PCOS have been variously decribed as low, normal, or elevated (1). The explanation for these disparities lies partly in the clinical characteristics of the populations of patients studied and partly in the method of blood collection. Specifically, GH secretion is impaired in the presence of obesity, and random blood sampling will give different and less reliable results from those obtained from multiple sampling during 24 h (1). If these variables are borne in mind, the evidence for a distinct disturbance of circulating GH concentrations is not compelling. There is more consistency in the reports of circulating IGF-I concentrations that, with one or two exceptions, suggest that these concentrations are normal in women with PCOS (1). This is illustrated in Figure 18.2, which demonstrates the overall median values derived from 24-h profiles of IGFBP-1, *sex hormone binding globulin* (SHBG), and insulin, as well as IGF-I.

Concentrations of IGFBP-1 have a marked diurnal variation, but are significantly lower throughout the day in PCOS individuals than those in age- and weight-matched controls. Levels of SHBG are also lower in PCOS subjects than in controls. At the same time, concentrations of insulin are strikingly higher than normal and show an inverse correlation with levels of both IGFBP-1 and SHBG. These data are in keeping with previous findings in women with PCOS (10–12) and are consistent with the hypothesis that suppression of IGFBP-1 (and of SHBG) is due to concordant hyperinsulinemia. This concept is further supported by the observation that insulin exerts a direct inhibitory effect on the production of both IGFBP-1 and SHBG by human hepatoma cells in vitro (13).

Intraovarian IGF System in Normal and Polycystic Ovaries

Recent studies suggest that the human ovary itself can synthesize as well as utilize IGFs and IGFBPs. We have detected, by specific radioimmunoassay, both IGF-I and IGF-II in conditioned medium from incubation of

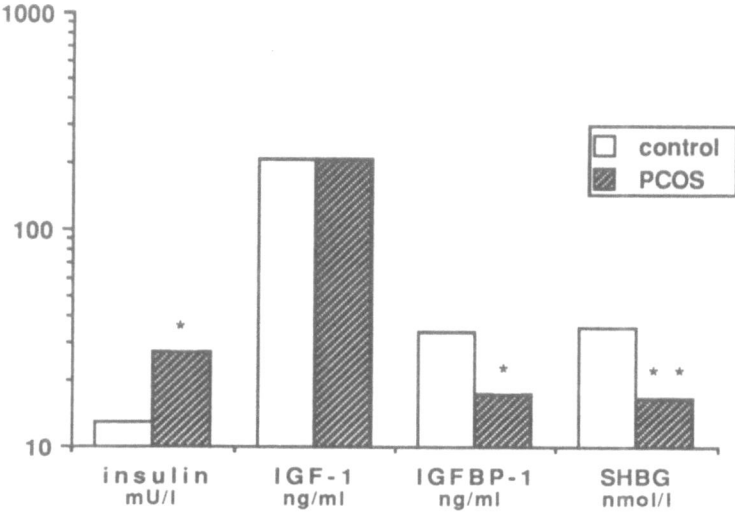

FIGURE 18.2. The 24-h median concentrations of insulin, IGF-I, IGFBP-1, and SHBG derived from samples measured at 2-h intervals in 10 women with anovulation and PCOS and 10 weight-matched controls with normal ovaries and regular cycles. (*$P < 0.01$; **$P < 0.007$; Mann-Whitney U-test.)

human ovarian theca and stroma (14). There were, however, no differences in IGF production between normal and *polycystic ovaries* (PCO). Similarly, analysis of IGFBPs, by either immunoassay or Western ligand blotting, revealed no difference in the expression of IGFBP-1, -2, and -4 by interstitial tissue of PCO compared with normal ovaries (14, 15). In granulosa cell cultures IGFBP-1 production by PCO follicles was also similar to that in normal ovaries (7) (Fig. 18.3). Furthermore, in a study of 45 *follicular fluid* (FF) samples, we found no difference between PCO and non-PCO follicles across a broad size range (7) (Fig. 18.4). The data regarding intraovarian IGFBP-1 concentrations are of particular interest since it may be inferred that the reduced circulating concentrations of this binding protein are not reflected by a similar change at ovarian level. In other words, low serum levels of IGFBP-1 cannot be assumed to have an impact on intraovarian IGF action.

San Roman and Magoffin have made the fascinating observation that FF samples aspirated from follicles of PCO contain much higher than normal concentrations of IGFBP-2 and -4 and the 29,000 binding protein (16). These findings, however, appear explicable largely on the basis of the higher number of atretic follicles aspirated from PCO. Indeed, Cataldo and Giudice have suggested that these and other binding proteins may be functionally important in the regulation of folliculogenesis and atresia

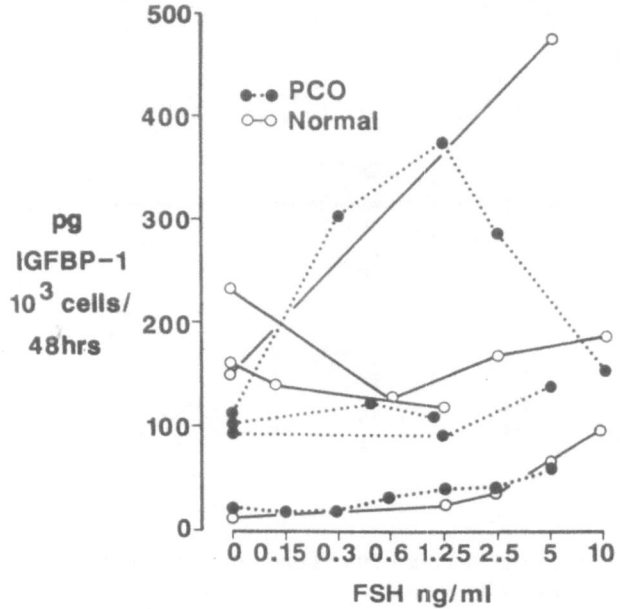

FIGURE 18.3. IGFBP-1 responses to FSH in granulosa cells from 4 PCO (solid circles) and 4 normal (open circles) ovaries. The means are shown; the error bars, which are small throughout, have been omitted for the sake of clarity. Reprinted with permission from Mason, Margara, Winston, Seppala, Koistinen, and Franks (7), © The Endocrine Society, 1993.

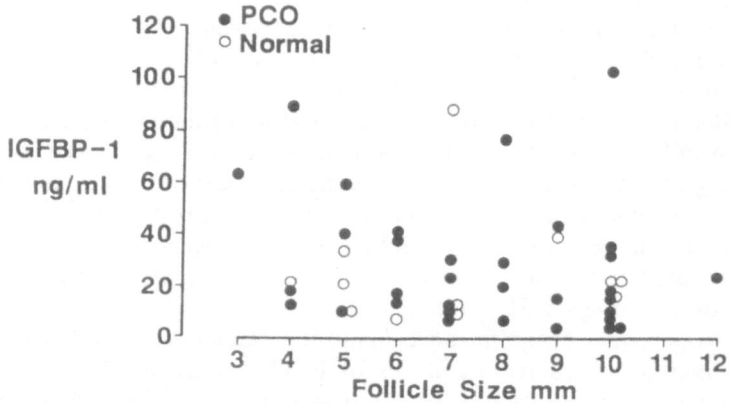

FIGURE 18.4. IGFBP-1 concentrations in FF from individual follicles from either PCO (solid circles) or normal (open circles) ovaries.

(17). The IGFBPs are, in turn, regulated by gonadotropins and growth factors (1, 16), and it seems unlikely that abnormalities of binding protein production represent the primary abnormality in the disordered folliculogenesis of PCOS.

Hyperinsulinemia and Anovulation in PCOS

The association between hyperinsulinemia and PCOS is now very well recognized (18). Hyperinsulinemia appears to reflect peripheral insulin resistance, but it has been unclear whether the ovary is resistant to insulin action or whether it is "reading" the supraphysiological levels of circulating insulin. Preliminary data from our own laboratory suggest that granulosa cells from women with PCOS retain responsiveness to insulin in vitro (19). Clinical studies have revealed that hyperinsulinemia is a feature of anovulatory women with PCOS, but it does not occur in equally hyperandrogenemic weight-matched subjects who have PCO, but regular cycles (20) (Fig. 18.5). These data suggest the possibility that hyperinsulinemia and/or insulin resistance are causally related to the mechanism of anovulation.

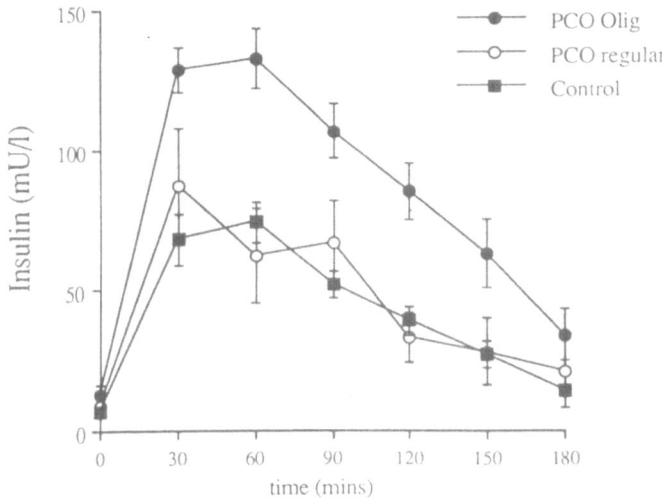

FIGURE 18.5. Insulin responses to 75-g oral glucose in women with PCO ($n = 72$) and weight-matched controls ($n = 30$). PCO subjects are subdivided into those with oligomenorrhea (solid circles) and those with regular cycles matched for weight and serum androgens (open circles). Controls are represented by solid squares. Reprinted with permission from Robinson, Kiddy, Gelding, et al. (20), © Blackwell Scientific Publications, 1993.

How can excessive insulin concentrations be implicated in the cause of anovulation in PCOS when the observed effect of insulin on the ovary is to *stimulate* steroidogenesis? The answer may be found in the interaction of insulin with *luteinizing hormone* (LH) in the granulosa cells of maturing follicles. In the preovulatory follicle LH has disparate effects on steroidogenesis (stimulation) and cell proliferation (inhibition) (21). The inhibitory effect of LH on cell growth is dose dependent and, in the normal cycle, is probably of physiological significance only at the onset of the LH surge. Women with PCOS typically have hypersecretion of LH, and recent data from our group indicate that in vitro, insulin preincubation will greatly enhance the steroidogenic response to LH in granulosa cell cultures (22). Our hypothesis is that in PCOS the combination of elevated levels of circulating LH and increased responsiveness to its action on the granulosa cell results in an LH "signal" that may enhance steroidogenesis, but prematurely arrests follicular maturation. Clearly, more work is required to test this hypothesis, but we would submit that hyperinsulinemia, together with hypersecretion of LH, could account for the mechanism of anovulation without the need to implicate derangement of the GH-IGF-I system.

Conclusion

GH, IGFs, and IGFBPs have all been demonstrated to affect steroidogenesis, and in some cases mitogenesis, in the human ovary. The human ovary has an intrinsic IGF system, and it seems reasonable to conclude that this is of functional importance in modulating the effects of gonadotropins on folliculogenesis. There is, however, no clear evidence of abnormalities of intraovarian production or function of IGFs and IGFBPs that specifically distinguish PCO from normal ovaries. Anovulatory women with PCOS are characterized by hypersecretion of LH and of insulin. A mechanism has been proposed by which interaction of these two endocrine factors could lead to the arrest of follicular maturation and, therefore, anovulation.

References

1. Katz E, Ricciarelli E, Adashi EY. Review: the potential relevance of growth hormone to female reproductive physiology and pathophysiology. Fertil Steril 1993;59:8–34.
2. Homburg R, West C, Torresani T, Jacobs HS. Cotreatment with human growth hormone and gonadotropins for induction of ovulation: a controlled clinical trial. Fertil Steril 1990;53:354–60.
3. Burger HG, Kovacs GT, Polson DW, et al. Ovarian sensitization to gonadotrophins by human growth hormone: persistance of the effect beyond the treated cycle. Clin Endocrinol (Oxf) 1991;35:119–22.

4. Mason HD, Martikainen H, Beard RW, Anyaoku V, Franks S. Direct gonadotrophic effect of growth hormone on oestradiol production by human granulosa cells. J Endocrinol 1990;126:R1–4.
5. Carlsson B, Bergh C, Bentham J, et al. Expression of functional growth hormone receptors in human granolosa cells. Hum Reprod 1992;7: 1205–9.
6. Erickson G, Garzo GV, Magoffin D. Insulin-like growth factor-1 regulates aromatase activity in human granulosa and granulosa-luteal cells. J Clin Endocrinol Metab 1989;69:716–24.
7. Mason HD, Margara R, Winston RML, Seppala M, Koistinen R, Franks S. Insulin-like growth factor-1 (IGF-1) inhibits production of IGF-binding protein-1 whilst stimulating estradiol secretion in granulosa cells from normal and polycystic ovaries. J Clin Endocrinol Metab 1993;76:1275–9.
8. Mason HD, Willis D, Holly JMP, Cwyfan-Hughes SC, Seppala M, Franks S. Inhibitory effects of insulin-like growth factor-binding proteins on steroidogenesis by human granulosa cells in culture. Mol Cell Endocrinol 1992;89: R1–4.
9. Giudice LC. Insulin-like growth factors and ovarian follicular development. Endocr Rev 1992;13:641–69.
10. Suikkari A-M, Ruutiainen K, Erkkola R. Low levels of low molecular weight insulin-like growth factor-binding protein in patients with polycystic ovarian disease. Hum Reprod 1989;4:136–9.
11. Kiddy DS, Hamilton-Fairley D, Seppala M, et al. Diet-induced changes in sex hormone-binding globulin and free testosterone in women with normal and polycystic ovaries: correlation with serum insulin and insulin-like growth factor I. Clin Endocrinol (Oxf) 1989;31:757–63.
12. Conway GS, Jacobs HS, Holly JMP, Wass JAH. Effects of luteinizing hormone, insulin-like growth factor-1 and insulin-like growth factor binding protein 1 in the polycystic ovary syndrome. Clin Endocrinol (Oxf) 1990;33: 593–603.
13. Singh A, Hamilton-Fairley D, Koistinen R, et al. Effect of insulin-like growth factor type 1 and insulin on the secretion of sex hormone-binding globulin and IGF-1 binding protein 1 by human hepatoma cells. J Endocrinol 1990; 124:R1–3.
14. Mason HD, Davies SC, Franks S, Holly JMP. Insulin-like growth factors 1 & 2 (IGFs 1 & 2) and IGF binding proteins are produced by normal and polycystic human ovaries [Abstract 26]. J Endocrinol 1991;131(suppl).
15. Mason HD, Willis D, Seppala M, Franks S. Insulin augments estradiol and inhibits IGFBP-1 secretion by normal and polycystic ovaries [Abstract 573]. Proc 75th annual meet Endocr Soc, Las Vegas, NV, 1993.
16. San Roman G, Magoffin DA. Insulin-like growth factor-binding proteins in ovarian follicles from women with polycystic ovarian disease: cellular source and levels in follicular fluid. J Clin Endocrinol Metab 1992;75:1010–6.
17. Cataldo NA, Giudice LC. Insulin-like growth factor-binding protein profiles in human ovarian follicular fluid correlate with follicular functional status. J Clin Endocrinol Metab 1992;74:821–9.
18. Dunaif A. Insulin resistance and ovarian dysfunction. In: Moller D, ed. Insulin resistance. New York: John Wiley, 1993:301–25.
19. Willis D, Mason HD, Franks S. Unpublished data.

20. Robinson S, Kiddy D, Gelding SV, et al. The relationship of insulin sensitivity to menstrual pattern in women with hyperandrogenism and polycystic ovaries. Clin Endocrinol (Oxf) 1993;39:351–5.
21. Yong EL, Baird DT, Hillier SG. Mediation of gonadotrophin-stimulated growth and differentiation by human granulosa cells by adenosine-3',5'-monophosphate: one molecule, two messages. Clin Endocrinol (Oxf) 1992; 37:51–8.
22. Willis D, Mason HD, Gilling-Smith C, Franks S. Insulin increases LH responsiveness of human granulosa cells. J Endocrinol 1993;139(suppl: abstract O29).

Part IV

Potential Utility of Growth Hormone in Clinical Reproductive Medicine

19

Diagnosing Growth Hormone Deficiency in Adults

PETER SÖNKSEN, ANDREW J. WEISSBERGER, AND KATHERINE VERIKIOU

Effects of GH and the Consequent Effects of GH Deficiency in Humans

It has long been recognized that *growth hormone* (GH) has potent effects both in vitro and in vivo. Until recently, however, it has only been considered as a form of hormone replacement for children.

Effects of GH in Humans

- Stimulates longitudinal bone growth
- Anabolic
- Lipolytic
- Diabetogenic
- Antinatriuretic

Proof that GH plays an important physiological role after attainment of adult stature has only been possible in the last six years since sufficient biosynthetic hormone has become available to allow placebo-controlled trials of GH replacement in adults with established and unequivocal *GH deficiency* (GHD). The concordance in the findings in the first 2 of these trials has been remarkable (1, 2), establishing the greater than expected potency and importance of GH in adults. It has also become clear that adults with GHD are disadvantaged in many ways not previously appreciated. The key features are given in the next section.

Adult GHD Syndrome

- Abnormal body composition
- Impaired physical performance

- Impaired quality of life
- Reduced life expectancy

Lean body mass (LBM), bone density, and extracellular body water are reduced, and body fat is increased (1, 2). Physical performance is impaired more than muscle strength (3, 4) and accounts for much of the morbidity of GHD, expressed by the patients themselves (often only if they are asked) as a "lack of energy and vitality" as well as "social isolation," which in reality means "I haven't the energy to go out tonight." Many aspects of quality of life are subnormal, sometimes to the degree that they may rate as an organic psychiatric disability (5). Some, but not all, of these symptoms originate from the physical weakness and loss of stamina associated with low LBM.

A large epidemiological study in and around Gothenburg (Sweden) has shown that hypopituitary adults on conventional hormone replacement with everything except GH have a substantially reduced life expectancy (6). They found an approximate doubling in overall mortality rates, mainly due to premature mortality from cardiovascular disorders in both men and women. Although further studies are needed to confirm this observation in other populations, the consistent finding of an excess of cardiovascular risk factors in this group of patients in many unrelated studies supports the epidemiological data.

Cardiovascular Risk Factors Found in Adults with GHD

- Increased body fat
- Central obesity
- Increased waist : hip ratio
- Raised serum cholesterol (particularly LDL)
- Raised triglycerides?
- Hyperinsulinemia?

It is of considerable historical interest that Dr. Maurice Raben, who pioneered the early extraction, purification, and clinical use of GH, made what are now known to be very accurate and astute observations about this yet undescribed syndrome and its response to GH replacement as early as 1963:

"One patient, a 35-year-old female, was treated in addition with hGH 3 mg three times a week. After two months of hGH she noticed increased vigour, ambition and sense of well-being. Observations will need to be made in more cases to indicate whether the favourable effect was more than coincidental (7)."

Effects of Age on GH Secretion and Action in Normal People

GH secretion reaches a peak during puberty and, unlike cortisol production, falls with increasing age (Fig. 19.1) (8). There is a similar pattern of plasma IGF-I levels in relation to age (9, 10). It appears that it is mainly reductions in sleep-related peaks of GH secretion that account for these changes (11). Rudman (9) was the first to draw attention to the possible relevance of this age-related fall in GH secretion to the changes in body composition that occur with advancing age. He demonstrated that with advancing age, an increasingly large proportion of the population had plasma IGF-I values below the laboratory normal range (Fig. 19.2). It has long been known that LBM falls with age, while at the same time, fat mass increases (12).

The mechanism(s) behind this progressive fall in GH secretion with advancing age remains to be clarified, but there is evidence to indicate that a number of processes are operating together. The number of GH (but not prolactin)-producing cells in the pituitary decreases with age (13), but it is not known whether this is "primary" atrophy or due to lack of hypothalamic stimulation. GH secretion in elderly individuals can be restored (at least partially) with repeated injections of *GH releasing hormone* (GHRH) (14), and this may indicate that a lack of hypothalamic stimulation is important in regulating GH-secreting cell mass in the pituitary. GH secretion can also be enhanced by a number of pharmacological maneuvers, including the administration of pyridostigmine (Fig. 19.3) (15). This implies that impaired neurotransmitter regulation of

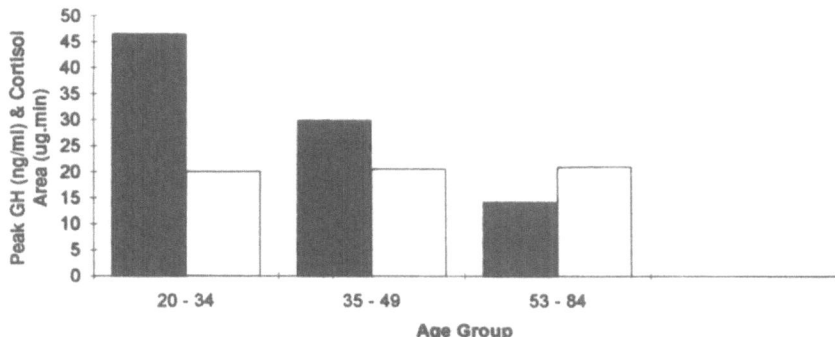

FIGURE 19.1. Effect of increasing age on GH and cortisol secretion in response to insulin-induced hypoglycemia. Redrawn with permission from Muggeo, Fedele, Tiengo, Molinari, and Crepaldi (8). Copyright © The Gerontological Society of America.

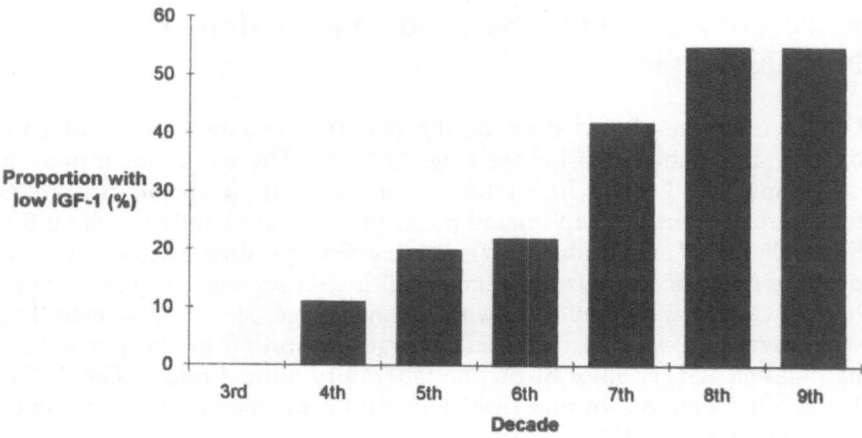

FIGURE 19.2. Effect of increasing age on the proportion of a normal population with plasma IGF-I below the range for young adults. Reproduced from the *Journal of Clinical Investigation*, 1981, 67:1361–1369 by copyright permission of the American Society for Clinical Investigation.

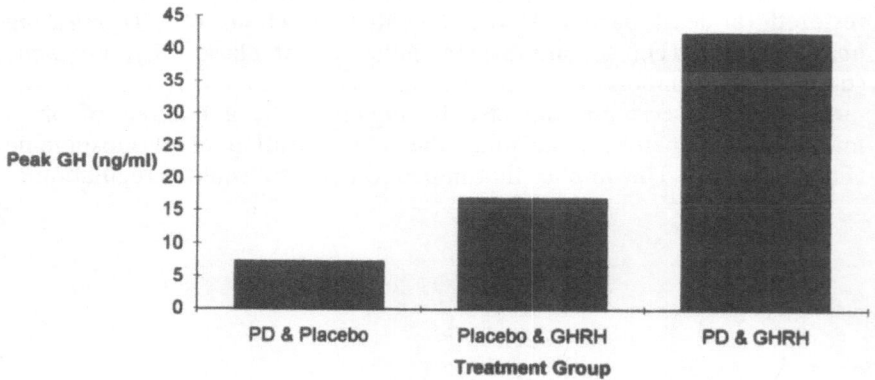

FIGURE 19.3. Effects of pyridostigmine and placebo on GHRH-induced GH secretion in elderly males. Redrawn with permission from Giusti, Marini, Sessarego, et al. (15).

GHRH, *somatostatin* (SRIH), or GH itself may also be important in the age-related decline in GH secretion.

The changes in LBM and in fat mass (and its distribution) and the pattern of increased plasma lipids associated with the fall in GH and IGF-I levels seen with increasing age are all remarkably similar to those seen in adults where the GHD results from pituitary disease or its treatment. Furthermore, Rudman and his colleagues have shown that in normal elderly individuals with low IGF-I levels, it is possible to reverse some of these changes by treatment with GH (16). There remains a distinct possibility that GH secretion fails with aging in a proportion of the population in a manner similar to that inevitably seen in the ovary, the so-called but controversial somatopause.

Adult-Onset Organic GHD

In order to investigate further the effects of normal aging and GH deficiency, we have examined a sequential group of adults with pituitary diseases who were referred to our unit for investigation and therapy. These cases have nearly all been investigated with an insulin hypoglycemia combined pituitary function test unless deemed unsafe because of epilepsy or ischemic heart disease. Their baseline measures of body composition and plasma IGF-I have been analyzed in relation to their GH response to hypoglycemia. These patients were all being investigated for a possible pituitary defect, and there was no control group of normal adults, although some of the patients' responses were normal by traditional criteria. There are, however, a number of points that arise from this analysis that we believe may contribute usefully to our understanding of the problems inherent in diagnosing GHD in adults where there is no clear pituitary pathology.

In all these patients hypoglycemia was induced by the *intravenous* (IV) injection of 0.1–0.15 U/kg of insulin under close, expert supervision in a specialist clinical investigation unit. In all cases included in the analysis, plasma glucose fell below 2 mmol/L, and symptomatic hypoglycemia developed. Plasma GH and IGF-I were measured in the routine hospital Chemical Pathology Department (Dr. M. Wheeler). The GH assay was a double monoclonal immunoradiometric assay with a sensitivity of 0.5 mU/L (0.25 ng/mL), while that for IGF-I was an extraction assay using the reagents kindly supplied by Drs. Teale and Marks (Guildford, UK) that was able to detect down to 5 nmol/L. The IGF-I values were interpreted in the context of an age-related normal range produced by Drs. Teale and Marks. The data reported were obtained over a 6-year period, 1988–1993, during which we routinely used IGF-I measurements in the investigation of pituitary disease. A summary of the underlying pathology of the patients is shown in Table 19.1.

Peak GH Response to Hypoglycemia

In Figure 19.4 patients have been categorized into 4 groups on the basis of peak GH to hypoglycemia. The majority of the patients (>60%) had a peak GH of less than 2 mU/L (1 ng/mL), a group we have categorized as

TABLE 19.1. Summary of underlying pathology in patients with pituitary disorders acquired in adulthood who underwent insulin tolerance testing at St. Thomas' Hospital, London, 1988–1993.

Chromophobe adenoma	23
Cushing's disease/Nelson syndrome	20
Macroprolactinoma	17
Craniopharyngioma	8
Sheehan syndrome	6
Pituitary apoplexy	5
Other	5
	84

Note: Excludes patients with acromegaly.

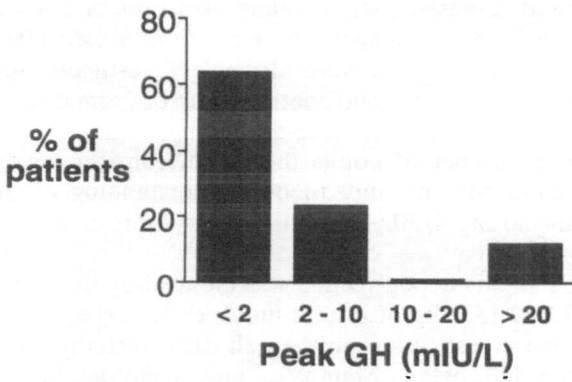

FIGURE 19.4. Proportion of patients (from Table 19.1) with TGHD (peak GH: <2 mU/L), PGHD (peak GH: 2–10 mU/L), gray zone (peak GH: 10–20 mU/L), and normal GH secretion (peak GH: >20 mU/L).

unequivocal *total GHD* (TGHD). Slightly more than 20% had a peak value of 2–10 mU/L (1–5 ng/mL), which we have accepted as *partial GHD* (PGHD). Only 1 patient had a result in the 10- to 20-mU/L (5- to 10-ng/mL) range, which we regard as the gray zone separating normal from abnormal. In order to eliminate any possibility that we are over-diagnosing GHD, we have classified this individual as not having GHD in the further analysis presented here. Examining the other baseline characteristics of those with TGHD and comparing them with all the rest, there was no difference in mean age (which is helpful in excluding significant age-related effects). BMI was slightly but significantly greater in the TGHD group (Fig. 19.5), but this may simply reflect the severity of the GHD. Although significantly lower, the mean IGF-I value of the TGHD

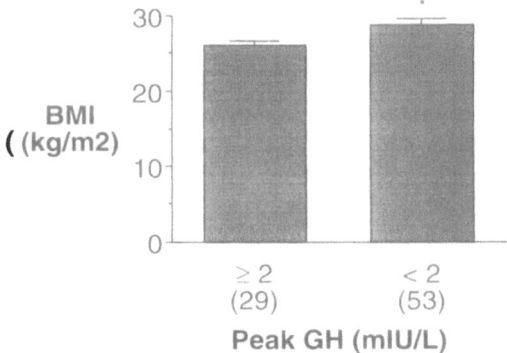

FIGURE 19.5. Mean (±SEM) BMI in those with TGHD compared with the rest of the group. (*P < 0.05.)

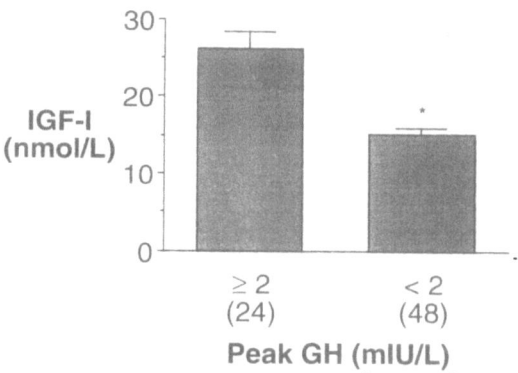

FIGURE 19.6. Mean (±SEM) serum IGF-I in the TGHD group compared with the rest of the group. (*P < 0.05.)

group was, rather surprisingly, still more than half that of the other patients (Fig. 19.6).

If the patients are subdivided into 2 groups on the basis of whether their IGF-I values fell above or below the normal age-related mean, it is clear that those with a low IGF-I have a very low peak GH (Fig. 19.7). If the subgroup with IGF-I below the normal age-related mean is in turn divided into 2 further subgroups—those still within the normal range (low-normal) and those with unequivocally low values—then the differences in peak GH are even more marked: None of the patients with a low IGF-I had a peak GH above 2 mU/L and are thus totally GH deficient (Fig. 19.8). Moreover, the vast majority of patients had IGF-I values in the lower half of the normal range, yet averaged a peak GH

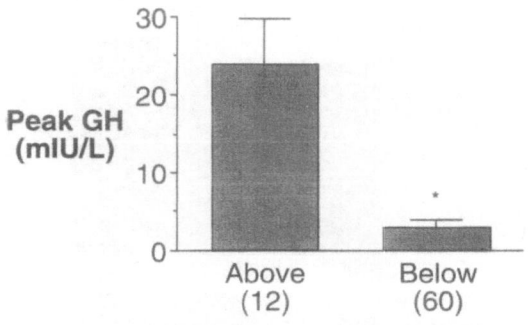

FIGURE 19.7. Mean (±SEM) peak GH during ITT in 2 subgroups selected on basis of their IGF-I being above or below the age-related mean. (*P < 0.05.)

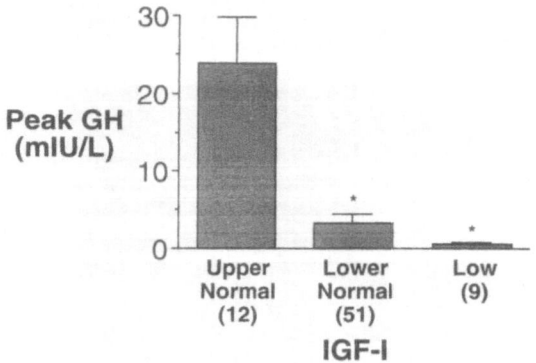

FIGURE 19.8. Mean (±SEM) peak GH in 3 subgroups selected on the basis of their serum IGF-I being above or below the age-related mean (but within the normal range) or actually below the lower limit of the age-related normal range. (*P < 0.05 for low and low-normal groups vs. upper-normal group.)

response of less than 5 mU/L (Fig. 19.8). Thus, in this series all those with a low IGF-I had GHD, but not all cases of GHD had low IGF-I.

We have also analyzed the data according to the current conventional hormone replacement therapy that the patients require and receive (which in each case has been determined by one of a team of specialist doctors working closely together, using similar criteria and good collaborative discussion and audit). A similar and consistent pattern of peak GH response is seen when the patients are subdivided on this basis (Fig. 19.9). The peak GH in those receiving replacement for 1 or 2 hormone deficiencies is intermediate between those with none and those with 3 or

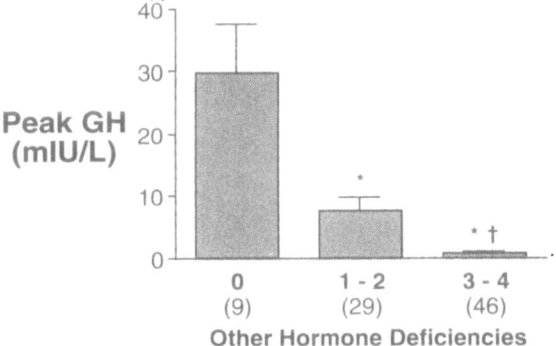

FIGURE 19.9. Mean (±SEM) peak GH during ITT in relation to the number of hormone deficiencies for which the patients are receiving replacement therapy. (*$P < 0.05$ vs. 0; [†]$P < 0.05$ vs. 1–2.)

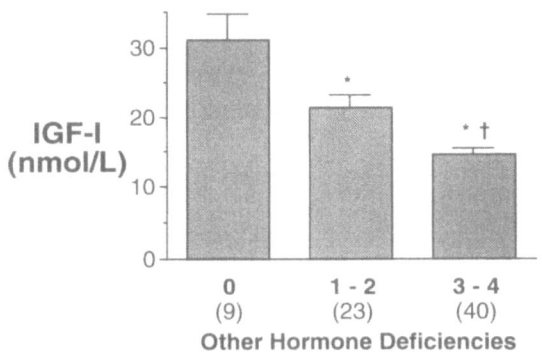

FIGURE 19.10. Mean (±SEM) serum IGF-I level in relation to the number of hormone deficiencies for which the patients are receiving replacement therapy. (*$P < 0.05$ vs. 0; [†]$P < 0.05$ vs. 1–2.)

4. There is also a graded fall in IGF-I with an increasing number of hormone deficiencies (Fig. 19.10).

When further examining the relationship between the number of pituitary hormone deficiencies and peak GH, there are at one end of the scale a few patients with true, isolated GHD (either total or partial) and otherwise completely normal pituitary function, while virtually all of those with 3–4 deficiencies are totally GHD, and the remainder are partially GHD (Fig. 19.11). A rather different pattern is seen in the distribution of patients in relation to the number of hormone deficiencies and their serum IGF-I (Fig. 19.12). None of the patients with either intact

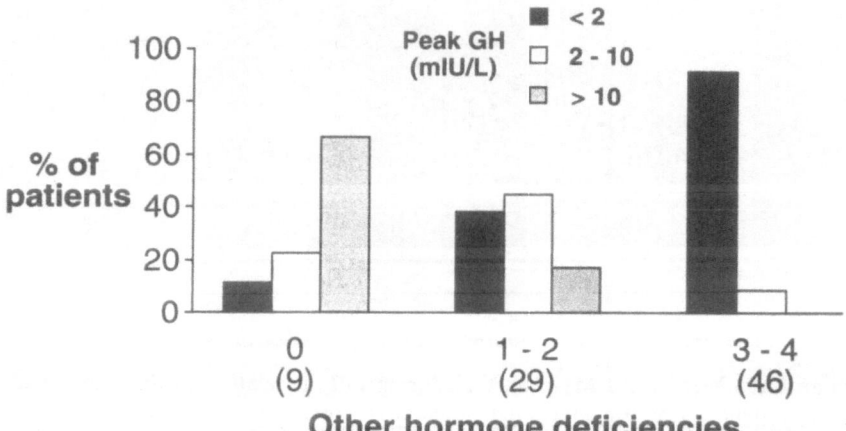

FIGURE 19.11. Distribution of patients within subgroups, defined by the number of coexisting hormone deficiencies receiving therapy, according to peak GH response to hypoglycemia.

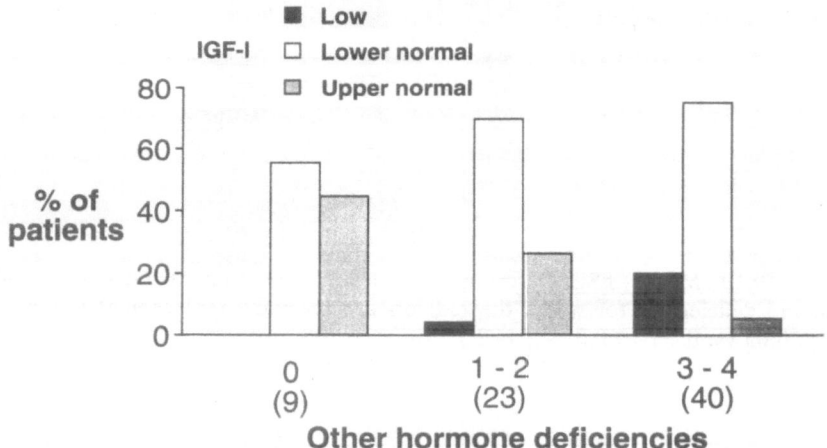

FIGURE 19.12. Distribution of patients within subgroups, defined by the number of coexisting hormone deficiencies receiving therapy, according to the serum IGF-I position within or below the age-related normal range.

pituitary function or isolated GHD had IGF-I values below the lower limit of the age-related normal range; likewise, the majority of those with 3–4 pituitary hormone deficiencies had IGF-I values in the normal range, albeit almost exclusively below the mean.

Making a Diagnosis of GHD in an Adult

In clinical endocrine practice it is now much easier to make a diagnosis of GHD, as we are now more aware of the syndrome and the benefits that can be achieved through GH replacement therapy. It was, however, only through our initial clinical trial and the fact that we incorporated several quality-of-life questionnaires into the trial protocol that we ever became aware of the extent of the loss of well-being these patients had been suffering in silence. In retrospect, most were so grateful to have retained their vision (or part of it) and to have had the "tumor" dealt with by fairly traumatic surgery, often followed by a long course of radiotherapy, that they seldom if ever complained. The clinical features are frequently obtained only by direct questioning (not leading questions, but just those targeted at the major problem areas), but when they do appear, they can be quite dramatic. As an illustration, there was one man who came with his wife and said that he was "really as well as could be expected," but when questioned directly about depression, admitted that he was actively thinking about suicide, and the last time this thought had come into his head was in the waiting room only 10 minutes before the consultation. His wife was quite unaware of this and was, not surprisingly, devastated. We need, therefore, to be alert to the syndrome and to seek actively to establish a diagnosis.

Clinical Features Suggesting GHD in Adults

- History
- Symptoms
- Signs
- Test results

The *history* of an adult with possible GHD should include any (i) known pituitary pathology/treatment and/or (ii) full conventional and appropriate hormone replacement. Possible *symptoms* are (i) impaired psychological well-being, including poor general health, impaired emotional reaction, depressed mood, impaired self control, anxiety, reduced vitality and energy, and increased social isolation; (ii) increased abdominal adiposity; and (iii) reduced strength and exercise capacity. Along with these symptoms, the following *signs* may suggest GHD: (i) mixed truncal/generalized obesity; (ii) increased waist:hip ratio; (iii) thin, dry skin, cool peripheries, and poor venous access; (iv) mild to moderate reduction in muscle strength; (v) moderate reduction in exercise performance; and (vi) a psychological state characterized by low, labile mood. Finally, in conjunction with the above, the following *test results* would suggest the possibility of adult GHD: (i) a stimulated GH level below 10 mU/L; (ii) low or low-normal IGF-I; (iii) hyperlipidemia ($\uparrow$ LDL cholesterol and $\uparrow$ TG); (iv) low

TABLE 19.2. Long-term (>6 months) follow-up of adult GHD patients at St. Thomas' Hospital, London ($N = 41$).

	Mean	Range
Age	47	22–67
Duration of treatment (months)	31	8–73
Dose (U/day)	2.2	0.7–5.3
Dose (U/kg/day)	0.029	0.008–0.079

glomerular filtration rate and renal plasma flow; (v) reduced LBI/increased fat mass; (vi) reduced basal metabolic rate; and (vii) reduced bone density.

St. Thomas' Hospital Longer-Term Experience of GH Replacement in Adults with GHD

We now have 41 patients on long-term GH replacement therapy (more than 6 months). All patients are taking a daily bedtime *subcutaneous* (sc) injection, and most are using the Kabi Pen, which makes the administration easier than with other methods. GH is more difficult to prepare, store, and inject than insulin, and it is vital that patients are trained by a suitably experienced nurse. Persuading and training patients to inject themselves have not been issues of significance, although many have initially said that they couldn't possibly ever do it.

Many, but by no means all, of these patients started GH in a formal placebo-controlled trial and then wished to continue with the treatment on completion of the trial. Some have been ineligible for formal trials through age or associated conditions and have been commenced on GH on an individual basis. We now start patients on a very low dose, usually 0.0125 U/kg per day, increasing the dose at intervals of 2–3 weeks, and monitoring their symptoms and general well-being, weight, body composition (impedance method), blood pressure, and serum IGF-I. We aim to keep the latter within the age-related normal range. A summary of the ages of our patients, doses, and duration of GH treatment is given in Table 19.2.

Adverse Reactions

If treatment is started with a low dose, such as that recommended above, there are infrequent but reported effects that include sodium retention with edema, hypertension, carpal tunnel syndrome, and benign intracranial hypertension. "Growing pains" with arthralgias and muscle aches are common. Occasionally patients have developed significant hyperglycemia and biochemical and clinical thyrotoxicosis. All appear to be dose related and remit on dose reduction.

The early trials of GH replacement in adults were associated with frequent side effects, mostly relating to initial sodium retention (which is probably linked to the biological action of GH in stimulating protein synthesis). This reflects the difficulty we initially had in deciding what dose should be used. We did not anticipate that GH would be as effective and potent as it turned out to be. In retrospect, although the dose used was comparable to that used in children, we should have been more cautious (as we have learned to be in hypothyroidism). Individual sensitivity varies enormously, and doses need monitoring and adjusting on an individual basis. In our patients on long-term treatment, the dose now averages 0.029 U/kg/day (Table 19.2), but there is a 10-fold range of sensitivity. Caution should be used in placing too much weight on serum IGF-I measurements, particularly if there is any underlying condition that may impair hepatic IGF-I generation (pancreatic or liver disease, malabsorption, insulin-dependent diabetes, etc.) since paracrine and autocrine IGF-I action could still play a role in the development of side effects.

Caution should also be taken regarding thyroid hormone replacement. GH activates the conversion of thyroxine to triiodothyronine, and some patients who have been stable on a given dose of thyroxine for many years develop frank thyrotoxicosis that in this age group may be masked, declaring itself through such cardiac manifestations as atrial fibrillation. There is clinical anecdotal evidence that a similar process may operate with people on testosterone replacement with depot injections, where after a long period of stability, greasy and spotty skin, increased libido, and priapism have occurred and disappeared after reduction in testosterone dose. These effects are dose related and not true side effects, as they are just an overexpression of GH's normal physiological role. However, as in the case of starting thyroxine in someone with longstanding hypothyroidism, it is imperative that we proceed slowly and with great care.

Suggestions for the Future

From our experience we believe that there are many people with adult GHD who are at risk as a result of their untreated GHD. In the UK, we are fortunate in that cost is the only restriction on the prescription of GH. Providing the GH replacement can be justified medically, it is available through the National Health Service at no direct cost to the patient. In order to justify the cost, the specialist has to work in close collaboration with the patient's primary care doctor (general practitioner), who is able to prescribe the GH providing he or she can convince local managers (Family Health Services Authority) that the treatment is effective and appropriate. At the present time, the UK National Health Service is (like most others) under extreme pressure to contain or cut expenditure. Even in that climate, we have found most GPs positive about GH replacement because we involve them from the outset in making the decision as to whether or not the patient has benefited from the additional hormone replacement.

Our current policy that we recommend others to consider is

1. Identify at-risk patient: Address and assess quality-of-life issues and assess mood (look actively for depression).
2. Look for clinical features.
3. Undertake baseline tests: blood pressure, body composition (height/weight/impedance), IGF-I, and quality-of-life assessment (NHP ± other similar questionnaires).
4. Perform provocative test (if results of recent test not available): insulin hypoglycemia (unless contraindicated), GHRH, arginine, and clonidine.
5. Use clinical judgment: Liaise with family doctor.
6. Consider a 3- to 6-month trial of therapy: Liaise with family doctor and with partner/family.
7. Evaluate in conjunction with partner/family and family doctor.

Conclusion

The diagnosis of GHD in adults with pituitary pathology is relatively straightforward, but many patients are lost to follow-up and may benefit enormously from therapy. Therapy is effective in the majority of patients, restores their perceived quality of life to normal, and improves vigor and productivity. Further research is needed on the longer-term benefits/risks of GH replacement in these people. The side effects of treatment are dose related and avoidable in the majority of cases by starting at a low dose and increasing it gradually. The issue of GHD as a component of the normal aging process is real and requires much further research.

References

1. Jørgensen JOL, Pedersen SA, Thuesen L, et al. Beneficial effects of growth hormone treatment in GH-deficient adults. Lancet 1989;i:1221–5.
2. Salomon FS, Cuneo RC, Hesp R, Sönksen PH. The effects of treatment with recombinant human growth hormone on body composition and metabolism in adults with growth hormone deficiency. N Engl J Med 1989;321:1797–803.
3. Cuneo RC, Salomon FS, Wiles CM, Hesp R, Sönksen PH. Growth hormone treatment in growth hormone-deficient adults, I. Effects on muscle mass and strength. J Appl Physiol 1991;70:688–94.

4. Cuneo RC, Salomon FS, Wiles CM, Hesp R, Sönksen PH. Growth hormone treatment in growth hormone-deficient adults, II. Effects on exercise performance. J Appl Physiol 1991;70:695–700.
5. McGauley GA, Cuneo RC, Salomon FS, Sönksen PH. Psychological well-being before and after growth hormone treatment in adults with growth hormone deficiency. Horm Res 1990;33(suppl 4):52–4.
6. Rósen T, Bengtsson B-Å. Premature mortality due to cardiovascular disease in hypopituitarism. Lancet 1990;336:285–8.
7. Raben M. Clinical use of human growth hormone. N Engl J Med 1962;266: 82–6.
8. Muggeo M, Fedele D, Tiengo A, Molinari M, Crepaldi G. Human growth hormone and cortisol response to insulin stimulation in ageing. J Gerontol 1975;30:546–51.
9. Rudman D, Kutner MH, Rogers MC, Lubin MF, Fleming GA, Bain RP. Impaired growth hormone secretion in the adult population. J Clin Invest 1981;67:1361–9.
10. Clemmons DR, Van Wyk JJ. Factors controlling blood concentration of somatomedin C. Clin Endocrinol Metab 1984;13:113–43.
11. Prinz PN, Weitzmann ED, Cunningham GR, Karacan I. Plasma growth hormone during sleep in young and aged men. J Gerontol 1983;38:519–24.
12. Forbes GB, Reina JC. Adult lean body mass declines with age: some longitudinal observations. Metabolism 1970;19:653–63.
13. Sun YK, Xi YP, Fenoglio CM, et al. Effect of age on the number of pituitary cells immunoreactive to growth hormone and prolactin. Hum Pathol 1984;15: 169–80.
14. Iovino M, Monteleone P, Steardo L. Repetitive growth hormone-releasing hormone administration restores the attenuated growth hormone (GH) response to GH-releasing hormone testing in normal ageing. J Clin Endocrinol Metab 1989;69:910–2.
15. Giusti M, Marini G, Sessarego P, et al. Growth hormone secretion in ageing: effect of pyridostigmine on growth hormone responsiveness to growth hormone-releasing hormone. Recenti Prog Med 1991;82:665–8.
16. Rudman D, Feller AG, Nagraj HS, et al. Effects of human growth hormone in men over 60 years old. N Engl J Med 1990;323:1–6.

20

Growth Hormone Delivery, Dosing, and Pharmacokinetics in Adults

Marc R. Blackman

GH Deficiency in Nonelderly Adults

Until the early 1980s, the suggestion that *growth hormone* (GH) might exert clinically significant effects past the time of adolescence was not accepted among most endocrinologists. This was because the only important physiologic action of GH was thought to be stimulation of the growth of long bones, while its actions on other tissues were appreciated mainly as pathological phenomena in acromegaly. Also, it was generally thought that adult patients with either isolated GH deficiency or pan-hypopituitarism—the latter typically treated with thyroid, adrenal, and sex hormones—did relatively well. Nonetheless, physicians treating appreciable numbers of such patients noted that many complained of a lack of energy and strength, low levels of libido, and a "plump" physique, with fat depots relatively resistant to exercise and dieting. The relationship of the latter problems to GH deficiency remained moot because the supply of human pituitary-extracted GH was only sufficient to treat GH-deficient short children.

More recently, body composition analyses of nonelderly adults with GH deficiency has revealed them to have increased whole-body and intraabdominal fat and reduced *lean body mass* (LBM) and bone density (1–3). Other studies have shown that cardiac output and reserve capacity are reduced (4, 5), skeletal muscle strength is diminished (6, 7), and glucose and lipid metabolism are altered unfavorably (1, 8) in such patients. Not surprisingly, the life expectancy of GH-deficient patients appears to be shorter than that of normal adults (9).

With the current, widespread availability of *recombinant human GH* (rhGH), it has become possible to carefully assess the utility of treating GH-deficient adults. Short-term (4–6 months) studies have demonstrated that GH increases nitrogen retention, protein synthesis, and LBM and decreases percent body fat (1, 6, 10–13) in these patients. There are also improvements in calcium retention and bone mass (14), as well as favorable effects of rhGH on skeletal muscle strength (6, 7, 11), cardiac function (5, 15), plasma lipid profiles (1), and, importantly, on measures of quality of life (16). Preliminary data suggest that long-term (e.g., 2 years) intervention with rhGH may be similarly beneficial in GH-deficient adults (17). Because GH exerts potent anabolic effects on so many tissues, substantial current research effort is under way to define the possible clinical utility of GH administration in adults suffering from a number of catabolic disease states. These include patients with burns, hip fractures, cancer cachexia, AIDS, gastrointestinal malabsorptive or short bowel syndromes, perioperative or critical care stress, and neurodegenerative diseases, to name a few. Further research will be necessary to determine the safety, efficacy, and cost benefit of GH administration in nonelderly adults with panhypopituitarism or isolated GH deficiency, as well as in these other patient groups.

Aging Effects on Body Composition and Metabolism and on GH

Elderly adults who do not have pituitary disease per se may also benefit from GH replacement. Aging is associated with a loss of LBM (especially skeletal muscle) and bone mass and an increase in percent body fat (18–20), changes similar to those occurring in adult GH deficiency. Moreover, with aging there is a redistribution of body fat to the intra-abdominal fat depots (19). A centripetal fat distribution is associated with reduced glucose tolerance and increased triglyceride and LDL cholesterol levels (21), changes also seen in old persons (22). Finally, altered cardiac (4, 23, 24) and immune (25–27) function are concomitants of both aging and GH deficiency.

Numerous studies reveal that GH secretion and plasma levels of *insulin-like growth factor I* (IGF-I) decline with age in healthy men and women (28). Endogenous GH is secreted episodically, with barely detectible basal levels prevailing during interpulse intervals. GH pulse amplitude increases substantially during slow wave sleep, leading to a diurnal rhythmicity in which most GH is secreted at night. Thus, frequent diurnal (or 12-h nocturnal) sampling is necessary to assess the magnitude of GH secretion. Numerous studies comparing spontaneous GH secretion in young versus old men and women have revealed decreases in GH pulse amplitude and 24-h mean integrated levels of GH with age (29–34). Mean reductions in integrated GH secretion range from 50% to 70% in persons over 65 years of age, and many older individuals secrete as little GH as patients with clinical GH deficiency. The decrease in 24-h GH secretion rates with age appears to be related to lower pulse amplitude, especially during the night, rather than to diminished pulse frequency (29, 35).

Several short-term studies have assessed the effects of GH administration on body composition and function in healthy elderly persons. In a placebo-controlled study (36) in which rhGH was administered to 12 older men in a VA hospital clinic setting, increases in LBM and reductions in percent body fat were observed, but increases in bone density were not convincingly demonstrated. Other studies of GH replacement in older persons have confirmed these findings, as well as demonstrated GH treatment to produce positive nitrogen balance, calcium retention, reductions in cholesterol, and increased utilization of fatty acids for energy metabolism (37–40). To put these studies in perspective, however, it must be pointed out that (i) all have been relatively small, (ii) none have looked for (or found) truly functional or clinical benefits of GH replacement, and (iii) the incidence of adverse effects of exogenous GH, especially nerve-trapping syndromes and fluid retention, appeared to be substantial. Thus, larger, more clinically oriented studies will be needed to assess the risk:benefit ratio of GH replacement in the elderly.

Pharmacokinetics of GH

Endogenous *human pituitary growth hormone* (hGH) is composed of several molecular species, including 22-kd, 20-kd, dimeric, and oligomeric forms. By contrast, biosynthetic hGH and rhGH consist of monomeric 22-kd GH. Some biosynthetic hGH preparations contain an additional methionine residue at the N-terminus, and early preparations were immunogenic because of contaminating *E. coli* proteins acting as adjuvants. The current availability of rhGH has evoked increased interest in investigating wider indications for GH administration, particularly in adults. Consequently, it is necessary to define the pharmacokinetic properties of rhGH and optimize the route, frequency, timing, dose, and duration of rhGH therapy for specific uses in men and women of various ages.

Early studies suggested that there are no major pharmacokinetic differences between endogenous (pituitary) and exogenous (rhGH) GH. In a recent review of this topic (41), Jorgensen noted considerable variation in the reported values for the MCR ($93.5-235 \, L/day/m^2$), distribution space ($46.7-84.5 \, mL/kg \, BW$), and serum half-time ($T_{1/2}$) of GH. Endogenous GH is cleared according to its molecular size (42), and elimination is considered to follow first-order kinetics (42–44), although a slower phase of disappearance (43) occurs at very low levels of GH. The volume of distribution of GH is intermediate between that of plasma volume and extracellular water.

Variation in reported pharmacokinetic properties of GH can be explained, in part, by the fact that earlier studies were generally conducted using injections of ^{125}I-labeled pituitary extracts, disappearance of GH after removal of pituitary tumors in acromegalic patients, or infusions to equilibrium of exogenous GH. Moreover, in most of these investigations, the possible confounding effects of endogenous GH, or alterations in diurnal rhythmicity of GH secretion and/or disposition, *GH binding proteins* (GHBPs), body fat mass and distribution, renal function, gonadal steroids, age, diet, state of nutriture, or level of physical activity were not completely eliminated. More recent studies have attempted to account for many of these potential confounders and have employed suppression of endogenous GH (with *somatostatin* [SRIH] or its analogs), frequent venous blood sampling techniques, ultrasensitive (e.g., IRMA or chemiluminescence) GH assays, and sophisticated computer algorithmic assessments (e.g., deconvolution analysis) to ascertain the pharmacokinetic properties of rhGH under diverse physiological and pathophysiological conditions.

In one recent investigation (45) conducted in healthy young men, the rates of endogenous and exogenous (rhGH) GH disappearance from the circulation ($T_{1/2}$) were compared in the morning versus night. The $T_{1/2}$ for endogenous GH was 23 ± 1.1 min (mean $\pm$ SE) in the morning versus 26 ± 1.7 min in the evening ($P < 0.02$), whereas the corresponding values for monomeric 22,000 rhGH were 14 ± 1.0 versus 19 ± 1.0 ($P < 0.01$). For endogenous GH the $T_{1/2}$ correlated negatively with estradiol ($P < 0.01$) and positively with *sex hormone binding globulin* (SHBG) ($P < 0.03$), whereas the $T_{1/2}$ for rhGH was inversely related to both height ($P < 0.05$) and weight ($P < 0.01$). Thus, in young men, the $T_{1/2}$ appears to be faster for monomeric rhGH than for endogenous GH and faster in the morning than at night. Moreover, sex steroid and body size contribute to the variability of the $T_{1/2}$ of GH.

Obesity and age exert independent negative effects on characteristics of GH secretion and clearance in men. In an earlier study Veldhuis et al. (46), using deconvolution analysis, reported that obese, compared with nonobese, men exhibited reductions in both the $T_{1/2}$ of endogenous GH (11.7 ± 1.6 vs. 15.5 ± 0.8 min; $P < 0.01$) and in GH secretory burst frequency (3.2 ± 0.5 vs. 9.7 ± 0.7/day; $P < 0.01$), but not in other measured parameters of GH secretion. There were significant negative correlations between the degree of obesity, as assessed by the *body mass index* (BMI), and both daily GH secretion and GH secretory burst frequency. As a consequence, the daily GH production rate in obese men was reduced to 25% of that in control men.

More recently, in a study of healthy nonobese men of different ages, these same investigators reported that age was associated with significant reductions in the $T_{1/2}$ of endogenous GH and GH secretory burst frequency, whereas the BMI was associated with significant decreases in the $T_{1/2}$ of GH and in GH secretory burst amplitude (47). There were independent negative correlations between the daily GH secretory rate and both age ($P < 0.004$) and BMI ($P < 0.03$), which together accounted for about 60% of the variability in daily GH production rates ($P <$ 0.0006). Although serum testosterone concentrations were inversely related to both age and BMI ($P < 0.02$), multiple regression analysis revealed that the negative effects of age on GH secretory burst frequency were independent of both BMI and testosterone level. Because of the greater effects of aging than of obesity, the daily GH production rate in men was estimated to decrease by about 14% for each advancing decade and by about 6% for each incremental unit of BMI.

In another investigation, healthy older versus younger men exhibited a 3- to 4-fold greater renal clearance rate of rhGH, whereas the metabolic clearance rate/body weight and $T_{1/2}$ for GH did not differ with age (48). These data suggest that renal tubular reabsorption of GH is diminished, whereas the disappearance of GH is not altered, with advancing age. More recently, Haffner et al. examined the effects of GH plasma concentration and *glomerular filtration rate* (GFR) on the pharmacokinetics of rhGH in a group of healthy and uremic males and females of different ages (49). In all subjects increasing plasma GH levels were associated with a decreased MCR and an increased $T_{1/2}$ for GH. In patients with chronic renal failure compared with control subjects, the MCR of GH was reduced by about 50%, and the $T_{1/2}$ for GH was increased by about 25%–50%. The estimated renal fraction of total MCR was reduced from 25%–53% in normal subjects to 4%–15% in uremic patients. The apparent volume of GH distribution was similar in normal and uremic subjects. Thus, the MCR of GH appears to be a function of both the plasma GH concentration and the GFR. In addition, extrarenal elimination of GH, presumably via (hepatic) *GH receptor* (GH-R)-mediated mechanisms, may be saturable at high physiological concentrations of GH, whereas renal MCR is independent of plasma GH levels.

Several implications derive from the latter study. First, the data may help resolve prior contradictory results with regard to age- or sex-related differences in the MCR of GH. Second, they suggest that computer algorithms, such as deconvolution analysis, need to account for the dependence of $T_{1/2}$ for GH on the instantaneous GH concentration. Third, at any given dose of rhGH adjusted for body-surface area, the average GH concentrations to which target tissues are exposed will be inversely related to the GFR.

Route of GH Administration

The advent of *subcutaneous* (sc) treatment regimens has been prompted by the recognition that sc versus *intramuscular* (I.M.) administration of GH is characterized by slower appearance (T_{max}) and disappearance phases in serum; lower, more physiological peak values (C_{max}); and, possibly, lesser bioavailability due to local degradation (41). There is also a positive relationship between injection volume and absorption rate, maximum concentration (C_{max}), and bioavailability (50), perhaps because of enhanced hormone exposure to the sc capillary bed and decreased degradation. Subcutaneous injections do not lead to more local side effects or antibody formation and are associated with improved patient compliance. It remains to be proved whether the anabolic, endocrine-metabolic, or other tissue responses to GH in younger or older adults are better after nightly sc GH injections versus after less frequent I.M. administration. The pharmacokinetic characteristics of GH when administered by injection pen devices are similar to those after routine administration with a needle and syringe, although patient compliance with the former regimen may be greater (51).

Intranasal GH administration has also been investigated. In one pharmacokinetic study (52) conducted in GH-deficient children, various doses of rhGH, in combination with a membrane permeation enhancer, were compared with single sc injections of GH. For all 3 intranasal doses, C_{max} values occurred at 20–30 min, with a decline to baseline at 2–3 h. The C_{max} values of the intranasal doses varied from 5.7%–15.8%— whereas the uptakes (AUCs) were only 1.6%–3%—of the corresponding values for the sc dose. The C_{max} values after intranasal GH were similar to endogenous physiological peak GH levels. No significant short-term side effects other than transient nasal irritation were noted.

Frequency and Timing of GH Administration

GH is normally secreted in rhythmic pulses, with highest frequency and amplitude associated with stages III to IV sleep. Using sensitive, but conventional, immunoradiometric assays for GH, it has not yet been possible to completely characterize and quantify daytime spontaneous GH secretory patterns, although the advent of ultrasensitive chemiluminescent assays for GH should allow for resolution of this difficulty. The pattern of GH release is influenced by a number of factors, including age, gonadal steroids, food intake, state of nutriture, and physical activity. Prior treatment regimens utilizing I.M. or sc GH injections 2–3 times weekly cannot approximate the normal endogenous diurnal rhythmicity of GH secretion. In GH-deficient children the change to daily sc injections of GH was associated with improved growth and metabolic responses in some (53–56), but not other (57), studies. In one short-term study in GH-deficient adults (58), the same cumulative dose of *intravenously* (IV) administered GH, given as 8 bolus injections or as a continuous infusion, led to higher IGF-I levels and integrated GH release profiles than did GH given as 2 bolus injections. To date, there have been no long-term studies comparing the effects of administering GH by daily sc or I.M. injections with those achieved by continuous sc versus IV infusions.

The relative efficacy of nocturnal versus daytime administration of GH in altering anabolic and other metabolic outcomes remains a subject of controversy. In one 7-day study in GH-deficient children, nitrogen balance increased to a greater extent after nighttime versus daytime GH injections (59), whereas in another study in which GH was given 3 times weekly for 4 months, no difference in linear growth response was observed after daytime versus nighttime treatment (60). By contrast, increased long-term growth velocity has been reported after changing from less frequent daytime injections, given for an average of 5 years, to nightly treatment with GH (53–56). In a single, randomized study of nonelderly adults with GH deficiency, comparing 4-week periods of evening, daytime, and no GH injections, more physiological nocturnal GH secretory patterns, lipid profiles, and levels of the gluconeogenic precursors alanine and lactate were observed after nighttime versus daytime hormone administration (61). To date, there have been no reports comparing the short- or long-term effects of evening versus morning administration of GH in other nonelderly or elderly adult populations with GH-deficient states.

Dose of GH Treatment

Assessment of the production rate of GH is often used as the basis for selection of the GH dose necessary for replacement therapy. The *production rate* (PR) can be calculated by using the following formula: PR = MCR × endogenous GH concentration (62). Use of the more sophisticated statistical technique of deconvolution affords the opportunity to determine the PR or secretory rate at any given point on the serum GH profile, assuming the MCR is known (63). Most data on the PR of GH derive from short-term studies in normal, nonelderly adults and reveal values of about $0.3–0.5 \, mg/24 \, h/m^2$ body-surface area (64, 65). As noted above, recent studies emphasize that the MCR of GH is dependent on both the serum concentration and the GFR of GH (49), thus explaining some of the previously noted variations in GH MCR based on age, sex, body composition, and state of nutriture.

Ultimately, assignment of the replacement dose of GH must be based on the clinical response to that dosage of GH, which will vary depending on the specific outcome measure assessed, and the physiological or pathophysiological state of the subject. As recently reviewed (28), there is at present insufficient information regarding optimal dosimetry of GH in the short- or long-term treatment of GH-deficient adults. Because the side effects of GH administration (e.g., salt and water retention, carpal tunnel syndrome, arthralgias, and glucose intolerance) appear to increase in frequency with advancing patient age, dosage regimens of GH for nonelderly and elderly adults have been lowered, and most patients now being treated with nightly sc doses of rhGH receive 0.05–0.15 μg/kg/day. Undoubtedly, these recommendations will be modified in the future.

Duration of GH Treatment

Most studies of GH replacement therapy in GH-deficient children have revealed diminished growth responsivity after long-term versus short-term treatment (41). By contrast, there is relatively little information regarding optimal duration of hormone treatment in GH-deficient nonelderly or elderly adults. In part, this is because the time necessary to elicit the diverse anabolic and metabolic effects of GH in adults is unknown and likely to differ depending on the target tissue or system being assessed (e.g., skeletal muscle, bone, immune, and psychobehavioral systems) and the (patho)physiological state of the subject. Not surprisingly, studies in GH-deficient adults have revealed that biochemical and radiographic changes resulting from GH treatment precede clinically and functionally relevant effects, with the former occurring after days or a few weeks and the latter occurring after several months (28). A recent preliminary report suggests that GH treatment of such patients may continue to exert clinically beneficial effects for periods of 2 years (17).

References

1. Salomon F, Cuneo RC, Hesp R, Sönksen PH. The effects of treatment with recombinant human growth hormone on body composition and metabolism in adults with growth hormone deficiency. N Engl J Med 1989;321:1797–803.
2. Christiansen JS, Jørgensen JOL. Beneficial effects of GH replacement therapy in adults. Acta Endocrinol (Copenh) 1991;125:7–13.
3. Jørgensen JOL, Blum WF, Møller N, Ranke MB, Christiansen JS. Short-term changes in serum insulin-like growth factors (IGF) and binding protein 3 after different modes of intravenous growth hormone (GH) exposure in GH-deficient patients. J Clin Endocrinol Metab 1991;72:582–7.
4. Merola BAC, Cittadini A, Colao A, et al. Cardiac structural and functional abnormalities in adult patients with growth hormone deficiency. J Clin Endocrinol Metab 1993;77:1658–61.
5. Amato G, Carella C, Fazio S, et al. Body composition, bone metabolism, and heart structure and function in growth hormone (GH)-deficient adults before and after GH replacement therapy at low doses. J Clin Endocrinol Metab 1993;77:1671–6.
6. Jørgensen JOL, Thuesen L, Ingemann-Hansen T, et al. Beneficial effects of growth hormone treatment in GH-deficient adults. Lancet 1989;1:1221–5.
7. Cuneo RC, Salomon F, Wiles CM, Sönksen PH. Skeletal muscle performance in adults with growth hormone deficiency. Horm Metab Res 1990;33(suppl 4):55–60.

8. Bonnet F, Lodeweyckx MV, Eeckels R, Malvaux P. Subcutaneous adipose tissue and lipids in blood in growth hormone deficiency before and after treatment with human growth hormone. Pediatr Res 1974;8:800–5.

9. Rosen T, Bengtsson BA. Premature mortality due to cardiovascular disease in hypopituitarism. Lancet 1990;336:285–8.

10. Crist DM, Peake GT, Mackinnon LT, Sibbit WL, Kraner JC. Exogenous growth hormone treatment alters body composition and increases natural killer cell activity in women with impaired endogenous growth hormone secretion. Metabolism 1987;36:1115–7.

11. Jørgensen JOL, Pedersen SA, Thuesen L, et al. Long-term growth hormone treatment in the growth hormone deficient adults. Acta Endocrinol (Copenh) 1991;125:449–53.

12. Crist DM, Peake GT, Loftfield RB, Kraner JC, Egan PA. Supplemental growth hormone alters body composition, muscle protein metabolism and serum lipids in fit adult: characterization of dose-dependent and dose-recovery effects. Mech Ageing Dev 1991;58:191–205.

13. Cuneo R, Salomon F, Wiles CM, Hesp R, Sönksen PH. Growth hormone treatment in growth hormone-deficient adults, II. Effects on exercise performance. J Appl Physiol 1991;70:695–700.

14. Van der Veen E, Netelenbos JC. Growth hormone (replacement) therapy in adults: bone and calcium metabolism. Horm Res 1990;33(suppl 4):65–8.

15. Cuneo RC, Salomon F, Wilmshurst P, Byrne C, Al E. Cardiovascular effects of growth hormone treatment in growth-hormone-deficient adults: stimulation of the renin-aldosterone system. Clin Sci 1991;81:587–92.

16. McGauley GA, Cuneo RC, Salomon F, Sönksen PH. Psychological well-being before and after growth hormone treatment in adults with growth hormone deficiency. Horm Res 1990;33(suppl 4):52–4.

17. Johannsson G, Rosen T, Bosaeus I, Bengtsson BA. The effects of 2 years treatment with recombinant human growth hormone in growth hormone deficient adults [Abstract #18]. Proc 76th annu meet Endocr Soc, Anaheim, CA, June 15–18, 1994.

18. Forbes GB, Reina JC. Adult lean body mass declines with age: some longitudinal observations. Metabolism 1970;19:653–63.

19. Forbes G. The adult decline in lean body mass. Hum Biol 1976;48:161–73.

20. Chon S, Vartsky D, Yasamura S. Compartmental body composition based on total body nitrogen, potassium and calcium. Am J Physiol 1980;239:E5524–30.

21. Kalkhoff RK, Hartz AH, Rupley D, Kissebah AH, Kelber S. Relationship of fat distribution to blood pressure, carbohydrate tolerance, and plasma lipids in healthy obese women. J Lab Clin Med 1983;102:621–7.

22. Heiss G, Tamir I, Davis CE, et al. Lipoprotein-cholesterol distributions in selected North American populations: the Lipid Research Clinics Program Prevalence Study. Circulation 1980;61:302–15.

23. Lakatta E, Goldberg AP, Fleg JL. Reduced cardiovascular and metabolic reserve in older persons: disuse, disease, aging? In: Lipschitz DA, Chernoff R, eds. Nutrition and aging II: health promotion and disease prevention in the elderly. New York: Raven Press, 1988.

24. Fleg JL, Lakatta EG. Role of muscle loss in the age-associated reduction in VO2max in older men. J Appl Physiol 1990;68:329–33.

25. Gupta S, Fikrig SM, Noval MS. Immunological studies in patients with isolated growth hormone deficiency. Clin Exp Immunol 1983;19:87–90.
26. Abbassi V, Bellanti JA. Humoral and cell-mediated immunity in growth hormone-deficient children: effect of therapy with growth hormone. Pediatr Res 1985;19:299–301.
27. Kelley KW. The role of growth hormone in modulation of the immune response. Ann NY Acad Sci 1990;594:95–103.
28. Corpas EC, Harman SM, Blackman MR. Human growth hormone and human aging. Endocr Rev 1993;14:20–39.
29. Ho KY, Evans WS, Blizzard RM, et al. Effects of sex and age on 24-hour profile of growth hormone secretion in men: importance of endogenous estradiol concentrations. J Clin Endocrinol Metab 1987;64:51–8.
30. Prinz PN, Weitzman ED, Cunningham GR, Karacan I. Plasma growth hormone during sleep in young and aged men. J Gerontol 1983;38:519–24.
31. Zadik Z, Chalew SA, McCarter RJ, Meistas M, Kowarski AA. The influence of age on the 24-hour integrated concentration of growth hormone in normal individuals. J Clin Endocrinol Metab 1985;60:513–6.
32. Kelijman M. Age-related alterations of the growth hormone/insulin-like-growth-factor I axis. J Am Geriatr Soc 1991;39:295–307.
33. Rudman D, Vintner MH, Rogers CM, Lubin MF, Fleming GH, Raymond PB. Impaired growth hormone secretion in the adult population: relation to age and adiposity. J Clin Invest 1981;67:1361–9.
34. Vermeulen A. Nyctohemoral growth hormone profiles in young and aged men: correlations with somatomedin-C levels. J Clin Endocrinol Metab 1987;64:884–8.
35. Iranmanesh A, Lizarralde G, Veldhuis JD. Age and relative adiposity are specific negative determinants of the frequency and amplitude of growth hormone (GH) secretory bursts and the half-life of endogenous GH in healthy men. J Clin Endocrinol Metab 1991;73:1081–8.
36. Rudman D, Feller AG, Nagraj HS, et al. Effect of human growth hormone in men over 60 years old. N Engl J Med 1990;323:1–6.
37. Binnerts A, Wilson JHP, Lamberts SWJ. The effects of human growth hormone administration in elderly adults with recent weight loss. J Clin Endocrinol Metab 1988;67:1312–6.
38. Ponting GA, Halliday D, Teale JD, Sim AJW. Postoperative positive nitrogen balance with intravenous hyponutrition on growth hormone. Lancet 1988;2:438–9.
39. Kaiser FE, Silver AJ, Morley JE. The effect of recombinant human growth hormone on malnourished older individuals. J Am Geriatr Soc 1991;39:235–40.
40. Marcus R, Butterfield G, Holloway L, et al. Effects of short term administration of recombinant human growth hormone to elderly people. J Clin Endocrinol Metab 1990;70:519–27.
41. Jorgensen JOL. Human growth hormone replacement therapy: pharmacological and clinical aspects. Endocr Rev 1991;12:189–207.
42. Hendricks CM, Eastman RC, Takeda S, Asakawa K, Gorden P. Plasma clearance of intravenously administered pituitary human growth hormone: gel filtration studies of heterogeneous components. J Clin Endocrinol Metab 1985;60:864–7.

43. Parker ML, Utiger RD, Daughaday WH. Studies on human growth hormone, II. The physiological disposition and metabolic fate of human growth hormone in man. J Clin Invest 1962;41:262–70.
44. Refetoff S, Sonksen PH. Disappearance rate of endogenous and exogenous human growth hormone in man. J Clin Endocrinol Metab 1970;30:386–91.
45. Holl RW, Schwarz U, Schauwecker P, Benz R, Veldhuis J, Heinze E. Diurnal variation in the elimination rate of human growth hormone (GH): the half-life of serum GH is prolonged in the evening, and affected by the source of the hormone, as well as by body size and serum estradiol. J Clin Endocrinol Metab 1991;77:216–20.
46. Veldhuis J, Iranmanesh A, Ho KKY, Waters MJ, Johnson ML, Lizarralde G. Dual defects in pulsatile human growth hormone secretion and clearance subserve the hyposomatotropism of obesity in man. J Clin Endocrinol Metab 1991;72:51–9.
47. Iranmanesh A, Lizarralde G, Veldhuis J. Age and relative adiposity are specific negative determinants of the frequency and amplitude of growth hormone (GH) secretory bursts and the half-life of endogenous GH in healthy men. J Clin Endocrinol Metab 1991;73:1081–8.
48. Sohmiya M, Kato Y. Renal clearance, metabolic clearance rate, and half-life of human growth hormone in young and aged subjects. J Clin Endocrinol Metab 1992;75:1487–90.
49. Haffner D, Schaefer F, Girard J, Ritz E, Mehls O. Metabolic clearance of recombinant human growth hormone in health and chronic renal failure. J Clin Invest 1994;93:1163–71.
50. Chantelau E, Sonnenberg GE, Rajab A, Romisch J, Berger M. Absorption of subcutaneously administered regular human and porcine insulin in different concentrations. Diabetes Metab 1985;11:106–11.
51. Jorgensen JOL, Moller J, Jensen FS, Jorgensen JT, Christiansen JS. Growth hormone administration by means of an injection pen. Pharmacol Toxicol 1989;65:96.
52. Hedin L, Olsson B, Diczfalusy M, et al. Intranasal administration of human growth hormone (hGH) in combination with a membrane permeation enhancer in patients with GH deficiency: a pharmacokinetic study. J Clin Endocrinol Metab 1993;76:962–7.
53. Kastrup KW, Christiansen JS, Koch Anderson J, Orskov H. Increased growth rate following transfer to daily sc administration from three weekly im injections of hGH in growth hormone deficient children. Acta Endocrinol (Copenh) 1983;104:148–52.
54. Hermanussen M, Geiger-Benoit K, Sippel WG. Catch-up growth following transfer from three times weekly im to daily sc administration of hGH in GH deficient patients, monitored by kinemometry. Acta Endocrinol (Copenh) 1985;109:163–8.
55. Albertsson-Wikland K, Westphal O, Westgren U. Daily subcutaneous administration of human growth hormone in growth deficient children. Acta Pediatr Scand 1986;75:89–97.
56. Kikuchi, Masakatsu S, Miyamoto A, Ohie T, Mori C, Mikawa H. Growth response to daily subcutaneous administration of growth hormone. Acta Pediatr Jpn Overseas Ed 1988;30:557–63.

57. Soyka LF, Bode HH, Crawford JD, Flynn FJ. Effectiveness of long-term human growth hormone therapy for short stature in children with growth hormone deficiency. J Clin Endocrinol Metab 1970;30:1–14.
58. Jorgensen JOL, Moller N, Lauritzen T, Christiansen JS. Pulsatile versus continuous intravenous administration of growth hormone (GH) in GH deficient patients: effects on circulating insulin-like growth factor I and metabolic indices. J Clin Endocrinol Metab 1990;70:1616–23.
59. Rudman D, Friedes D, Patterson JH, Gibbas DL. Diurnal variation in the responsiveness of human subjects to human growth hormone. J Clin Invest 1973;52:912–8.
60. Matustik MC, Furlanetto RW, Meyer WJ. Chronobiologic considerations in human growth hormone therapy. J Pediatr 1983;103:543–6.
61. Jorgensen JOL, Moller N, Lauritzen T, Alberti KGM, Orskov H, Christiansen JS. Evening versus morning injections of growth hormone (GH) in GH0-deficient patients: effects on 24-hour patterns of circulating hormones and metabolites. J Clin Endocrinol Metab 1990;70:207–14.
62. Tait JF. Review: the use of isotopic steroids for the measurement of production rates in vivo. J Clin Endocrinol Metab 1963;23:1285–97.
63. Albertsson-Wikland K, Rosberg S, Libre E, Lundberg L-O, Groth T. Growth hormone secretory rates in children as estimated by deconvolution analysis of 24-h plasma concentration profiles. Am J Physiol 1989;257:E809–15.
64. Taylor AL, Finster JL, Mintz DH. Metabolic clearance and production rates of human growth hormone. J Clin Invest 1969;48:2349–58.
65. Kowarski A, Thompson RG, Migeon CJ, Blizzard RM. Determination of integrated plasma concentrations and true secretory rates of human growth hormone. J Clin Endocrinol Metab 1971;32:356–60.

21

Role of Growth Hormone Therapy in Turner Syndrome

E. Kirk Neely and Ron G. Rosenfeld

The classic phenotypic appearance associated with the 45,X karyotype was first described by Otto Ullrich in 1930 and by Henry Turner in 1938 (1, 2). The syndrome is appropriately called Ullrich-Turner in Germany, but simply *Turner syndrome* (TS) in the United States. (We shall use the latter term by convention.) In the 1940s, the infantilism described by Turner was related to rudimentary development of the ovaries; thus the term *gonadal dysgenesis* (3). In 1959, not long after the correct number of human chromosomes was elucidated, the syndrome was attributed to loss of an X chromosome (4). Properly, it should be stated that X chromosome monosomy results from the loss of the second sex chromosome, whether X or Y, presumably secondary to a nondisjunctional event in an initial mitotic division. Further cytogenetic studies in the 1960s determined that TS due to structural abnormality of the second X chromosome or to various mosaicisms may be clinically indistinguishable from X chromosome monosomy.

Natural History of Growth in TS

Several descriptions of growth resulting from X chromosome loss have identified relatively distinct phases in the development of short stature. The paper of Ranke et al. (5), for example, described the natural history of growth in the absence of hormonal therapy in 150 German girls with TS. In their study intrauterine growth was frequently retarded, with a mean reduction of 540 g in birth weight and 2.8 cm in birth length. Nevertheless, TS is usually diagnosed at birth only when lymphedema or neck webbing are apparent (primarily in 45,X girls). Growth velocity in the first 3 years of life is relatively normal, but a more marked, progressive decline occurs relative to the height velocity of normal females during the rest of childhood.

The 5th percentile of the normal female curve and the 95th percentile of the TS curve diverge at about 9 years of age, but the actual age when this occurs is dependent on the familial target height. Growth. failure becomes even more clinically apparent in adolescence due to the absence of a pubertal growth spurt, and the height nadir relative to normal female growth curves is typically found at 14 years of age. Girls with TS exhibit delayed skeletal maturation prior to adolescence, and epiphyseal fusion is markedly delayed, with typical closure at 18–20 years. Paradoxically, continued slow growth during late adolescence allows for effective growth therapy over a prolonged period.

The German data, as well as the more recent growth standards of Naeraa and Nielsen (6), have provided a basis for evaluating therapeutic regimens designed to augment growth. Lyon et al. (7) combined Ranke's data with 3 other studies of growth in untreated girls with TS and constructed a standard growth curve called the *Lyon curve*. These standard plots permit an accurate projection of adult height from the current-height *standard deviation scores* (SDS), demonstrating a correlation coefficient of 0.95 between first-measured height at 3–12 years of age and final adult height (8). Accumulated experience with several hundred TS children and adolescents in the United State have confirmed the accuracy and utility of the standardized TS growth curves developed in Europe.

Interpretation of final-height data is complicated by many factors, including karyotype, parental heights, and earlier hormonal treatment with estrogens and androgens. Because TS is less likely to be diagnosed in the more phenotypically normal patients, particularly those with deletions of the long arm of the X chromosome or with mosaicisms, reported adult heights may even be slightly underestimated due to ascertainment bias. Final heights may be affected by ethnic diversity as well, but the growth patterns of patients with TS have proven to be fairly consistent from one country to another. Mean final height from the Lyon curve is 142.9 ± 7.3 cm, a figure quite comparable to the untreated final height reported from multiple European and U.S. studies. However, reported final heights in Japan are about 138 cm, while adult heights of approximately 147 cm have been reported in 2 northern European studies, including the Danish group (6), and in the Seattle population described by Sybert (9) (where a higher frequency of mosaicism is noted in her series). Resolution of the latter discrepancies is important for assessing the benefits of growth therapies since studies with *human growth hormone* (hGH) have not used untreated control groups after the first year, with the exception of the ongoing Canadian study (10).

The growth failure associated with TS results from the contributions of skeletal dysplasia, estrogen deficiency, and, possibly, GH secretory dysfunction, but essentially the cause remains unclear. There are no conclusive data concerning the histologic appearance of bone and cartilage in children with TS, and increased bone resorption and defective osteoblast renewal and differentiation have both been reported (11). In theory, pathological bone development could represent either an intrinsic defect in ossification due to loss of critical genes on the X chromosome or a secondary effect of intrauterine edema in a manner comparable to the postulated development of cardiac anomalies. Preadolescent growth failure probably evolves secondary to this underlying defect in ossification since growth failure begins before the age at which reduction in GH secretion has generally been detected.

Evidence of disordered GH secretion later in childhood and in adolescence is fairly convincing. Although many patients with TS are not GH deficient as defined by provocative GH testing (i.e., they have maximal serum GH levels of >7–10 ng/mL), most investigators have documented relatively low peak stimulated GH levels or actual GH deficiency in their TS populations. TS adolescents, but not younger children, are reported to have reduced mean 24-h GH levels, peak amplitudes, and peak frequencies when compared with age- or bone age-matched controls (12, 13). Serum levels of *insulin-like growth factor I* (IGF-I), which mediates most actions of GH, are only about half of normal levels during adolescence, and levels are in the lower half of the normal range in childhood (14). Failure of the normal pubertal rise in GH and IGF-I secretion in TS adolescents may be related either to a primary hypothalamic-pituitary abnormality or to estrogen deficiency, and the latter argument is supported by an increase in GH concentrations in response to low-dose estrogen therapy (15). Thus, growth failure is an evolving process involving putative defects in bone formation in childhood and hormonal insufficiency in adolescence.

Failure of Androgens or Estrogens to Influence Final Height in TS

Estrogens and anabolic steroids were widely utilized for growth therapy in TS in the era before *recombinant human GH* (rhGH), but their use has declined in recent years, as it has become clear that the short-term growth stimulation does not result in significant increments in final height. Oxandrolone, the most widely studied anabolic agent, is available in the United States on a limited clinical research basis and has been employed in combination with hGH in several ongoing studies. The anabolic effects of oxandrolone had been observed in the absence of any significant increase in serum IGF-I levels, suggesting that the growth-promoting actions are mediated by a direct effect on the epiphysis rather than via the GH-IGF-I axis, thus raising the possibility that the clinical benefits of hGH and androgens could be additive.

Many studies reported modest increases in predicted or final adult height with androgen therapy alone. However, most of the early reports were retrospective or uncontrolled, often utilizing a comparison of androgen and estrogen treatment. Recently, Naeraa et al. (16) have reported a 3- to 4-cm improvement in adult height when oxandrolone therapy was initiated prior to a bone age of 13 years. Similarly, Crock et al. (17) in Australia documented an increase in height velocity in 35 subjects treated with oxandrolone, with the greatest responses in younger girls, and an increase of 4–5 cm in final height over earlier predicted height. In contrast, Sybert (10) reported in 66 patients that the mean adult height of patients given either oxandrolone or fluoxymesterone (148 ± 4.7 cm) did not differ significantly from the height of untreated patients (146.3 ± 5.5 cm). The data from most of these studies suggest that anabolic steroids accelerate growth over several years of treatment and may result in slight improvement of adult height.

Comparably, trials of oral estrogens for growth have confirmed a significant short-term acceleration in height velocity, but have failed to document an improvement in predicted or final height. Indeed, many studies have reported that estrogens, even in very low doses, result in an accelerated skeletal maturation (18) that is particularly evident in younger girls. As a result, no consistent improvement in final height has been demonstrated with estrogen alone, and estrogen provides no additional growth benefit in conjunction with hGH therapy. Estrogen therapy is naturally accompanied by some degree of secondary sexual development, rendering treatment in young girls even less palatable. Thus, estrogen therapy is contraindicated in preadolescent patients and plays no role in growth therapy in TS, but should be introduced at the appropriate time and dose for feminization of adolescents with TS.

GH Therapy

Turner himself in 1938 undertook the first documented administration of exogenous GH (2). He treated 3 girls for several months and was unsuccessful in altering their growth rates, presumably administering injections of bovine pituitary extract, as hGH was not available until the 1950s. In 1960, Escamilla et al. (19) gave hGH to a patient with TS, resulting in an incremental increase in growth velocity from 3.8 to 7.5 cm/ year over a brief treatment period. Later trials of pituitary-derived hormone, necessarily small-scale, supported the efficacy of hGH in the treatment of short stature in TS, but did not establish definitive final-height data. With the development of rhGH, TS offered an ideal group for early large-scale prospective clinical trials. Consequently, all of the large industrialized countries have initiated countrywide collaborative trials that by now have proceeded to 2 or more years, and it can be stated that hGH trials in TS have unequivocally demonstrated a marked increase in growth velocity. However, it should be emphasized that long-term treatment data and final-height data are just appearing in the literature.

The oldest and most influential trial in TS began in the United States in 1983 (20). Seventy girls between 4.7 and 12.4 years of age, all with a normal provocative GH testing (GH: >7 µg/L), were randomly assigned to a control group (no treatment), oxandrolone (0.125 mg/kg/day), hGH (0.125 mg/kg/3 times a week), or a combination of oxandrolone and hGH. After 12–20 months all groups except hGH alone were collapsed into the combination group at a reduced oxandrolone dose (due to virilization in 30% of subjects). After year 3 of treatment, most subjects received hGH injections daily instead of 3 times a week.

In the most recent report, 50 subjects had finished 5 years of therapy. In subjects receiving hGH alone, growth velocity increased from 4.5 cm/year in the pretreatment period to 6.6 and 5.4 cm/year in years, 1 and 2, respectively. Growth rates of 5.2, 5.1, and 4.3 cm/year in years 4–6 of hGH treatment were still greater than untreated growth rates derived from the Lyon curve, which progressively decline in the absence of the pubertal growth spurt. The increase in growth rate was greater in the combination group than with hGH alone. Mean growth rates in the first 3 years in subjects started on hGH and oxandrolone (at the lower dose of 0.0625 mg/kg/day) were 8.3, 6.7, and 6.3 cm/year, and year 1 growth rate with hGH and the higher oxandrolone dose was 10 cm/year. These data are presented in Figure 21.1.

Over 6 years of therapy, subjects receiving hGH alone grew an average of 31.2 cm, and those in the combination group grew 37.3 cm, in comparison with an expected total growth of 19 cm in untreated subjects. However, 6 years of combination therapy resulted in a skeletal age advance of 6.3 ± 1.6 years, while the mean bone age advance with hGH alone was 5.2 ± 1.4 years. (The expected bone age advance in TS is approximately 0.8 year per year of chronologic age.) As a result of bone age acceleration in the combination group, the relative decline in growth velocity SDS in later treatment years has been more dramatic than with hGH alone, and the current 6.2-cm excess in cumulative growth in the combination group (vs. hGH alone) cannot be anticipated to remain when final-height data are complete. Results from this study, consistent with earlier, smaller studies of oxandrolone alone, suggest that the addition of oxandrolone to hGH increases growth velocity more dramatically than hGH alone, but offers little additional long-term benefit.

Augmentation of final height by hGH is the ultimate measure of therapeutic success. At this point, 14/17 (82%) subjects in the hGH group and 41/45 (91%) in the combination arms, with many still growing, have already exceeded projected adult heights derived from the Lyon growth curve. Mean current height for the 42 patients who have terminated therapy is 151.9 cm, which is significantly greater than the initially pro-

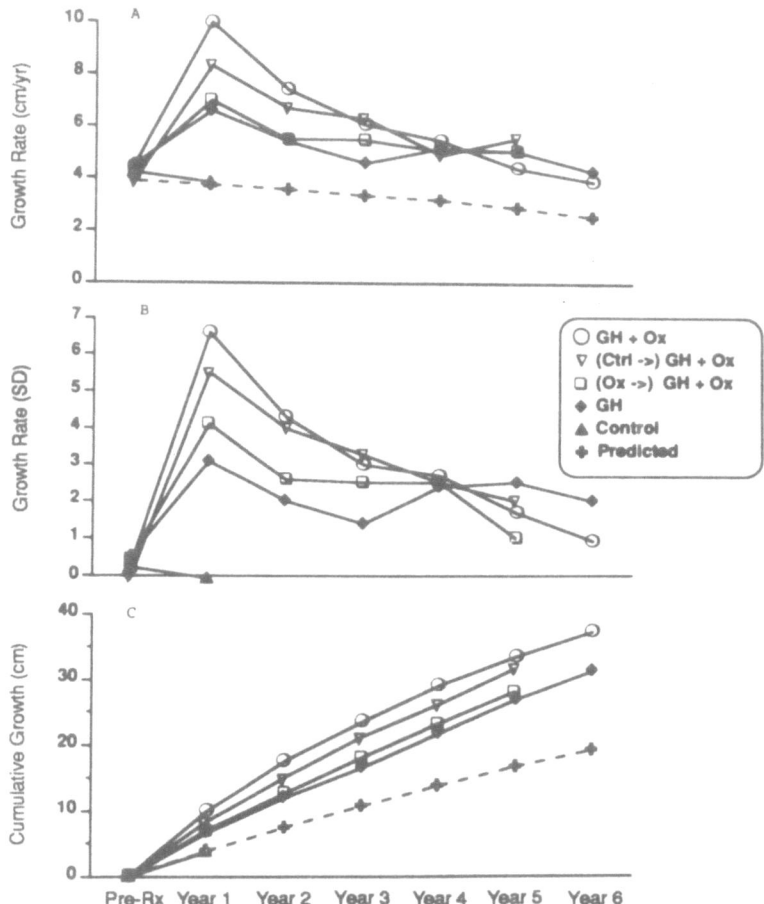

FIGURE 21.1. Annual growth rate for each U.S. TS study arm as indicated. In *A* the growth rate is expressed in cm/year and compared to predicted untreated growth rate; in *B* it is expressed as SDS; and in *C* it is shown as cumulative growth in cm/year in excess of cumulative growth for untreated patients. Reprinted with permission from Rosenfeld, Frane, Attie, et al. (20).

jected mean adult height of 143.8 cm. The mean gain over projected adult height for all groups is 8.1 cm, despite a mean age of 9.3 years at initiation and thrice-weekly administration.

Many international studies have corroborated the results from the initial years of the U.S. study. A German collaborative study utilized a similar randomization to hGH alone or hGH in combination with oxandrolone (21). In year 2, the initial dose of oxandrolone was reduced to 0.05 mg/kg/day in the combination group, and hGH was increased to

3.4 IU/m^2/day in the hGH group. Height velocities in the hGH-treated group were 4.0, 6.3, and 5.3 cm/year at baseline and in years 1 and 2, respectively, and predicted adult heights increased by approximately 4 cm. The increment in growth velocity was even more pronounced in the combination group. Similarly, Nilsson et al. (22) treated Swedish TS patients with a daily combination of hGH and oxandrolone, and the annual growth rate increased from 3.9 cm/year to 9.4 and 6.8 cm/year in years 1 and 2, respectively.

Takano et al. have reported the Japanese experience with different dosages and preparations of hGH in TS. In their treatment of 203 girls for 1 year, hGH alone increased the mean annual growth rate from 3.7 to 5.1 cm/year, while hGH with stanazolol (0.04 mg/kg/day) enhanced the mean growth velocity to 6.9 cm/year (23). In 46 children, the effects of hGH at 0.05 or 0.1 IU/kg/week have been compared over 3 years (24). Growth rate in the group receiving the higher dose was significantly greater in year 1 only; after 3 years relative advances in bone age were 2.5 and 2.9 years in the 2 groups, respectively. Thus, it is unclear whether final height would be significantly improved with the higher dose. However, data are convincing that daily dosing is preferable to less frequent dosing. This was suggested by the increase in growth rate in the U.S. study following a change from thrice-weekly to daily dosing in year 3. Additionally, the Dutch group randomized 52 girls to 24 IU/m^2/week, given either 3 or 6 times weekly. After 2 years the total-height increment was 8.6 and 11.3 cm in the 2 groups, respectively. Although bone age advance was slightly greater with more frequent hGH dosing, the change in height SDS for bone age was still 0.4 SDS greater from the daily dosing regimen (25).

In the Belgian experience (26), 41 girls were treated with hGH alone or in combination with a very low dose of ethinyl estradiol (25 ng/kg/day). Increased skeletal maturation was noted in estrogen-treated patients under 11 years old, potentially resulting in loss in final height. In the group receiving hGH alone, growth velocity increased from 4.0 cm/year at baseline to 7.4, 5.8, 5.0, and 3.7 cm/year in each of 4 years of therapy. The mean proximate final height of 20 girls with a bone age of >13 years was 149.3 cm after 4 years of treatment, with no discernible differences between the hGH alone and hGH plus ethinyl estradiol groups.

In summary, all of the many international studies utilizing hGH alone or in combination with androgens or estrogens have demonstrated an accelerated height velocity and increase in height SDS for bone age in the first year of therapy. As in hGH therapy of GH deficiency or idiopathic short stature, height velocity gradually declines in subsequent years of therapy, but remains above the expected growth velocity for untreated girls. Partial final-height data have been accumulated in several studies, with most suggesting a mean final height of 150 cm or more in the subjects with relatively advanced age at the initiation of therapy.

Human GH and Bone Mineral Density in TS

Although troubled by major methodological problems, reports of osteopenia in TS have been widely cited. Bone mineral status is of particular concern in TS because ovarian failure and the relative GH deficiency may both contribute markedly to insufficient acquisition of bone mass during the crucial adolescent years and a similar inability to maintain bone mass during adulthood.

Early reports of osteopenia with TS employed skeletal X-rays and computerized tomographic scans, but more recently, single and dual photon absorptiometry and *dual-energy X-ray absorptiometry* (DEXA) have been utilized to report *bone mineral content* (BMC), measured in g, and *bone mineral density* (BMD), measured in g/cm^2, the latter measure compensating for the surface area of bones. In adults with TS early but limited studies documented a markedly reduced BMC in some patients, with less reduction in those on estrogen replacement (27, 28). The evidence for osteopenia in children and adolescents with TS is less convincing, and an increased incidence of fractures has never been documented. Furthermore, the reported decreases in mean radial BMC in children with TS have little significance since BMC is highly dependent on body and bone size. Studies measuring BMD have reported contradictory results (29, 30) and have not effectively determined when relative demineralization begins. Our group has recently reported that adolescents with TS receiving GH therapy (as virtually all our patients do) exhibit normal bone mineral properties (31). Thus, it is not clear when estrogen replacement is required for bone accretion, but it is likely that hGH therapy and timely estrogen treatment improve the chances that individuals with TS will develop normal bone density. Whether continued hGH therapy in TS during adulthood might contribute to maintenance of bone density remains to be tested.

Conclusion

On the assumption, still speculative, that earlier treatment with hGH may benefit final height, we offer hGH therapy, at 0.05 mg/kg/day as soon as the patient falls below the 5th percentile for height, an event that commonly occurs in the early school years. We do not routinely perform provocative GH testing in girls with TS because most subjects will respond positively to hGH therapy in the face of adequate stimulated GH levels. Despite the apparent efficacy of oxandrolone added to hGH in immediate growth stimulation, its routine use is not justifiable. Neither the optimal duration of therapy nor the appropriate time to discontinue hGH has been established; without any data to support more definitive criteria, we generally use a growth rate of <2.5 cm/year together with a bone age of >14 years. The optimum age and the dose of estrogen for initiation of feminization remain controversial, although we generally introduce low-dose estrogen at an age of 12–15 years, followed by full replacement and progesterone within 2–3 years.

The efficacy of hGH in short-term growth stimulation is no longer in question, and the proportion of subjects in the U.S. trial who have achieved 150 cm has convinced most investigators of the long-term benefit. The most prominent dissent comes from investigators who believe that untreated adult heights in TS are substantially greater than the 143-cm standard from the Lyon curve and other studies. At this time, most pediatric endocrinologists believe that the provisional U.S. study benefit, describing a mean 8- to 10-cm gain from 6 years of therapy (with thrice-weekly, then daily injections) and a final height in excess of 150 cm, will hold up in the final analysis. Furthermore, earlier therapy may lead to even greater height gains, although there is no evidence to date.

Justification of any therapy for the short stature inherent in TS must be considered in the context of broader ethical arguments regarding the use of hGH therapy in other conditions, such as GH deficiency and idiopathic short stature. While final-height data from the initial TS hGH trials are not yet complete, a general consensus has been reached that TS is an appropriate use for hGH.

References

1. Ullrich P. Turner's syndrome and status Bonnevie-Ullrich. Am J Hum Genet 1949;1:179–202.
2. Turner HH. A syndrome of infantilism, congenital webbed neck, and cubitus valgus. Endocrinology 1938;23:566–74.
3. Grumbach MM, Van Wyk JJ, Wilkins L. Chromosomal sex in gonadal dysgenesis: relationship to male pseudohermaphroditism and theories of human sex differentiation. J Clin Endocrinol Metab 1955;15:1161–93.
4. Ford CE, Miller OJ, Polani PE, de Almeida JC, Briggs JH. A sex chromosome anomaly in a case of gonadal dysgenesis (Turner's syndrome). Lancet 1959;1:711–3.
5. Ranke MB, Pfluger H, Rosendahl W, et al. Turner syndrome: spontaneous growth in 150 cases and review of the literature. Eur J Pediatr 1983;141:81–8.
6. Naeraa RW, Nielsen J. Standards for growth and final height in Turner's syndrome. Acta Paediatr Scand 1990;79:182–90.
7. Lyon AJ, Preece MA, Grant DB. Growth curve for girls with Turner syndrome. Arch Dis Child 1985;60:932–5.
8. Frane JW, Sherman BM, Genentech Collaborative Group. Predicted adult height in Turner syndrome. In: Rosenfeld RG, Grumbach MM, eds. Turner syndrome. New York: Marcel Dekker, 1990:405–19.
9. Holland J, Brnjac L, Alexander D, et al. Turner syndrome and final adult stature: a randomized controlled trial using human growth hormone and low

dose ethinyl estradiol. In: Ranke MB, Rosenfeld RG, eds. Turner syndrome: growth promoting therapies. Amsterdam: Excerpta Medica, 1991:195–200.

10. Sybert VP. Adult height in Turner syndrome with and without androgen therapy. J Pediatr 1984;104:365–9.

11. Rosenfeld RG, Grumbach MM, eds. Turner syndrome. New York: Marcel Dekker, 1990.

12. Ross JL, Long LM, Loriaux DL, Cutler GB Jr. Growth hormone secretory dynamics in Turner syndrome. J Pediatr 1985;106:202–6.

13. Massarano AA, Brook CGD, Hindmarsh PC, et al. Growth hormone secretion in Turner's syndrome and influence of oxandrolone and ethinyl estradiol. Arch Dis Child 1989;64:587–92.

14. Cuttler L, Van Vliet G, Conte FA, Kaplan SL, Grumbach MM. Somatomedin-C levels in children and adolescents with gonadal dysgenesis: differences from age-matched normal females and effect of chronic estrogen replacement. J Clin Endocrinol Metab 1985;60:1087–92.

15. Mauras N, Rogol AD, Veldhuis JD. Increased hGH production rate after low-dose estrogen therapy in prepubertal girls with Turner's syndrome. Pediatr Res 1990;28:626–30.

16. Naeraa RW, Nielsen J, Pedersen IL, Sorensen K. Effect of oxandrolone on growth and final height in Turner's syndrome. Acta Paediatr Scand 1990;79:784–9.

17. Crock P, Werther GA, Wettenhall HNB. Oxandrolone increases final height in Turner syndrome. J Paediatr Child Health 1990;26:221–4.

18. Vanderschueren-Lodeweyckx, et al. Growth-promoting effect of growth hormone and low-dose ethinyl estradiol in girls with Turner's syndrome. J Clin Endocrinol Metab 1990;70:122–6.

19. Escamilla RF, Hutchings JJ, Deamer WC, Li CH. Clinical experiences with human growth hormone (LI) in pituitary infantilism and in gonadal dysgenesis. Acta Endocrinol Suppl (Copenh) 1960;51:253A.

20. Rosenfeld RG, Frane J, Attie KM, et al. Six-year results of a randomized prospective trial of human growth hormone and oxandrolone in Turner syndrome. J Pediatr 1992;121:49–55.

21. Stahnke N, Stubbe P, Keller E. Recombinant human growth hormone and oxandrolone in treatment of short stature in girls with Turner syndrome. Horm Res 1992;37(suppl 2):37–46.

22. Nilsson KO, Swedish Paediatric Study Group for Growth Hormone Treatment. The Swedish somatonorm Turner trial: two-year results. Acta Paediatr Scand Suppl 1989;356:160A.

23. Takano K, Shizume K, Hibi I. Turner's syndrome: treatment of 203 patients with recombinant human growth hormone for one year. Acta Endocrinol (Copenh) 1989;120:559–68.

24. Takano K, Shizume K, Hibi I. Treatment of 46 patients with Turner's syndrome with recombinant human growth hormone for three years: a multicentre study. Acta Endocrinol (Copenh) 1992;126:296–302.

25. Rongen-Westerlaken C, van Es A, Wit J-M, et al. Growth hormone therapy in Turner's syndrome. Am J Dis Child 1992;146:817–20.

26. Massa G, Vanderschueren-Lodeweyckx M. Growth promoting effect of growth hormone and low dose ethinyl estradiol in girls with Turner syndrome: 4-year

results. In: Hibi I, Takano K, eds. Basic and clinical approach to Turner syndrome. Elsevier Science, 1993:327–32.
27. Smith MA, Wilson J, Price WH. Bone demineralisation in patients with Turner's syndrome. J Med Genet 1982;19:100–3.
28. Naeraa RW, Brixen K, Hansen RM, et al. Skeletal size and bone mineral content in Turner's syndrome: relation to karyotype, estrogen treatment, physical fitness, and bone turnover. Calcif Tissue Int 1991;49:77–83.
29. Kirkland RT, Lin T-H, LeBlanc AD, Kirkland JL, Evans HJ. Effects of hormonal therapy on bone mineral density in Turner syndrome. In: Rosenfeld RG, Grumbach MM, eds. Turner syndrome. New York: Marcel Dekker, 1990:319–25.
30. Ross JL, Long LM, Feuillan P, Cassorla F, Cutler GB. Normal bone density of the wrist and spine and increased wrist fractures in girls with Turner's syndrome. J Clin Endocrinol Metab 1991;73:355–9.
31. Neely EK, Marcus R, Rosenfeld RG, Bachrach LK. Turner syndrome adolescents receiving growth hormone are not osteopenic. J Clin Endocrinol Metab 1993;76:861–6.

22

Role of Growth Hormone in the Treatment of Postmenopausal Osteoporosis

John G. Haddad

Postmenopausal osteoporosis is a common disorder that can lead to acute and chronic morbidity, affecting many in the population (1, 2). Surely, prophylactic and therapeutic strategies that prevent or improve this condition are welcome. In recent years much attention has been paid to *secondary osteoporosis* (known disorders and medications or such influences as diet, activity, and alcohol that negatively influence the skeleton), as well as to agents that inhibit osteolysis (gonadal steroids, *calcitonin* [CT], and the bisphosphonate compounds) (3). Increasingly, however, agents associated with more robust bone formation, such as low-dose, intermittent *parathyroid hormone* (PTH), sodium fluoride, *growth hormone* (GH), and various growth factors, are being considered as potential additions to our therapeutic armamentarium.

GH effects on bone are best recognized for promoting longitudinal growth during modeling, as well as acral growth during remodeling, of the skeleton. Our concern here is to sort out the impacts on bone density effected by GH deficiency, replacement therapy, and excess. Because of the limited availability of hGH until recently, this chapter concentrates on data available over the past 20 years.

GH Deficiency and the Skeleton

Considerable evidence supports the observation that GH release is attenuated in aging adults (4, 5). Rudman has reviewed some of the changes associated with aging (6)—GH deficiency and excess—and these are adapted for display in Table 22.1. Major factors that could have an impact on bone are (i) a direct skeletal effect of GH, and (ii) an indirect

269

TABLE 22.1. Musculoskeletal effects of GH replacement and excess.

Muscle protein synthesis	Increase
Linear skeletal growth	Increase (children)
Acral skeletal growth	Increase (adults)
Bone density, spine	Increase or decrease
Bone density, forearm	Increase

effect of GH mediated by *insulin-like growth factor I* (IGF-I) and/or the anabolic effect of GH on skeletal muscle that would increase skeletal strain. In GH-deficient children, forearm bone density is low, even when corrected for bone age.

Direct GH effects on skeletal cells are clearly recognized in epiphyseal cartilage, but the bone cell actions of GH have received less study. Since GH could affect bone directly or by IGF-I mediation, its effects are complex. Its ability to increase bone turnover is recognized (7), and it can increase calciuria, renal phosphorus reclamation, and intestinal absorption of calcium. With GH treatment biochemical markers of bone formation (serum osteocalcin and procollagen peptides) and bone resorption (urine hydroxyproline and deoxypyridinoline) increased (8–10) when given to GH-deficient adults. A 5% increase in spinal *bone mineral density* (BMD) was seen after 6 months of therapy, and mid- and distal forearm BMD increased by 4.3% and 4.1%, respectively (11). One year of GH therapy to children with hypopituitarism did not change forearm BMD in an uncontrolled study (12).

Bone Status in Acromegaly

Studies reveal accelerated skeletal remodeling in acromegaly. Both increased cortical bone formation (periosteal apposition) and bone resorption at endosteal surfaces (especially trabecular or cancellous bone) are recognized in this condition (7, 13–15). Since other pituitary hormones (estrogen and testosterone) can be deficient with GH-producing pituitary adenomas, one must be mindful that their absence has a role in the skeletal effects seen. Functional increase in skeletal muscle mass could have a favorable effect on bone mass and density since osteoblasts repond positively to increased strain.

Recent studies have delineated the BMD alterations seen in acromegaly (15, 16). It appears that the appendicular (mostly compact or cortical bone) BMD and axial (mostly trabecular or cancellous bone) BMD change in opposite directions after chronic GH excess. Significantly reduced spine BMD values did correlate with these patients' gonadal status, however (16). At the forearm, acromegalic patients had higher BMD, regardless of their gonadal status. With treatment of the acromegaly, interesting findings were seen. With the return of IGF-I levels to normal, appendicular BMD decreased at a rate of 1% per year. In patients with persistently elevated IGF-I levels in serum, forearm BMD increased at an annual rate of 1.5%.

GH Treatment in Osteoporosis and in the Elderly

Aloia and colleagues have made early and important studies of the effects of GH in osteoporosis (17–19) (Table 22.2). In osteoporotic patients, a 2-dose schedule at 6 months each did not reveal changes until the higher dosage. Forearm BMD decreased, and whole-body calcium (neutron activation analysis) did not change. Some complications of therapy (hyperglycemia and carpal tunnel syndrome) were seen. In studies by this group, comparisons of GH and *salmon calcitonin* (SCT) therapy were made in osteoporotic patients (18, 19). Moderate increases in whole-body calcium were seen after 2 years of SCT treatment. The addition of GH to the SCT treatment, concurrently or in a coherence regimen, did not cause significant changes. With whole-body analyses of calcium, it is possible that opposite regional changes in trabecular and cortical bone may have offset each other.

Studies of the effects of *recombinant human GH* (rhGH) over 7 days revealed changes in biochemical markers of skeletal metabolism (8). Increased osteocalcin (29%) and hydroxyproline (100%) were accompanied by increased PTH, 1,25-$(OH)_2$D, and urine calcium. In a study of 12 healthy elderly men over 1 year, Rudman analyzed parameters during a 6-month baseline period and during a 6-month GH treatment period (0.03-mg rhGH/kg/thrice weekly). In spite of an 8.8% increase in *lean body mass* (LBM), only a 1.6% increase in lumbar BMD was observed, and no changes in forearm or proximal femur BMD were noted (20).

TABLE 22.2. Skeletal effects of GH treatment in osteoporosis and in the elderly.

Year	Author	Duration of study	Bone change
1976	Aloia	6 months	Iliac crest; bone biopsy showed increased resorption; no change in whole-body calcium and forearm BMD
1987	Aloia	24 months (coherence)	No augmentation of CT effect
1990	Rudman	6 months	Increase in spine BMD; no change in hip and arm BMD
1990	Marcus	1 week	Increase in osteocalcin s; increase in hydroxyproline u

In contrast to the consistently positive effects of GH treatment on nitrogen balance and muscle mass, such therapy has less consistent and less marked effects on skeletal BMD in elderly and osteoporotic patients. The positive skeletal effects seen in proven GH-deficient patients after GH treatment are not as convincing in GH-treated elderly or osteoporotic patients without GH deficiency. Most studies do not reveal osteoporotic women to have lower GH secretion than their nonosteoporotic peers.

Experimental Studies in Animals

A relevant group of studies in animals indicates the skeletal effects of GH. In tibiae from rats single doses of GH increased protein synthesis in the bone, but the compartments were not studied separately (21). In a 3-month treatment of adult dogs of both sexes, new bone formation increased strikingly in 63% of the skeletal regions studied, but no specific description of the axial or trabecular-rich bones was offered (22). Since only 20% of adult bone is trabecular, the influences of GH on this skeletal region can only be detected by analyses of cancellous bone, better represented in the vertebrae, proximal femur, and distal radius. In postmenopausal osteoporosis, bone morbidity is heavily expressed in these areas by vertebral, hip, and wrist fractures.

An interesting report described the effects of GH on bone loss in female monkeys made hypogonadal with a *gonadotropin releasing hormone* (GnRH) agonist (23). GH treatment 3 times a week appeared to reduce the loss of spine BMD over 10 months. Curiously, serum osteocalcin levels increased in the animals who received the GnRH analog alone, as well as in those who received GH along with the analog. Since vehicle-treated monkeys gained significantly higher spine BMD, the GH treatment was only partially helpful in reducing the loss of bone density. A disturbing effect of GH was observed on closed tibial fracture healing in rats (24). Although GH-treated animals had earlier external callus formation, the callus was loosely structured and not removed by normal modeling/remodeling events. By 40 days control animals showed healing, but GH-treated animals had not healed by 80 days, and a large marrow invasion of the callus was apparent.

Factors to Consider

In order to come to some conclusions about whether or not GH has a role in the treatment of osteoporosis, several issues warrant our attention. First, GH has anabolic effects on cortical bone, but catabolic effects on trabecular bone when it is given to GH-sufficient subjects or when it has continuously perfused the body at high titers for long periods (acromegaly).

Second, in GH-deficient adults, however, anabolic skeletal effects appear to occur as the result of GH treatment. Third, since GH preparations and dosage schedules have differed in many of the studies aimed at detecting skeletal changes, it is difficult to be confident about the results of such studies, especially in comparisons. Fourth, although a primate study suggests a role for GH in reducing hypogonadism-associated spinal bone loss, no such human data are available, and several other agents (gonadal steroids, CT, and bisphosphonates) are known to be effective.

Future Studies

With rhGH providing a standardized source material and with improved availability, future studies should address the important issues concerning the effect of GH in postmenopausal and senescent osteoporosis in individuals without clear-cut GH deficiency. Perhaps, the senescent (ages 65 and older) form of osteoporosis would be a more attractive target since bone mass augmentation is needed then more than the prevention of bone loss. Other augmentation agents, such as sodium fluoride and intermittent low-dose PTH, are currently being further examined for reduced morbidity associated with bone density gain (25).

Recently, IGF-I has been examined for its skeletal effects in animals and influences on skeletal biochemical markers in humans. M. Drezner at Duke University (personal communication) has observed IGF-I to increase bone volume and bone formation rates in dogs, with increased BMD at the spine and hip. Clinical trials were slated to start in late 1993.

Since chronic GH treatment can reduce trabecular BMD and increase compact bone, its regional effects are the opposite of those reported for low-dose, intermittent PTH treatment (25, 26). Some of the anabolic effects of PTH on bone have been observed to be mediated by IGF-I. If so, the present data indicate that treatments with GH and PTH are different by their own effects and systemic versus local IGF-I effects on skeletal regions. R. Lindsay at Helen Hayes Hospital, NY (personal communication), has recently observed good results when osteoporotic women were treated with PTH and estrogen. A combined GH therapy with anti-bone-resorbing agents, therefore, might be an attractive possibility.

References

1. Riggs BL, Melton LJ. Involutional osteoporosis. N Eng J Med 1986;314: 1676–86.
2. Consensus conference: osteoporosis. JAMA 1984; 252:799–80.
3. Recker RR. Current therapy for osteoporosis. J Clin Endocrinol Metab 1993;76:14–6.

4. Rudman D, Kutner MH, Rogers CM, Lubin MF, Fleming GA, Bain RP. Impaired growth hormone secretion in the adult population. J Clin Invest 1981;67:1361–9.

5. Thorner MO, Vance ML, Horvath E, Kovacs K. The anterior pituitary. In: Wilson JD, Foster DW, eds. Williams textbook of endocrinology. 8th ed. Philadelphia: WB Saunders, 1992:229–34.

6. Rudman D. Growth hormone, body composition and aging. J Am Geriatr Soc 1985;33:800–7.

7. Parfitt AM. Growth hormone and adult bone remodeling. Clin Endocrinol (Oxf) 1991;35:467–70.

8. Marcus, Butterfield G, Holloway L, et al. Effects of short term administration of recombinant hGH to elderly people. J Clin Endocrinol Metab 1990;70: 519–27.

9. Bengtsson B, Eden S, Lonn L, et al. Treatment of adults with growth hormone deficiency with recombinant human GH. J Clin Endocrinol Metab 1993;76:309–17.

10. Schlemmer A, Johansen J, Pedersen S, Jargensen J, Hassager C, Christiansen C. The effect of growth hormone (GH) therapy on urinary pyridinoline cross-links in GH-deficient adults. Clin Endocrinol (Oxf) 1991;35:471–6.

11. O'Halloran DJ, Tsatsoulis A, Whitehouse RW, Holmes SJ, Adams JE, Shalet SM. Increased bone density after recombinant human growth hormone (GH) therapy in adults with isolated GH deficiency. J Clin Endocrinol Metab 1993;76:1344–8.

12. Shore RM, Chesney RW, Mazess RB, Rose PG, Bergman GJ. Bone mineral status in growth hormone deficiency. J Pediatr 1980;96:393–6.

13. Aloia JF, Roginsky MS, Jowsey J, Dombrowski CS, Shukla K, Cohn SH. Skeletal metabolism and body composition in acromegaly. J Clin Endocrinol Metab 1972;35:543–51.

14. Ezzat S, Melmed S, Endres D, Eyre DR, Singer F. Biochemical assessment of bone formation and resorption in acromegaly. J Clin Endocrinol Metab 1993;76:1452–7.

15. Seeman E, Wohner HW, Offord KP. Differential effects of endocrine dysfunction on the axial and appendicular skeleton. J Clin Invest 1982;69: 1302–9.

16. Diamond T, Nery L, Posen S. Spinal and peripheral bone mineral densities in acromegaly: the effects of excess growth hormone and hypogonadism. Ann Intern Med 1989;111:567–73.

17. Aloia JF, Zanzi I, Ellis K, et al. Effects of growth hormone in osteoporosis. J Clin Endocrinol Metab 1976;43:992–9.

18. Aloia JF, Vasani A, Meunier P, et al. Coherence treatment of postmenopausal osteoporosis with growth hormone and calcitonin. Calcif Tissue Int 1987;40:253–9.

19. Aloia JF, Vawani A, Kapoor A, Yeh JK, Cohn SH. Treatment of osteoporosis with calcitonin with and without growth hormone. Metabolism 1985; 34:124–9.

20. Rudman D, Feller AG, Nagraj HS, et al. Effects of human growth hormone in men over 60 years old. N Eng J Med 1990;323:1–6.

21. Martinez JA, DelBarrio AS, Larralde J. Evidence of a short term effect of growth hormone on in vivo bone protein synthesis in normal rats. BBA 1991;1093:111–3.

22. Harris WH, Heaney RP. Effect of growth hormone on skeletal mass in adult dogs. Nature 1969;223:403–4.
23. Mann DR, Rudman CG, Akinbami MA, Gould KG. Preservation of bone mass in hypogonadal female monkeys with recombinant human growth hormone administration. J Clin Endocrinol Metab 1992;74:1263–9.
24. Mosekilde L, Bak B. The effects of growth hormone on fracture healing in rats: a histological description. Bone 1993;14:19–27.
25. Riggs BL, Melton LJ. The prevention and treatment of osteoporosis. N Engl J Med 1992;327:620–7.
26. Neer R, Slovik D, Daly M, Lo C, Potts JT, Nussbaum S. Treatment of postmenopausal osteoporosis with daily PTH plus calcitriol. In: Christiansen C, Overgaard K, eds. Osteoporosis. Copenhagen: Osteopress ApS, 1990: 1314–7.

Part V

Potential Utility of Growth Hormone in Ovulation Induction

23

Evidence for the Utility of Growth Hormone in the Enhancement of Ovulation

HOWARD S. JACOBS

Two lines of evidence suggested a role for *growth hormone* (GH) in the reproductive process. The first was the observation that lowering GH levels in female rats delays puberty and decreases ovarian steroidogenesis in response to stimulation by gonadotropins (1). It was also observed that puberty is delayed and prolonged in children with Laron-type dwarfism (2), which is resistant to the action of GH, whereas puberty and gonadal maturation are induced by the administration of GH to hypopysectomized animals (3).

The second line of evidence suggesting a role for GH in the reproductive process was more deductive. Thus, in reviewing the endocrine events of the human ovulation cycle, one notes during the follicular phase a striking contrast between the explosive increase in *estradiol* (E_2) secretion, reflecting the exponential increase in follicular growth, and the only modest increase in the concentrations of immunoreactive *follicle stimulating hormone* (FSH). This difference in the patterns of E_2 and FSH suggested the possibility of ovarian paracrine factors that amplify the effect of FSH on follicle growth. It is now clear that in addition to the gonadotropins that regulate granulosa cell proliferation, intraovarian peptides also play a part in the control of follicular development (4) and that the *insulin-like growth factors* (IGFs) are particularly relevant in this matter (5).

At about the time the results of these in vitro studies were appearing in the literature, we noted that a number of our patients who were undergoing induction of ovulation were resistant to treatment with gonadotropins. We hypothesized that by cotreatment with GH we might increase endogenous IGF-I concentrations and thereby sensitize the ovaries to stimulation by gonadotropins. At that time, we had in mind that the putative action of GH would operate through a paracrine mechanism—that is to say, that treatment with GH would increase production of IGF-I within the ovarian follicle itself.

Accordingly, we initiated a series of clinical studies: The first was open, but thereafter we moved to randomized, placebo-controlled studies for induction of ovulation for in vivo and *in vitro fertilization* (IVF). The latter study was designed in order to gain access to *follicular fluid* (FF) from GH-treated cycles, but we also used the experimental setup to determine whether, for the same dose of gonadotropins, women who had been cotreated with GH would yield more follicles and more oocytes than women who received placebo. A detailed account of all of these studies has recently been published (6); this chapter is confined to those studies that were randomized and prospective.

Controlled Clinical Trial of Cotreatment with Human GH and Gonadotropins for Induction of Ovulation in Anovulatory Patients

Study Design and Patients

We studied 16 patients with amenorrhoea who required *human menopausal gonadotropin* (hMG) for induction of ovulation and who had previously been treated for at least 1 cycle in which a minimum of 25 hMG ampoules were needed. Five of the patients had had pituitary surgery (for macroprolactinoma, pituitary-dependent Cushing disease, chromophobe adenoma, and craniopharyngioma). Five patients suffered from isolated hypogonadotropic hypogonadism, 1 had Kallmann syndrome, and 5 patients had ultrasound-diagnosed *polycystic ovaries* (PCO).

Patients were assigned to treatment with either GH or placebo using a blinded randomization procedure. Eight patients received GH and hMG, and 8 patients received placebo with hMG. The assignment code was broken at the end of each treatment cycle. Those who received GH were considered to have completed the study, whereas those who had received placebo were then offered cotreatment with GH in a subsequent cycle.

In all treatment cycles hMG was administered according to the individually adjusted dose scheme, starting with 1 ampoule per day for 5 days and increasing the dose with 1 ampoule per day until a sufficient ovarian response was obtained as detected by ultrasound. The *daily effective dose* was then continued until 1 or more follicles attained a size of >17 mm in diameter. *Human chorionic gonadotropin* (hCG), 10,000 IU, was administered in order to induce ovulation. GH, 24 IU, was administered *intramuscularly* (I.M.) on alternate days to a total dose of 144 IU or until the day of hCG administration, starting on day 1 of the hMG therapy. Patients who needed treatment for more than 12 days continued with hMG alone, as the GH treatment was stopped after 6 injections. Placebo was administered in the same way as the GH.

The mean ages of patients in the GH and the placebo group were 31.9 (range: 23–40 years) and 30.1 (range: 26–34 years), respectively, with no statistically significant difference. The mean ($\pm$ SD) serum *luteinizing hormone* (LH), FSH, and E_2 concentrations were 2.7 $\pm$ 1.7 IU/L, 1.6 $\pm$ 1.5 IU/L, and 62.5 $\pm$ 35 pmol/L, respectively. The various clinical diagnoses were equally distributed between the 2 treatment groups, and, in particular, there were 2 hypophysectomized patients in the placebo group and 3 in the GH-treated group.

Results

Three patients conceived in their first study cycle: 1 in the placebo group and 2 in the GH group. During the GH treatment the number of ampoules of hMG required to induce follicular development was reduced ($P = 0.008$), the duration of treatment was shorter ($P = 0.011$), and the daily effective dose was lower ($P = 0.035$) compared with the placebo cycles (Table 23.1). The number of follicles that were induced to a diameter of >17 mm was marginally higher in the placebo (2.4 $\pm$ 1.1) than in the GH group (1.4 $\pm$ 0.7) ($P = 0.05$). The number of cohort follicles (14–16 mm in diameter) was not significantly different in the placebo (2.9 $\pm$ 2.2) and GH groups (2.4 $\pm$ 1.4) ($P = 0.58$). The range of serum E_2 concentrations was 850 to 7340 pmol/L, and there was no significant difference between the groups. All placebo cycles and all but 1 of the GH-treated cycles were ovulatory, as detected by ultrasound scanning and by midluteal serum progesterone concentrations. Serum IGF-I concentrations rose during GH treatment and peaked between the 2nd and 3rd injection at a mean

TABLE 23.1. Comparison of the number of ampoules, days of treatment, and daily effective dose of hMG in placebo and GH groups.

Variable	Group	No. of cycles	Prestudy cycle	Treatment cycle
No. of hMG ampoules	Placebo	8	43.6 $\pm$ 13.1	42.5 $\pm$ 13.1
	GH	8	34.4 $\pm$ 7.8	24.5 $\pm$ 9.7
			$P = 0.110$	$P = 0.008$
Days of treatment	Placebo	8	19.3 $\pm$ 3.7	18.5 $\pm$ 3.8
	GH	8	17.3 $\pm$ 2.3	13.4 $\pm$ 3.2
			$P = 0.220$	$P = 0.011$
Daily effective dose of hMG	Placebo	8	3.13 $\pm$ 0.83	3.38 $\pm$ 0.74
	GH	8	2.88 $\pm$ 0.64	2.50 $\pm$ 0.76
			$P = 0.510$	$P = 0.035$

Note: Values are mean $\pm$ SD.

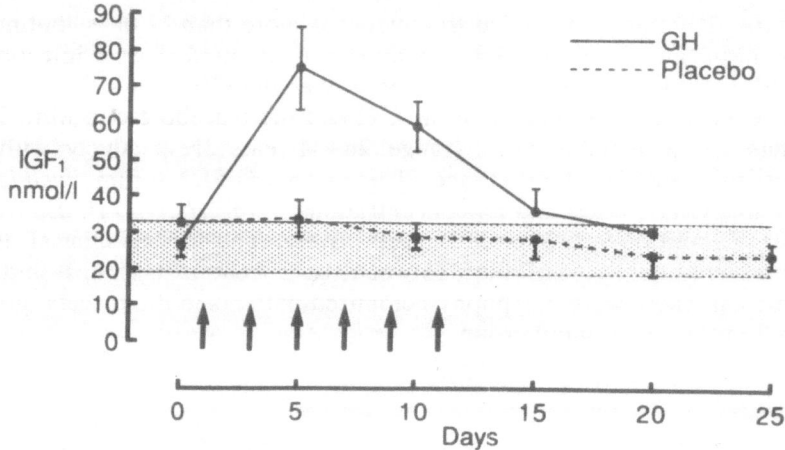

FIGURE 23.1. Serum IGF-I concentrations in hMG + GH and hMG + placebo treatment cycles. The arrows indicate the timing of the injections of GH and placebo. The shaded areas indicate the normal range.

of more than twice the upper limit of normal, falling back into normal range within 1 week of the last GH injection (Fig. 23.1). There was no change in serum IGF-II concentrations. Neither of the growth factor concentrations changed significantly during placebo cycles.

Comments

The results of this prospective, double-blind, randomized study confirmed the findings in the initial open trial (7) that cotreatment with GH sensitized the human ovary to stimulation by gonadotropins. The results did, however, raise an important question—whether the effect of GH that we observed was exerted directly on the ovary or mediated through IGF-I. In order to address this question, we studied patients undergoing treatment by IVF and *embryo transfer* (ET), as the IVF protocol offered us access to human FF.

A Prospective, Randomized, Double-Blind, Placebo-Controlled Trial in Patients Undergoing IVF-ET Following Pituitary Suppression

Study Design and Patients

At Hallam Medical Centre, London, between January and December 1989, we studied patients undergoing IVF-ET who were less than 38 years of age and who had undergone 1 or more IVF-ET cycles in which ovarian stimulation had been carried out using the combined regimen of *LH releasing hormone agonist* (LHRHa) and hMG and in which the response had been considered suboptimal. A *suboptimal response* was defined as one in which fewer than 6 oocytes were collected, from which fewer than 4 embryos developed. Twenty-five patients were recruited into a randomized, double-blind trial of cotreatment with GH compared both with the results of previous treatment and with cotreatment with placebo.

All patients had a pretreatment ultrasound scan of the ovaries to determine ovarian morphology. According to the ovarian scan, the patients were divided into 2 groups: those with normal ovaries and those with ultrasound-diagnosed PCO. This distinction was based on the criteria of Adams et al. (8). Patients in both groups were then randomized to receive GH or placebo in addition, of course, to their standard treatment for IVF. Thirteen patients were allocated to receive GH (24 U per injection given I.M.), and 12 were allocated to receive placebo injections, starting on day 1 of hMG treatment. The GH or placebo was given on alternate days until the administration of hCG or for a maximum period of 2 weeks. At the completion of the cycle of treatment, the assignment code was broken. Those who had received GH were considered to have completed the study; those who had received placebo entered an open study in which they received GH in order that they not be deprived of a potentially beneficial therapy. In all cases, an interval of 2 months was allowed to elapse between cycles of treatment.

All patients had had at least 1 previous IVF-ET attempt using pituitary gonadotropin suppression (buserelin, Suprefact, Hoechst, Hounslow, UK), 200 µg *subcutaneously* (sc). In all treatment cycles the analog was administered daily from day 1 of the menstrual period for a minimum of >14 days (Fig. 23.2). When ovarian suppression was confirmed (serum E_2 concentrations: <150 pmol/L), treatment with hMG was started, and treatment with this dose of buserelin continued until the day of hCG administration. For induction of follicular growth, treatment was commenced with 3 ampoules of hMG (225 IU of FSH and 225 IU of LH) daily for at least 6 days. Further dosages were individualized on the basis of ultrasound examinations and E_2 levels until an adequate ovarian response was obtained.

Human CG, 5000 IU, was administered when 3 follicles of 14 mm in diameter were detected on ultrasound scan of the ovaries—with at least 1 being >17 mm in diameter—in the presence of serum E_2 concentration >1500 pmol/L. Oocyte recovery was performed 35 h later using transvaginal ultrasound-directed follicle aspiration. The technique of IVF, culture of oocytes and embryos, fertilization, and ET were as described by Owen et al. (9).

The FF for analysis of IGF-I was collected by carefully selecting the clear portion of the aspirate using 3 separate tubes. The FF was sub-

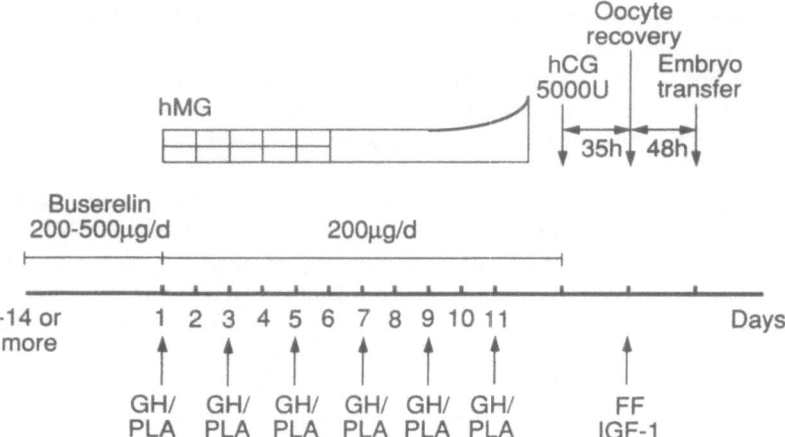

FIGURE 23.2. Schedule of ovarian stimulation following suppression with LHRH analog and cotreatment with GH or placebo. (FF = follicular fluid; PLA = placebo.)

sequently centrifuged and separated from cellular particles, and the supernatant was stored frozen at −20°C. Only FF obtained from follicles containing an oocyte were used for measurement of IGF-I concentrations. In total, 44 FF were analyzed: 26 from GH-augmented cycles and 18 from placebo cycles.

During the period of this study, up to 4 embryos that had shown evidence of normal cleavage were transferred to the uterus after 48 h of culture. Remaining embryos of sufficient quality were cryopreserved. Luteal phase support was given to all patients in the form of hCG (5000 IU) on the day of ET and again 3 days later.

The mean (± SD) age of the patients was 32.4 ± 3.0 years in the GH group, which was not significantly different from that in the placebo group (33.5 ± 2.6 years). The mean *body mass index* (BMI) was also not significantly different (22.6 ± 4.6 in the GH group; 22.0 ± 1.8 in the placebo group) between the 2 groups. The various clinical diagnoses were equally distributed between the 2 groups: 11 patients had tubal damage, 7 had unexplained infertility, 3 had oligospermia, 2 had antisperm antibodies, and 2 failed donor insemination.

Of the 25 patients, 18 were diagnosed as having PCO based solely on ovarian ultrasound morphology (7). Pretreatment serum LH concentrations were not significantly different between the patients with ultrasound-diagnosed PCO (median: 6.2 IU/L; range: 3.1–25.8) and those with ultrasound findings of normal ovaries (median: 5.5 IU/L range 3.7–6.8).

Results

Comparing the results between the 2 treatment groups, in those women who were randomized to receive GH, the total dose of hMG used was significantly less (median: 28 ampoules; range: 20–85) than in the group receiving placebo (median: 36 ampoules; range: 24–100) ($P < 0.05$). No significant differences were found when comparing the number of follicles >14 mm in diameter on the day of hCG administration between the 2 groups and when comparing the number of oocytes collected, yet more oocytes were fertilized from the patients receiving GH than from those receiving placebo ($P < 0.04$).

When we analyzed the subgroup of 18 patients with ultrasound-diagnosed PCO, a similar pattern was observed. Group analysis of the 8 PCO patients receiving placebo and the 10 receiving GH showed that despite using a significantly smaller dose of hMG ($P < 0.01$) in the patients receiving GH, more follicles developed ($P < 0.05$), more oocytes were collected ($P < 0.03$), and more of those were fertilized ($P < 0.004$) and cleaved ($P < 0.02$). There was, however, no significant difference in the maximum serum E_2 concentrations between the treatment groups. Comparing the number of follicles that developed in the prestudy cycles with the number that developed in the subsequent cycles involving the 2 different treatment modalities, a significant effect of placebo, as well as GH, was found (Fig. 23.3). There was, however, a significantly greater improvement in those who received GH treatment ($P < 0.04$).

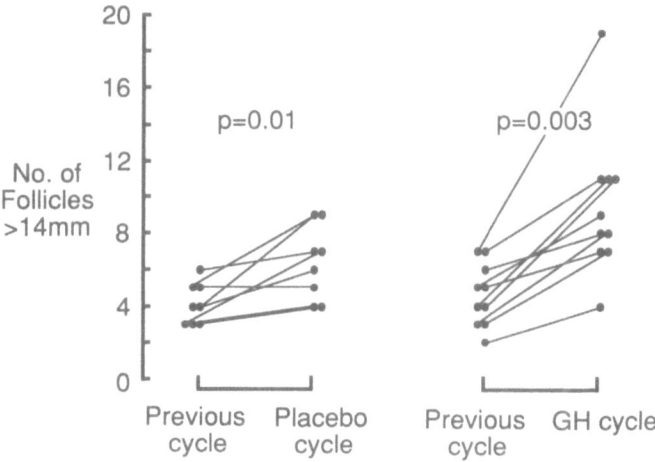

FIGURE 23.3. Comparison of the difference between the number of follicles developed in the prestudy cycles and the subsequent study cycles in the 2 different treatment modalities. In applying the Mann-Whitney test to compare the improvement between the 2 groups, a greater degree of improvement was found in those who received GH treatment ($P < 0.04$).

Serum IGF-I concentrations rose significantly during the GH treatment cycles and peaked between the 2nd and 3rd injections, falling back into the normal range by the day of egg collection. No change in IGF-I concentrations was noted in the placebo group. Although there was no significant difference in mean serum IGF-I concentrations on the day of oocyte collection (23.6 ± 7.0 nmol/L for the placebo group vs. 28.9 ± 8.2 nmol/L for the GH group), FF IGF-I concentrations were significantly higher in the patients treated with GH than in those treated with placebo (24.9 ± 6.9 vs. 19.6 ± 3.3 nmol/L, respectively; $P < 0.04$). In both groups FF IGF-I concentrations were significantly lower compared with serum concentrations (19.6 ± 3.3 vs. 26.3 ± 7.0 nmol/L, respectively, for the placebo group; $n = 10$, $P < 0.0002$; and 24.9 ± 6.9 vs. 28.9 ± 8.2 nmol/L, respectively, for the GH group; $n = 13$; $P < 0.01$).

Four women conceived after their treatment cycle with GH and delivered 2 sets of twins and 2 singletons (a rate of 46% live births per aspiration). One patient had an ectopic pregnancy following a cycle with placebo treatment.

Comments

The results of this prospective, randomized, double-blind, placebo-controlled study confirmed our earlier findings in studies of induction of ovulation and IVF-ET (10–12). Cotreatment with GH increased the ovarian response to stimulation by gonadotropin in patients who previously had been resistant to hMG. The present results demonstrated a 22.3% reduction in the total dose of gonadotropins required for ovarian stimulation, despite which, the ovarian performance improved, as judged by the increase in the number of follicles developing and of oocytes retrieved.

When we analyzed the results from the subgroup of patients with ultrasonically diagnosed PCO, we found an even greater reduction in the total dose of gonadotropins required for ovarian stimulation and an even greater increase in the number of follicles that developed. This result suggested that even though they had previously responded suboptimally to gonadotropin stimulation, these patients, when cotreated with GH, were potentially very sensitive to gonadotropin therapy. The cause of the initial lack of adequate response to treatment with LHRHa and gonado-tropins alone remains unknown; it might be caused by an inadequate intraovarian production of growth factors or to a relative GH deficiency caused by the analog-induced estrogen deficiency.

The data showed significantly higher concentrations of IGF-I in the FF of patients treated with GH compared with placebo, although there was no significant difference in serum IGF-I concentrations on the day of egg collection. These results differed from those published by Volpe et al.

(13), who found no significant difference in FF IGF-I concentrations between the GH-treated patients and the control group. However, their study group included only 4 patients. The presence of IGF-I in the FF in both groups in our study might suggest a role for this growth factor in the development of the follicle. The significant rise of the concentrations of IGF-I in the FF following GH treatment would be consistent with local production of IGF-I by granulosa cells, as has been demonstrated in in vitro experiments (14). However, it is known that quantitatively, IGF-I is synthesized mainly in the liver (15), and in view of the significantly lower levels of IGF-I in FF compared with serum found in our study and in that of Rabinovici et al. (16), an extraovarian source may be the main contributor to FF IGF-I.

Discussion

These and other studies (9–12, 17, 18) show that cotreatment with GH augments the gonadal response to stimulation by gonadotropins. The results of the studies that demonstrated a reduction in the total dose of gonadotropins required for inducing ovulation have obvious implications for programs of fertility treatment that require ovarian stimulation. They do, however, raise two important questions. The first is whether the effect we have observed is physiological and represents a form of replacement treatment or whether it represents a pharmacological effect. While in the initial studies we used very large doses of GH (about 6 times the dose used in the replacement treatment for short stature), our subsequent experience suggests that the effect can be obtained with much smaller amounts of GH. Thus, Homburg et al. (19) showed that a single injection of 24 IU of GH on day 1 of gonadotropin therapy was also effective in increasing ovarian sensitivity. A comparison of the results of a single dose of cotreatment with GH with the 6-dose GH protocol described above demonstrated an intermediate, albeit significant, reduction in the required number of ampoules, duration of treatment, and daily effective dose of hMG. This result strongly suggests a dose-dependent response, a conclusion endorsed by the results of a multicenter dose-ranging study organized by the Novo Nordisk company and presently being prepared for publication.

We therefore think that the results of our studies most likely represent the medical aspect of the essential physiological connection of growth and development. It is, after all, essential that the processes of growth and development are intimately coordinated so that the organism does not attempt to reproduce when it is physically too small for the mechanics to take place. Studies from our own group have shown that application of pulsatile LHRH to children with delayed puberty causes an increase in the rate of growth as one of the earliest phenotypic expressions of treatment (20, 21). In our opinion, therefore, the studies described above represent the clinical application of these physiological observations.

The second question raised by these observations is whether the effect of GH we have observed is exerted directly on the gonad or mediated through the IGFs. Since other chapters in this volume address this issue directly, it does not seem appropriate here to explore this aspect further, except to remark that our finding of higher concentrations of IGF-I in serum than in FF does not provide immediate support for the concept of GH-augmented ovarian synthesis of growth factor as the basis for the actions of GH on the ovary.

Finally, there is the practical question of which patients, in the light of present knowledge, should be considered for cotreatment with GH. So far as those with anovulatory infertility are concerned, it is those with gonadotropin and estrogen deficiency who seem to benefit most. What is certain is that women with incipient or manifest primary ovarian failure do not respond (22). So far as patients undergoing programs of assisted fertility are concerned, it seems that it is a relatively select group of women who benefit—that is, women with PCO who are undergoing ovarian stimulation using an LHRH analog and who prove resistant to stimulation by gonadotropin. Again, it is crucial to emphasize that cotreatment with GH of women with ovarian failure is doomed to failure. On the other hand, judicious use of this hormone may provide the clinician with an additional form of treatment for the otherwise-resistant case.

Acknowledgments. I thank Thomas Torresani, Ph.D., of Proteinhormon Labor, Kinderspital, Zurich, Switzerland, for measuring serum IGF-I concentrations. It is a pleasure to thank my colleagues Professor Roy Homburg, Dr. Elizabeth Owen, Dr. Zeev Shoham, Dr. Gerard Conway, and Dr. Hanne Ostergaard for their contributions to these studies. Novo Nordisk S/A, Gentofte, Denmark, supplied all of the growth hormone (Norditropin) we have used. I am particularly grateful to Dr. Anne-Marie Kappelgard for generous financial support at all stages of the project.

References

1. Advis JPS, White S, Ojeda SR. Activation of growth hormone short loop negative feedback delays puberty in the female. Endocrinology 1981;108: 1343–52.
2. Laron Z, Sarel R, Pertzelan A. Puberty in Laron type dwarfism. Eur J Pediatr 1980;134:79–83.
3. Shiekholislam BM, Stempfel RS Jr. Hereditary isolated somatotropin deficiency: effects of human growth hormone administration. Pediatrics 1972;49: 362–74.

4. Hammond JM. Peptide regulators in the ovarian follicle. Aust J Biol Sci 1981;34:491–504.
5. Adashi EY, Resnick CE, D'Ercole AJ. Insulin-like growth factors as intra-ovarian regulators of granulosa cell growth and function. Endocr Rev 1985;6:400–20.
6. Shoham Z, Homburg R, Owen EO, Conway GS, Ostergaard H, Jacobs HS. The role of treatment with growth hormone in infertile patients. Baillieres Clin Obstet Gynaecol 1992;6:267–81.
7. Homburg R, Eshel A, Abdalla HI, Jacobs HS. Growth hormone facilitates ovulation induction by gonadotrophins. Clin Endocrinol (Oxf) 1988;29:113–7.
8. Adams J, Franks S, Polson DW, et al. Multifollicular ovaries: clinical and endocrine features and response to pulsatile gonadotropin releasing hormone. Lancet 1985;2:1375–8.
9. Owen EJ, Davies MC, Kingsland CR, et al. The use of a short regimen of buserelin, a gonadotrophin-releasing hormone agonist, and human menopausal gonadotrophin in assisted conception cycles. Hum Reprod 1989;4:749–54.
10. Homburg R, Eshel A, Abdalla HI, et al. Growth hormone facilitates ovulation induction by gonadotrophins. Clin Endocrinol (Oxf) 1988;29:113–7.
11. Homburg R, West C, Torresani T, et al. Co-treatment with human growth hormone and gonadotrophins for induction of ovulation: a controlled clinical trial. Fertil Steril 1990;53:254–60.
12. Owen EJ, West C, Mason BA, et al. Co-treatment with growth hormone of sub-optimal responders in IVF-ET. Hum Reprod 1991;6:524–8.
13. Volpe A, Coukos G, Barreca A, et al. Ovarian response to combined growth hormone-gonadotropin treatment in patients resistant to induction of super-ovulation. Gynecol Endocrinol 1989;3:125–34.
14. Steinkampf MP, Mendelson CR, Simpson ER. Effects of epidermal growth factor and insulin-like growth factor-I on the levels of mRNA encoding aromatase cytochrome P-450 of human ovarian granulosa cells. Mol Cell Endocrinol 1988;59:93–7.
15. D'Ercole AJ, Stiles AD, Underwood LE. Tissue concentrations of so-matomedin C: further evidence for multiple sites of synthesis and paracrine or autocrine mechanisms of action. Proc Natl Acad Sci USA 1984;81:935–9.
16. Rabinovici J, Dandekar P, Angle MJ, et al. Insulin-like growth factor I (IGF-I) levels in follicular fluid from human preovulatory follicles: correlation with serum IGF-I levels. Fertil Steril 1990;54:428–33.
17. Owen EJ, Shoham Z, Mason BA, et al. Cotreatment with growth hormone, following pituitary suppression, for ovarian stimulation in in-vitro fertilization: a randomized, double-blind, placebo-control trial. Fertil Steril 1991;56:1104–10.
18. Blumenfeld Z, Lunenfeld B. The potentiating effect of growth hormone on follicle stimulation with human menopausal gonadotropins in a panhypo-pituitary patient. Fertil Steril 1989;52:328–31.
19. Homburg R, West C, Torresani T, et al. A comparative study of single-dose growth hormone therapy as an adjunct to gonadotrophin treatment for ovulation induction. Clin Endocrinol (Oxf) 1990b;32:781–5.
20. Stanhope R, Brook CGD, Pringle PJ, et al. Induction of puberty by pulsatile gonadotrophin releasing hormone. Lancet 1987;2:552–5.

21. Darendeliler F, Hindmarsh PC, Preece MA, et al. Growth hormone increases rate of pubertal maturation. Acta Endocrinol (Copenh) 1990;122/3:414–6.
22. Homburg R, West C, Ostergaard H, Jacobs HS. Combined growth hormone and gonadotropin treatment for ovulation induction in patients with non-responsive ovaries. Gynaecol Endocrinol 1991;5:33–6.

24

Evidence Against the Utility of Growth Hormone in the Enhancement of Ovulation

Marco Filicori

Ovulation induction is a commonly used procedure for the treatment of infertility. Traditionally, anovulatory women receive stimulatory drugs that result in follicular maturation, ovulation, and, hopefully, conception. Drugs used for ovulation induction include such antiestrogens as clomiphene citrate (1), *gonadotropin releasing hormone* (GnRH) (2), and exogenous gonadotropins. More recently, the introduction of assisted reproduction techniques prompted the use of *human menopausal gonadotropin* (hMG) in normal ovulatory women for the recruitment of multiple follicles; the availability of an elevated number of oocytes is essential for the optimal outcome of *in vitro fertilization* (IVF).

Recent evidence suggests that *growth hormone* (GH) directly or indirectly through the action of *insulin-like growth factors* (IGFs) can stimulate granulosa cell function (3) synergistically with *follicle stimulating hormone* (FSH). This finding has prompted the combined use of hMG and GH for clinical ovulation induction. The basic goals of GH supplementation in gonadotropin ovulation induction are listed in Table 24.1. Several recent studies have tested the feasibility of this approach; this chapter briefly reviews such studies and their impact on the pharmacology of ovulation induction.

hMG Supplementation with GH in Normal Ovulatory Women

Most women treated with hMG in assisted reproduction protocols have regular ovulatory menstrual cycles; exogenous gonadotropins are used to achieve multiple folliculogenesis and increase oocyte yield. The

TABLE 24.1. Goals of GH supplementation in hMG ovulation induction.

Improve treatment in suboptimal responders.
Shorten follicular stimulation.
Lower hMG requirements.
Increase follicle number.
Improve oocyte yield.

TABLE 24.2. GH supplementation in hMG ovulation induction: studies in normal women.

Author (ref.)	Type of study	Number of patients	GH dose (IU/cycle)
Shaker et al., 1992 (6)	GH vs. control cycles	10	144
Tapanainen et al., 1992 (7)	Randomized, placebo-controlled	54	>120
Younis et al., 1992 (8)	Randomized, placebo-controlled, double-blind	42	48

development of pharmacologic regimens that result in the maturation of a greater follicle number in a shorter time period and with the use of lesser amounts of hMG is considered highly desirable. Early results with GH supplementation (4, 5) suggested that at least some of these goals could be achieved. However, these early studies were mostly carried out in resistant or amenorrheic women. Only more recent extensive studies have addressed the issue of the efficacy of GH supplementation for the enhancement of ovulation induction procedures in normal women; a list of these studies is shown in Table 24.2.

Shaker et al. (6) studied 20 women undergoing IVF; 10 of them were normal women with regular menstrual cycles. Cycles in which GH supplementation was provided were compared to cycles with hMG only carried out in the same patients. The duration of stimulation, hMG dose, day of first response, preovulatory *estradiol* (E_2) levels, and number of follicles >14 mm in diameter were all unaffected by GH supplementation (Table 24.3). Furthermore, the number and quality of oocytes and embryos did not vary between the control and the GH-treated cycles. Similar clinical results were reported in the randomized, placebo-controlled study of Tapanainen et al. (7). However, serum E_2 and progesterone levels were lower, while *follicular fluid* (FF) testosterone concentrations were higher in the GH-treated subjects, suggesting that GH may somewhat affect ovarian steroidogenetic activity. Furthermore, IGF-I levels of the GH-treated cycles were increased in serum but not in FF, suggesting that FF IGF-I does not derive from local ovarian

TABLE 24.3. Effects of GH supplementation on hMG ovulation induction parameters.

	Author (ref.)		
	Shaker et al., 1992 (6)	Tapanainen et al., 1992 (7)	Younis et al., 1992 (8)
hMG dose	=[a]	=	=
FP duration[b]	=	=	=
Preovulatory E_2	=	=	=
No. of follicles	=	=	=
No. of oocytes	=	=	=

[a] Equals symbol (=) indicates no effect.
[b] FP duration = follicular phase duration.

production. Finally, Younis et al. (8) studied 42 normal ovulatory women undergoing IVF in a placebo-controlled, double-blind fashion (Table 24.2). In this study as well, the clinical parameters of ovarian stimulation and hMG dose requirements were not influenced by GH administration (Table 24.3), and the pregnancy rate was not different between the control and the GH-treated groups. Taken together, these results indicate that in normal women with regular ovulatory menstrual cycles, the addition of GH to hMG stimulatory regimens does not affect any major clinical parameter and does not result in a lowering of hMG dose requirements.

hMG Supplementation with GH in Anovulatory and Resistant Patients

While GH appears to be of limited efficacy in normal women, better results may derive from its use in women with deranged ovulatory function. Although a majority of women under 40 years promptly respond to hMG stimulation, some patients show an abnormal response to this drug in spite of intensive hMG administration. This feature is particularly prominent in panhypopituitary patients. *Polycystic ovary syndrome* (PCOS) patients may also respond abnormally to hMG, although an excessive ovarian response is often present; these patients are at greater risk of developing ovarian hyperstimulation.

The early study of Homburg et al. (4) in 7 patients resistant to hMG stimulation (Table 24.4) was the first clinical work indicating that GH may improve ovarian response to hMG; hMG dose requirements and the length of stimulation were lowered by GH, while the number of collected oocytes (in the 3 patients undergoing IVF) increased. However, the number of follicles detected at ultrasound was not affected (Table 24.5). The addition of GH appears to be particularly effective in hypopituitary patients, as clearly shown by Blumenfeld and Lunenfeld (9) in their early

TABLE 24.4. GH supplementation in hMG ovulation induction: studies in suboptimal responders.

Author (ref.)	Type of study	Type of patients	Number of patients	GH dose (IU/cycle)
Homburg et al., 1988 (4)	Prospective	Resistant	7	120
Blumenfeld et al., 1989 (9)	Case report	Hypopituitary	1	16–24
Homburg et al., 1990 (5)	Randomized, placebo-controlled	Hypopituitary, HH, PCO	16	144
Owen et al., 1991 (10)	Randomized, placebo-controlled, double-blind	PCO	18	24
Ibrahim et al., 1991 (11)	Prospective	Resistant	10	144
Shaker et al., 1992 (6)	GH vs. control cycles	Resistant	10	144

TABLE 24.5. Effects of GH supplementation on hMG ovulation induction parameters.

	Author (ref.)				
	Homburg et al., 1988 (4)	Homburg et al., 1990 (5)	Owen et al., 1991 (10)	Ibrahim et al., 1991 (11)	Shaker et al., 1992 (6)
hMG dose	↓ [a]	↓	↓	↓	= [b]
FP duration[c]	↓	↓	=	↓	=
Preovulatory E_2	NR [d]	=	=	↓	=
# follicles	=	=	↑	=	=
# oocytes	↑ [e]	NR	↑	↑	=

[a] Decrement.
[b] No effect.
[c] Follicular phase duration.
[d] Not reported.
[e] Increment.

case report and in the 1990 larger study of Homburg et al. (5). This latter study confirmed that hMG dose requirements and the period of stimulation were reduced in subjects with severe gonadotropin derangements, while preovulatory serum E_2 and the number of maturing follicles did not change (Table 24.5). Serum IGF-I levels were increased in GH-treated subjects.

The same group (10) later reported a double-blind, placebo-controlled study carried out in 25 hMG-resistant patients; 18 of these subjects had PCOS (Table 24.4). The interesting feature of this study is that unlike every other report, the number of maturing follicles in the PCOS group was found to be greater in GH-treated patients, suggesting that PCOS may be uniquely sensitive to this form of stimulation. Although both FF and serum levels of IGF-I were increased, FF concentrations were lower, again suggesting a nonovarian source of this hormone. Ibrahim et al. (11) also confirmed in their series of 10 resistant patients that GH supplementation may lower hMG dose requirements and improve oocyte yield (Table 24.5). Finally, the report of Shaker et al. (6) could not confirm in 10 resistant patients any of the positive results found in other studies; hMG dose and duration of administration, preovulatory E_2, and number of follicles and oocytes were unaffected by GH supplementation (Table 24.5).

Taken together, these results suggest that GH supplementation in resistant patients is probably effective in improving some of the parameters of hMG ovulation induction, such as hMG dose, duration of drug administration, and number of oocytes that can be recovered at IVF. PCOS patients appear to be particularly susceptible to this regimen.

Conclusions

The enthusiasm that GH coadministration for hMG ovulation induction initially raised has been mitigated by more recent reports suggesting limited, if any, effects on drug regimen and treatment outcome. No significant improvement derives from GH supplementation in normal ovulatory women. In the resistant patient, the benefits appear to be limited mostly to a reduction of hMG dose requirements and, possibly, oocyte yield. However, the high cost of recombinant GH (about U.S. $71.80 per 4 IU in Italy) more than offsets the economic benefit derived from the reduction in the number of hMG ampoules used in this procedure (costing around U.S. $16.20–$18.60 in Italy). Treatment improvement deriving from the use of these regimens in PCOS patients may be worth exploring. Nevertheless, it is unlikely that GH supplementation will find widespread application in ovulation induction procedures and in assisted reproduction.

Acknowledgment. We wish to thank Mrs. Silvia Arsento for excellent secretarial assistance.

References

1. Adashi EY. Clomiphene citrate: mechanism(s) and site(s) of action—a hypothesis revisited. Fertil Steril 1984;42:331–44.
2. Filicori M, Flamigni C, Meriggiola MC, et al. Ovulation induction with pulsatile gonadotropin-releasing hormone: technical modalities and clinical perspectives. Fertil Steril 1991;56:1–13.

3. Katz E, Ricciarelli E, Adashi EY. The potential relevance of growth hormone to female reproductive physiology and pathophysiology. Fertil Steril 1993;59: 8–34.

4. Homburg R, Eshel A, Abdalla HI, Jacobs HS. Growth hormone facilitates ovulation induction by gonadotrophins. Clin Endocrinol (Oxf) 1988;29:113–7.

5. Homburg R, West C, Torresani T, Jacobs HS. Cotreatment with human growth hormone and gonadotropins for induction of ovulation: a controlled clinical trial. Fertil Steril 1990;53:254–60.

6. Shaker AG, Fleming R, Jamieson ME, Yates RWS, Coutts JRT. Absence of effect of adjuvant growth hormone therapy on follicular responses to exogenous gonadotropins in women: normal and poor responders. Fertil Steril 1992;58:919–23.

7. Tapanainen J, Martikainen H, Voutilanein R, Orava M, Ruokonen A, Rönnberg L. Effect of growth hormone administration on human ovarian function and steroidogenetic gene expression in granulosa-luteal cells. Fertil Steril 1992;58:726–32.

8. Younis JS, Simon A, Koren R, Dorembus D, Schenker JG, Laufer N. The effect of growth hormone supplementation on in vitro fertilization outcome: a prospective randomized placebo-controlled double-blind study. Fertil Steril 1992;58:575–80.

9. Blumenfeld Z, Lunenfeld B. The potentiating effect of growth hormone on follicle stimulation with human menopausal gonadotropin in a panhypopituitary patient. Fertil Steril 1989;52:328–31.

10. Owen EJ, Shoham Z, Mason BA, Ostergaard H, Jacobs HS. Cotreatment with growth hormone, after pituitary suppression, for ovarian stimulation in in vitro fertilization: a randomized, double-blind, placebo-control trial. Fertil Steril 1991;56:1104–10.

11. Ibrahim ZH, Matson PL, Buck P, Lieberman BA. The use of biosynthetic human hormone to augment ovulation induction with buserelin acetate/human menopausal gonadotropin in women with a poor ovarian response. Fertil Steril 1991;55:202–4.

Author Index

Subject Index

PROCEEDINGS IN THE SERONO SYMPOSIA USA SERIES

Continued from page ii